MOSFET
Theory and Design

MOSFET
Theory and Design

R. M. Warner, Jr.

B. L. Grung

New York • Oxford
OXFORD UNIVERSITY PRESS
1999

Oxford University Press

Oxford New York
Athens Auckland Bangkok Bogotá Buenos Aires Calcutta
Cape Town Chennai Dar es Salaam Delhi Florence Hong Kong Istanbul
Karachi Kuala Lumpur Madrid Melbourne Mexico City Mumbai
Nairobi Paris São Paulo Singapore Taipei Tokyo Toronto Warsaw

and associated companies in
Berlin Ibadan

Published by Oxford University Press, Inc.
198 Madison Avenue, New York, New York 10016
http://www.oup-usa.org

Library of Congress Cataloging-in-Publication Data

Warner, R. M., Jr.
MOSFET theory and design / R.M. Warner, Jr., B.L. Grung.
p. cm.
Includes bibliographical references and index.
ISBN: 0–19–511642–9 (pbk.)
1. Metal-oxide-semiconductor field-effect transistors.
2. Device design. I. Grung, B. L. II. Title.
TK7871,95.W353 1998
621.3815'284—dc21 98–13184
CIP

Printing (last digit): 9 8 7 6 5 4 3 2 1

Printed in the United States of America
on acid-free paper

To our marvelous wives, Mae Warner and Sandi Grung,
who were also wonderfully supportive on this project.

RMW

BLG

Contents

Preface

MOSFET Theory and Design is a textbook developed for a one-semester course at the junior, senior, or graduate level of an engineering program. The material initially evolved in nineteen years of class testing at the University of Minnesota, and in several years of in-house course offerings for industrial firms at several locations throughout the country. The sole focus of this book—the MOSFET—is justified by the overwhelming dominance of this device in today's microelectronics technology, and by the efficiency that accompanies using a small book free of "extra" subject matter.

Achieving clarity was our primary aim in composing the book. Liberal use has been made of analogies and heuristic descriptions, and unnecessary jargon has been avoided. Perspective illustrations are employed in support of explanations wherein spatial relationships are important, replacing the primitive orthogonal-isometric drawings found in most texts. To increase first-reading comprehension, intermediate equations are given in most derivations. And consistent with the authors' deeply held pedagogical conviction, explanations proceed from the specific to the general.

We begin with elementary theory for the MOS capacitor and MOSFET, adding the complicating factors one by one. The interplay of MOS capacitor and *PN* junction that occurs in the MOSFET and the distinctions between these two constituent elements are treated physically and analytically, using some original tools. We endeavor to build explanations of the MOSFET from a simple level to complex real-life cases in carefully graduated steps extended to include advanced models. SPICE treatments of small-signal and large-signal problems follow, starting with the first-generation Level-1, -2, and -3 models, and ending with a discussion of the second and third generations of MOSFET models.

MOSFET Theory and Design includes analytic problems, computer problems, and design problems, and is accompanied by a carefully prepared solutions manual. Additional learning aids incorporated into the book include in-text exercises with accompanying solutions, and an end-of-book summary as well as a series of review questions that let a student know whether key points have been mastered. Finally, this book includes a comprehensive bibliography, not found in most competing books.

Minneapolis, Minnesota
November 1998

R. M. Warner, Jr.
B. L. Grung

Physical Constants

Quantity	Symbol	Value
Angstrom	Å	10^{-8} cm = 10^{-4} μm
Atmospheric pressure	—	1.013×10^{5} N/m^{2}
Avogadro's number	N	6.022×10^{23}/mol
Bohr magneton	—	1.165×10^{-29} V·s·m
Bohr radius	r_0	0.5292 Å
Boltzmann's constant	k	1.381×10^{-23} J/K = 8.618×10^{-5} eV/K
Charge of electron	q	1.602×10^{-19} C
Electron volt	eV	1.602×10^{-19} J
Gas constant	R	1.987 cal/mol·K
Ionization energy, hydrogen	ξ_{ion}	13.6 eV
Mass of electron	m	0.9110×10^{-30} kg
Mass of proton	—	1.673×10^{-27} kg
Micrometer	μm	10^{-6} m = 10^{-4} cm = 10^{4} Å
Permeability, free space	μ_0	$4\pi \times 10^{-7}$ V·s/A·m
Permittivity, free space	ϵ_0	8.854×10^{-14} F/cm = 8.854×10^{-12} F/m
Planck's constant	h	6.626×10^{-34} J·s
Planck's constant/2π	h	1.054×10^{-34} J·s
Speed of light in vacuum	c	2.998×10^{10} cm/s = 2.998×10^{8} m/s
Thermal voltage	kT/q	0.02586 V at 300 K = 0.02566 V at 297.8 K (= 24.8°C)
Thousandth of an inch	mil	25.4 μm

Properties of Silicon

Property	Value (at 300 K)	Value (at 297.8 K)	Unit
Atom density N_a	5.0×10^{22}		cm^{-3}
Atomic number Z	14		
Atomic weight	28.09		
Density	2.33		g/cm^3
Relative dielectric permittivity κ	11.7 (SiO_2:3.9)		
Absolute dielectric permittivity ϵ	1.035×10^{-12}		F/cm
Energy gap ξ_G	1.119	1.120	eV
Equivalent density of states			
Conduction band N_{CA}	3.22×10^{19}	3.18×10^{19}	cm^{-3}
Valence band N_{VA}	1.83×10^{19}	1.81×10^{19}	cm^{-3}
$N_C = N_V$ (for calculation)	3.04×10^{19}	3.01×10^{19}	cm^{-3}
Intrinsic carrier density n_i	1.20×10^{10}	1.00×10^{10}	cm^{-3}
Intrinsic Debye length L_{Di}	26.4	28.8	μm
Intrinsic mobilities			
Electron μ_n	1400		$cm^2/V\cdot s$
Hole μ_p	450		$cm^2/V\cdot s$
Lattice constant a_0	5.43		Å
Melting point	1415		C°
Specific heat	0.7		J/g·C°
Thermal conductivity	1.5		W/cm·C°
Thermal-expansion coefficient (linear)	2.6×10^{-6}		$(C°)^{-1}$

MOSFET
Theory and Design

Basic MOSFET Theory

A thin and tightly adherent layer of glass, mainly SiO_2, can be formed on the surface of a single-crystal silicon sample by heating it in an oxidizing atmosphere. The most common choices are oxygen and steam. Starting in the late 1950s, these procedures were moved along vigorously in the hope that they might provide at least a partial solution to the vexing problem of stabilizing surface conditions on silicon devices. Ultimately they permitted not only stabilization but also control of surface conditions; to accomplish this, however, enormous investments of time and resources were required to achieve the necessary understanding of the oxide–silicon system, and especially of the oxide–silicon interface.

At a relatively early stage, these procedures benefited the bipolar junction transistor (BJT) because its crucial phenomena occur mainly inside the silicon crystal. Further, even a limited grasp of oxide technology opened a whole new approach to BJT fabrication. But then came the MOSFET, an acronym that stands for metal–oxide–silicon field-effect transistor. By contrast with the BJT, the decisive phenomenon in the MOSFET occurs *at* the oxide–silicon interface. Consequently, the art and science of controlling interfacial properties required an order-of-magnitude improvement to make the MOSFET a truly practical device, a process that took approximately an additional ten years.

The central portion of the MOSFET is the *MOS capacitor*, a structure that received intensive study in its own right. When isolated, it is a one-dimensional "sandwich" consisting of the oxide with silicon on one side, and with an adherent field-plate electrode on the opposite side that is capable of manipulating potential at the oxide–silicon interface. With a favorable voltage polarity on the field plate it is possible to create a potential well in the silicon right at the interface, and in it minority carriers can spontaneously collect in a thin layer. Altering the voltage on the field plate in turn alters the depth of the potential well and, hence, its carrier population.

Creating an FET from the MOS capacitor involves making ohmic contact to opposite edges of the thin layer of minority carriers, now known as a *channel.* The field plate that is able to modulate the carrier population of the channel is known as the *gate.* Ohmic contact is achieved by placing regions whose conductivity type is opposite to that of the substrate in the MOS capacitor at each of the opposite edges. Applying a bias voltage from one of these contact regions to the other causes an ohmic flow of carriers through the channel between them, a flow that can be modulated by means of the gate. Thus this FET, like all FETs, is essentially a voltage-controlled resistor. Altering the control voltage changes the carrier population in the channel, which in turn alters the channel resistance.

There are several FET families, each having many members, and in each family there are differences of detail in how the control voltage interacts with the channel. Nonetheless, all of the FETs have similar gain properties, properties that are substantially inferior to those of the BJT through a wide current range. But, we should add quickly, the various families have offsetting advantages. In one case, that of the MOSFET, the sum of these advantages is so imposing that MOSFETs are the dominant solid-state devices in the world today.

1-1 Field-Effect Transistors

Well before the present solid-state-device era, intuitive thoughts were advanced on how to realize a solid-state amplifier; the structures proposed were essentially FETs [1, 2]. The serious postwar effort that led to the transistor also began with FETs [3, 4]. While that path was being explored, an unintended experimental event led to the discovery of the point-contact transistor [5], which found enough applications to be characterized as the first practical solid-state amplifier. But the point-contact transistor had serious shortcomings (and its operating principles remain obscure to this day). Consequently it was supplanted by the BJT within just a few years [6]. The events connected with these two devices (occurring from the mid-1940s to the mid-1950s) display a fascinating blend of scientific insight, art, engineering intuition, and happenstance [7].

A practical FET structure was described and analyzed in detail by Shockley in 1952 [8], a year *before* its reduction to practice by Foy and Wiegmann, both working under the direction of Dacey and Ross [9]. This was a case of having metaphorical lightning strike twice in the same place; the same inventor had accomplished the same feat just a few years earlier with the BJT! Shockley's FET employed a pair of *PN* junctions for defining the channel, and hence has become known as the junction FET, or JFET. His innovation placed the region of crucial properties well inside the semiconductor crystal and away from the troublesome and poorly understood surfaces. Varying degrees of reverse bias on these channel-defining junctions caused varying depletion-layer encroachment upon the channel, thus achieving the aim of altering the areal density (number per unit area) of carriers in the thin channel.

While the laboratory JFETs reported in 1953 [9] constituted a feasibility demonstration, they were not practical devices. The JFET requires a degree of control over areal impurity density in its channel that is about ten times better than that required, for example, in the base region of a BJT. Thus the JFET was a device awaiting new and more refined fabrication options. The first practical JFETs, made in 1960, incorporated channels produced by the epitaxial growth of silicon from a vapor [10, 11]. Today the technology of choice is usually ion implantation [12], which achieves unprecedented control of areal impurity density in thin layers.

In the discrete-device arena, JFETs have played a relatively small part. Unique properties, especially low noise, earned a "niche market" for them. At first JFETs did not play a role in integrated circuits either, because their peculiar requirements made them essentially incompatible with the most popular devices. Then it was found that a kind of hybrid JFET–MOSFET device could play an advantageous part in circuits dominated by the latter [13]. Today, JFETs are being incorporated into even bipolar integrated circuits, thanks to the almost commonplace availability of ion-implantation technology. In addition, the JFET may find application in monocrystalline three-dimensional integrated circuits [14].

The MOSFET [15] was the second important field-effect device to come along. The heart of its original embodiment was a metal–oxide–silicon sandwich, with the initial letters, as noted earlier, responsible for the acronym MOS, as in MOSFET and MOS technology. The sandwich was in fact a parallel-plate capacitor, with one metal plate and one silicon plate. Thus the structure indeed evokes early (and unsuccessful) experiments that endeavored to modulate the majority-carrier population in a thin semiconductor layer as a way of realizing a solid-state amplifier. But important new features were present in the MOSFET, and these spelled success. First, the new device exploited the infant oxide-growth technology that started with a polished single-crystal silicon sample. Significantly, one of the MOSFET inventors was also a pioneering developer of the oxide-growth technology and student of oxide–silicon interfaces [16]. Second, the MOSFET exploited the creation of an *inversion layer* in the silicon at the interface, rather than attempting to achieve a significant modulation of majority-carrier areal density in a thin sample. More specifically, the metal plate of the capacitor was used to create a potential well at the surface of the silicon, a well capable of retaining the desired carriers—electrons in a *P*-doped sample or holes in an *N*-doped sample. This constitutes an alternative description of the inversion layer at an insulator–semiconductor interface. As an isolated entity, this inversion layer is identical to that in a grossly asymmetric step junction at equilibrium. But the two cases are different in context. In the present case the carriers are confined within the potential well because they cannot penetrate the insulator on the other side of the interface, while in the *PN*-junction case, the inversion-layer carriers are in direct communication with the identical majority carriers on the heavily doped side of the junction. Because the MOSFET has overwhelming practical importance today, and because all field-effect devices involve similar principles, our focus in the balance of the book will be on the MOSFET.

The FET chronicle by no means ended with the MOSFET, however. A new family was proposed in 1966 wherein control was exercised through a metal–semiconductor junction, or *Schottky junction* [17], and reduction to practice was reported the following year [18]. The channel carriers in this device are confined between the depletion region associated with a Schottky junction and that associated with a *PN* junction. (Sometimes the *PN* junction is replaced by a junction between an extrinsic region and a semiinsulating region.) This structure was given the name MESFET, for metal–semiconductor field-effect transistor. It has been realized using a number of semiconductor materials, but GaAs is by far the most common. The MESFET, in fact, is the dominant compound-semiconductor transistor today.

A still more recent FET is the MODFET, which designates the modulation-doped FET. It creates a potential well for electrons in a region of extremely light doping, and as a result, the ionic scattering of drifting electrons is reduced to near zero, leading to extreme values of electron mobility, well above 10^4 cm^2/V·s at room temperature. (Hole mobilities are typically small in the GaAs family of materials.) By cooling the sample to reduce phonon activity, one can achieve electron mobilities exceeding 10^5

$cm^2/V \cdot s$. Such enhanced electron mobilities were first observed in 1978 [19], and were realized in an FET in 1980 [20]. This feature of the MODFET has given it a second name, HEMT, the acronym for high-electron-mobility transistor. The significance of high carrier mobility is high-speed operation of the resulting device. Still another name for the MODFET is TEGFET, for two-dimensional electron–gas FET.

The feat of separating channel electrons in the MODFET from the impurity atoms that contribute them is accomplished by means of a *heterojunction*, which is an interface between two semiconductor materials of differing energy gap. Two materials often used to form a heterojunction are the binary compound GaAs and the ternary compound AlGaAs. Heterojunctions are growing in importance in semiconductor technology, making possible proliferating device innovations. One of these is yet another FET, the HIGFET [21, 22], which stands for heterojunction insulated-gate field-effect transistor. Among its advantages is the ready achievement of complementary—or *N*-channel and *P*-channel—devices, which make possible superior performance.

A still further advance in the FET art uses heterojunctions to create a series of *quantum wells* or a *superlattice*, terms that describe a set of very thin, parallel layers of differing energy gap in a semiconductor crystal, usually involving binary and ternary compound semiconductors. Layer thickness must be appreciably less than the de Broglie wavelength and typically amounts to a few tens of angstroms. (A superlattice can also be created by doping variations, and even by light or sound waves!)

One kind of FET-like structure exploiting quantum-well layers uses them as multiple channels that are under the control of a gate electrode. Because carriers are able to move in the planes of such layers at great velocity, the resulting device is capable of high gain and fast switching [23]. An even more recent proposal would replace the two-dimensional quantum-well layers by one-dimensional quantum-well "wires" [24]. The transition from classical to quantum principles of operation opens new realms of device possibilities. A substantial number of variations on these and related themes already exist, and it is safe to predict that the future holds more.

1–2 MOSFET Definitions

The essential structure of a MOSFET is shown in Figure 1-1. For purposes of explanation one can imagine that this structure was cut out of an integrated circuit. The capacitorlike arrangement formed on and by the silicon substrate accounts for the acronym *MOS*, as noted both above and in the diagram. The metal top plate was made of aluminum in most early devices. But since about 1970, heavily doped polycrystalline silicon (*polysil*) has been favored. Many different insulating materials have been employed through the years, but thermally grown oxide formed *in situ* is currently and historically by far the most important option. This description of the silicon oxide means that the necessary silicon was supplied by the single-crystal substrate, and its growth was induced by heating the sample in an oxidizing atmosphere.

The metal top plate, or field plate, becomes the control electrode of the MOSFET and is known as the *gate*. Applying a sufficient voltage to the gate with respect to the substrate causes electrons (normally in the minority in the *P*-type substrate) to be at-

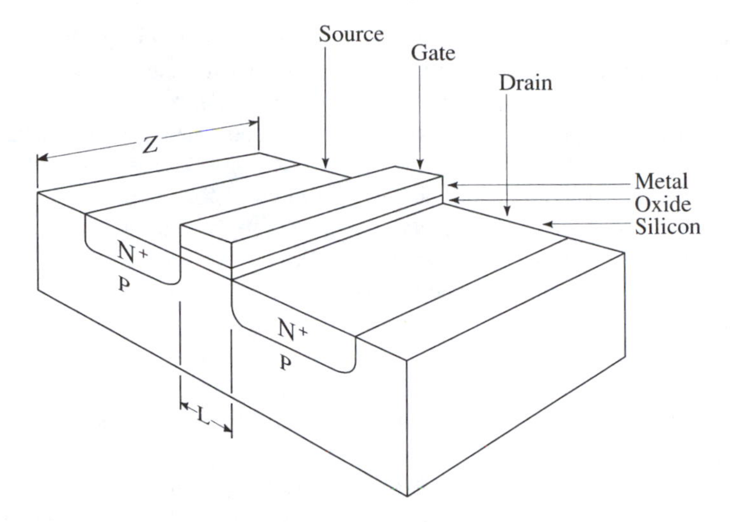

Figure 1–1 Essential structure of a MOSFET.

tracted to the interface, thus creating an inversion layer. This thin layer constitutes the resistor cited earlier and also the MOSFET *channel.* Ohmic contact is made to the ends of the channel by means of heavily doped *N*-type regions, usually formed by ion implantation or solid-phase diffusion, or some combination of the two.

It is evident in Figure 1-1 that the MOSFET exhibits bilateral symmetry. That is, channel carriers can be made to flow in either direction, depending on bias polarity. This symmetry is important in certain MOSFET applications. The two channel terminals have been named according to which one supplies carriers to the channel and which one receives carriers from it. These two terms are *source* and *drain*, respectively, and it is evident that bias polarity determines which is which. This choice of terms has been carried over from the JFET [8], and was made there to provide distinctive initial letters for subscript purposes. (The physicist's classically favored terms *source* and *sink* are obviously wanting with respect to this requirement.) To summarize, for the *N*-channel device under consideration, the more positive of the two channel terminals is the drain and the other is the source.

It is customary to refer all voltages in the MOSFET to the source voltage (just as in an earlier era, terminal voltages in a vacuum tube were referred to that of the cathode). Thus, two important bias values are the drain–source voltage V_{DS} and the gate–source voltage V_{GS}, using the double-subscript notation described in Appendix A. But the substrate constitutes a fourth terminal, and a bias from substrate to source indeed alters device properties; this is a situation that sometimes cannot be avoided, especially in integrated circuits. To deal with it, we need at the most simplistic level a distinctive subscript for "substrate" too, and so the homely term *bulk* has been adopted as an approximate synonym. It is intended to connote the interior of the substrate. In Chapter 4 we shall examine the effect of a bulk–source voltage V_{BS}. But for the initial description, let us assume that the source and substrate are electrically common, to serve as voltage reference. Further, for the moment let $V_{DS} = 0$.

There is a certain voltage V_{GS} at which the hole population immediately under the oxide–silicon interface will be identical to that deep within the bulk of the crystal. Under such conditions, a band diagram representing conditions from the interface to the interior of the P-type silicon crystal would have perfectly horizontal band edges. This magic gate-bias value is termed the *flatband voltage*, $V_{GS} = V_{FB}$, and is valuable as a conceptual starting point. When the gate terminal is now made more positive with respect to source (and substrate), the band edges begin to bend near the oxide–silicon interface. This matter will be examined in detail. At the moment, what is important is that potential ψ_S at the surface of the silicon crystal (at the oxide–silicon interface) is becoming more positive through the influence of the field plate. Hole density at the interface declines and electron density rises. This is dictated by the law of mass action, $pn = n_i^2$, because the silicon remains (for engineering purposes) at equilibrium. The reason for the caveat is that any leakage current through the oxide is accompanied by a net current in the silicon that violates equilibrium. But the superb insulating qualities of SiO_2—resistivity in excess of 10^{15} ohm·cm—make this a vanishing concern at present.

There exists another critical voltage at which the *electron* density right at the interface is equal to the hole density in the bulk. This condition is termed the *threshold of strong inversion* and is specified by the statement $V_{GS} = V_T$. The chosen term is apt, because channel conductance increases steeply with gate voltage for $V_{GS} > V_T$. At the threshold voltage and just below it, the electrons that form the incipient channel contribute a very small amount of drain–source conduction, leading to *subthreshold current*. For still smaller values of V_{GS}, there exists essentially an open circuit from drain to source because the applied voltage V_{DS} reverse biases the drain junction. Only a tiny fraction of V_{DS} appears as forward bias on the source junction. Thus to a good approximation, one can say that the channel conductance is "turned on" by V_{GS} at the threshold voltage V_T.

1-3 Rudimentary Analysis

Figure 1-2 shows explicitly the electrical commonality of source and substrate and their joint use as voltage reference. This time, however, we choose a value of V_{GS} well above V_T, as indicated by the symbol "++". But drain–source bias is small ("+"). Thus we have conditions in which there is only a small current in the channel, I_{CH}, consisting of electrons drifting from source to drain. The IR drop in the channel from drain to source amounts to V_{DS}, and by requiring the condition $V_{DS} \ll V_{GS}$ we can arrange to have the channel be nearly uniform in conductivity from one end to the other. That is, its areal density of electrons will be about the same at the source and drain ends.

To deal with the conducting properties of a thin and uniform sheet, it is useful to introduce a concept known as *sheet resistance*. This is especially so when the layer or sheet is nonuniform in the thickness direction, which, as Figure 1-2 shows, we take to be the x direction. Such is the case with the inversion layer that serves here as channel. Electron density peaks just under the oxide and then declines monotonically with increasing x. Having said this, we shall assume temporarily that the sheet of interest *is* uniform in the x direction, a restriction that will be relaxed after just two equations.

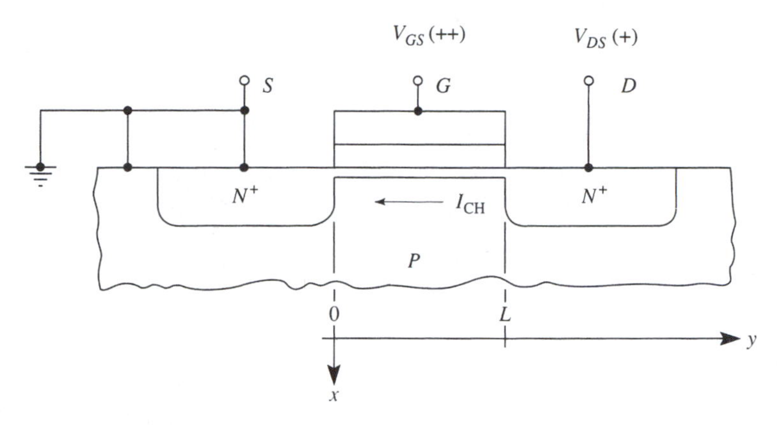

Figure 1–2 Cross-sectional diagram of a MOSFET under bias, showing axis assignments. Gate–source voltage is sufficient to produce an inversion layer; drain–source voltage is very small. The x direction is into the sample, with spatial origin at the silicon surface. The y direction is from source to drain, with origin at the source end of the channel. The z direction is out of the paper.

Further, let us use the symbol Z for the extent of the channel in the z direction, a common practice in MOSFET work.

Consider a uniform layer like that depicted in Figure 1-3(a) that has a resistivity ρ, a thickness X, and two other equal dimensions L. Since the cross-sectional area presented to the entering current is $A = XL$, it is evident from the discussion in Appendix B, and from Equation B1 in particular, that the resistance of this thin sample is

$$R = \rho \frac{L}{A} = \frac{\rho L}{XL} = \frac{\rho}{X}. \tag{1-1}$$

Thus it is evident that the resistance of a uniform square sample is independent of its lateral dimensions, depending only on resistivity and thickness. (It does not matter whether L is 1 micrometer or 1 mile!) This invariant quantity is designated sheet resistance R_S. It is measured in ohms. We sometimes specify it in "ohms per square" to label the quantity cited as a sheet resistance, but "square" has no dimensional-analysis significance. Because R_S is an intensive quantity, like bulk resistivity ρ, it is sometimes termed sheet *resistivity*. Either designation is correct.

EXERCISE 1–1 Explain qualitatively how it can be possible to have a 1-micrometer square exhibit the same side-to-side resistance as a 1-mile square having the same resistivity and thickness.

■ Increasing the size of the square places increasing amounts of material in the series path, but also in the parallel paths in a way that holds resistance constant.

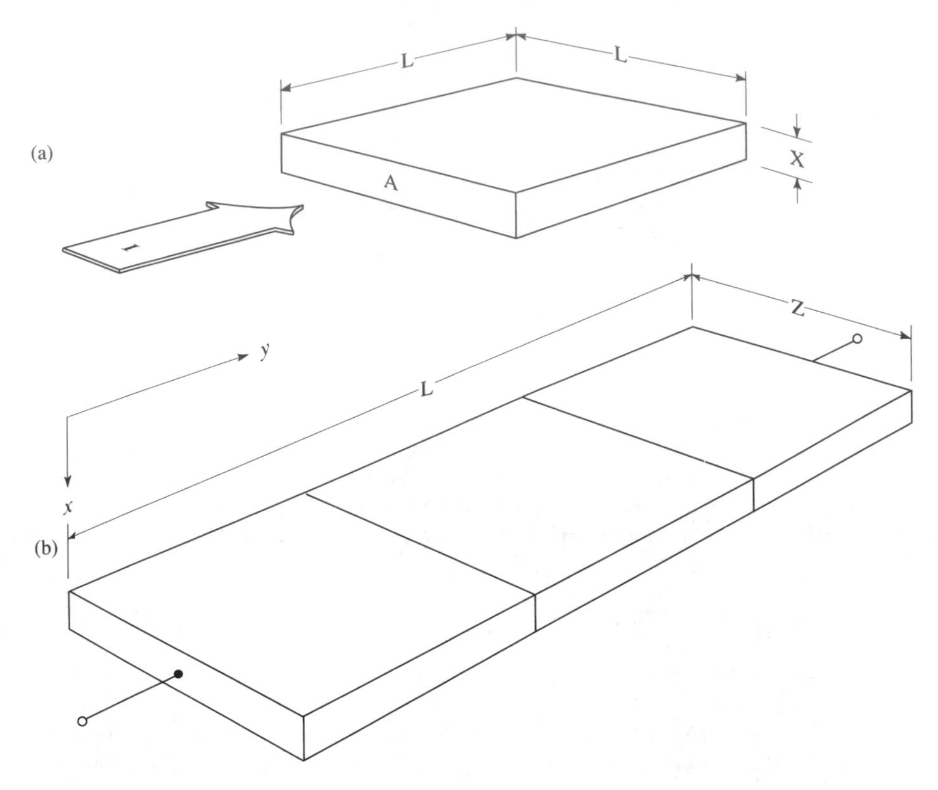

Figure 1–3 The sheet-resistance concept. (a) Current passes through a square layer of thickness X, entering through one edge of area $A = LX$. (b) The resistance of a rectangular sample is found by multiplying the sheet resistance by the aspect ratio of the sample L/Z.

Now turn to the rectangular resistor in Figure 1-3(b). Given $R_S \equiv \rho/X$, it is evident that this time we may write

$$R = R_S(L/Z). \tag{1-2}$$

In this example, three squares are connected in series, a number provided by the ratio L/Z, and the resistance of each square is R_S, so the resulting resistance is $R = 3R_S$. Had the squares been in parallel, we would have had $(L/Z) = 1/3$, and thus the resistance this time would have been $R = R_S/3$. The convenience of working with R_S in lieu of resistivity and dimensions is that *now* we need only be concerned about layer uniformity in the lateral directions. The function $\rho(x)$ is completely arbitrary; the properties of the layer need only be independent of y and z.

Now let us apply the sheet-resistance formulation to the problem posed by the MOSFET channel in Figure 1-2. Once again it is convenient to assume that we deal with a conducting layer of fully uniform properties, and then to relax the requirement later. Given a channel with a well-defined thickness X and a volumetric electron density n, we can write its areal charge density with due regard for algebraic sign as

$$Q_n = -qnX. \tag{1-3}$$

Then using $R_S \equiv \rho/X$ to eliminate X, we have

$$Q_n = \frac{-qn\rho}{R_S}. \tag{1-4}$$

But recalling that $\rho = 1/q\mu_n n$, we can rewrite Equation 1-4 as

$$Q_n = \frac{-qn}{qn\mu_n R_S} = \frac{-1}{\mu_n R_S}. \tag{1-5}$$

This relationship (admittedly nonobvious) eliminates all reference to channel resistivity and thickness; thus, realistic inversion-layer properties are now fully permitted. That is, the steep variation of $n(x)$ and hence $\rho(x)$ poses no problems.

Next we write a different expression for areal charge density in the inversion layer, one that takes advantage of the fact that the MOS sandwich structure in Figures 1-1 and 1-2 is very nearly a conventional parallel-plate capacitor. Its only modest departure from such description is that the inversion layer has a thickness of the order of a few hundred angstroms rather than the few angstroms that characterize the charge-layer thickness in a metal capacitor plate. Let us assume that inversion-layer formation commences abruptly at $V_{GS} = V_T$, which proves to be an excellent working approximation. Then the resulting expression is

$$Q_n = -C_{OX}(V_{GS} - V_T), \tag{1-6}$$

where C_{OX} is capacitance per unit area for the MOS capacitor. Finally, combining Equations 1-3 and 1-6 gives us an expression for channel sheet resistance in terms of MOSFET real physical properties and one terminal voltage, and independent of artificial quantities such as ρ and X that were temporarily assumed for the channel:

$$R_S = \frac{1}{\mu_n C_{OX}(V_{GS} - V_T)}. \tag{1-7}$$

EXERCISE 1–2 Why is there a minus sign on the right-hand side of Equation 1-6? The quantities C_{OX} and $(V_{GS} - V_T)$ are both positive, and the minus sign that accounts for negative electron charge is subsumed in Q_n, as Equation 1-3 shows.

■ For algebraic sign consistency in the capacitor law, $Q = CV$, it is necessary to examine the charge Q on the plate *to which* the voltage V is applied. But in Figure 1-2, we apply the voltage to the top plate, or gate, while the charge of interest is the inversion layer on the bottom plate. Thus the extra minus sign is needed for algebraic-sign consistency.

Next let us acknowledge that in the general case there *will* be a variation in channel properties from source to drain because of the unavoidable *IR* drop accompanying I_{CH}. This situation is represented in Figure 1-4. Note the polarities involved. Both V_{GS} and V_{DS} are positive, and $V_{DS} < V_{GS}$ but V_{DS} is no longer negligible. Because the drain end of the channel is more positive than the source end, and because the conducting channel plays the role of a capacitor plate, it follows that there is less voltage drop from gate to channel at the drain end than there is at the source end. However, in this instance the values of V_{GS} and V_{DS} are adjusted to ensure that above-threshold conditions exist even at the drain end of the channel. That is, the inversion layer is continuous from source to drain. Furthermore, the channel has been made long enough so that its properties (especially areal charge density) change slowly from source to drain. This structural choice is made so that the channel will exhibit near-one-dimensional properties—so that its variation in the y direction will be negligible compared to that in the x direction. (No variation exists in the z direction.) The analogous specification was made by Shockley in his approximate analysis of the JFET [8], and is known as his *gradual-channel* approximation. In recent times, the same approximation or assumption is often identified through the adjective *long-channel.*

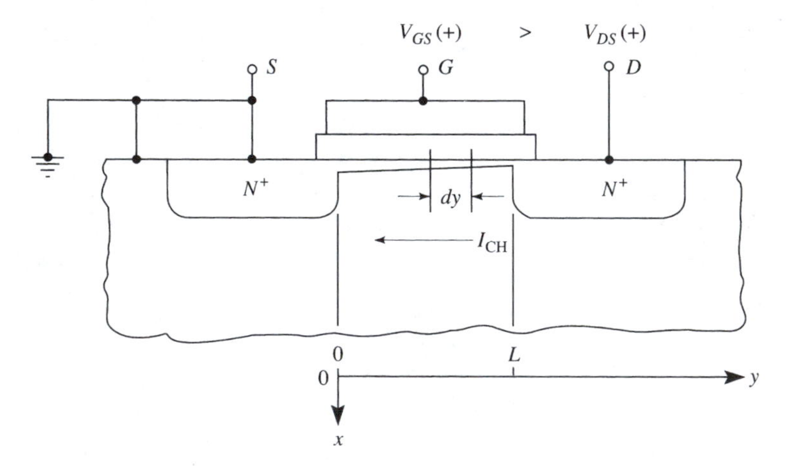

Figure 1-4 Cross-sectional representation of a long-channel MOSFET with significant gate–source voltage and drain–source voltage, but with $V_{GS} > V_{DS}$ sufficiently to ensure that an inversion layer exists at the drain end of the channel.

Now let the symbol $V(y)$ represent the "voltmeter voltage" in the channel, varying from $V(0) = 0$ to $V(L) = V_{DS}$ for the biasing conditions represented in Figure 1-4. Note, too, that the channel "thickness" variation shown there heuristically is to represent the monotonically declining charge density in the channel as the drain end is approached, a variation directly related to $V(y)$, which in turn stems from the aforementioned IR drop. The next step is to focus on a small channel element of length dy located at the arbitrary position y, as also shown in Figure 1-4. This element exhibits a sheet resistance $R_S(y)$, which also increases monotonically from source to drain. At the source end $R_S(y)$ is fixed by the voltage $(V_{GS} - V_T)$ and is given by Equation 1-7. At the drain end it is fixed by the smaller voltage $(V_{GS} - V_T - V_{DS})$. And at the arbitrary position y, $R_S(y)$ is fixed by the intermediate voltage $[V_{GS} - V_T - V(y)]$, so the appropriate modification of Equation 1-7 is

$$R_S(y) = \frac{1}{\mu_n C_{OX}[V_{GS} - V_T - V(y)]}. \tag{1-8}$$

The channel element of length dy exhibits a resistance dR that can be written immediately using the sheet-resistance formulation as

$$dR = \frac{dy}{Z} R_S(y) = \frac{dy}{Z\mu_n C_{OX}[V_{GS} - V_T - V(y)]}, \tag{1-9}$$

where Z, once more, is the dimension of the channel in the z or width direction. Now simplify the notation by letting $V(y) \equiv V$. Using Ohm's law we may write

$$I_{CH} = -\frac{dV}{dR}, \tag{1-10}$$

thus bringing us to the relationship sought between current and voltage.

EXERCISE 1–3 Why is there a minus sign in Equation 1-10?

■ The Ohm's-law convention, as noted in Appendix A, defines a voltage *drop* as positive. This is evident in the present example, wherein I_{CH} is negative (leftward) while V increases toward the right.

EXERCISE 1–4 Is I_{CH} a function of y, as are V and R_S?

■ No. The inversion-layer electrons that compose the channel are confined in a potential well from which they cannot escape except at a channel end.

1-4 Current-Voltage Equations

The desired expression for output current as a function of input and output voltages can now be derived by combining Equations 1-9 and 1-10, which yields

$$I_{\mathrm{CH}} = -\frac{Z\mu_n C_{\mathrm{OX}}(V_{GS} - V_T - V)\, dV}{dy}. \tag{1-11}$$

Separating variables and applying appropriate limits gives

$$I_{\mathrm{CH}}\int_0^L dy = -Z\mu_n C_{\mathrm{OX}}\int_0^{V_{DS}} (V_{GS} - V_T - V)\, dV, \tag{1-12}$$

and completing the integration yields

$$I_{\mathrm{CH}} = -\frac{\mu_n C_{\mathrm{OX}}}{2}\frac{Z}{L}[2(V_{GS} - V_T)V_{DS} - V_{DS}^2]. \tag{1-13}$$

Finally, then, after noting that $I_D = -I_{\mathrm{CH}}$, we have

$$I_D = \frac{\mu_n C_{\mathrm{OX}}}{2}\frac{Z}{L}[2(V_{GS} - V_T)V_{DS} - V_{DS}^2]. \tag{1-14}$$

EXERCISE 1–5 Justify the last change of algebraic sign.

■ The leftward (and hence negative) current I_{CH} results in an *inward*, and hence positive, terminal current I_D.

Letting V_{GS} be a parameter, we can use Equation 1-14 to plot in the MOSFET *output-plane* curves of drain current I_D versus drain–source voltage V_{DS}. The result is shown in Figure 1-5(a). Inspection of Equation 1-14 shows that each curve is a parabola with vertex displaced from the origin. Only the solid portions of the parabolas have physical meaning, and consequently only the shaded portion of the output plane is described by Equation 1-14. The relevant portion can of course be expanded by assigning still larger values to the parameter V_{GS}.

EXERCISE 1–6 Compare the three parabolas shown in Figure 1-5 with respect to their proportions.

■ Each is the *same* parabola, as Figure 1-5(a) suggests and Equation 1-14 confirms. The vertex is translated when V_{GS} is changed, so that more or less of the parabola has meaning.

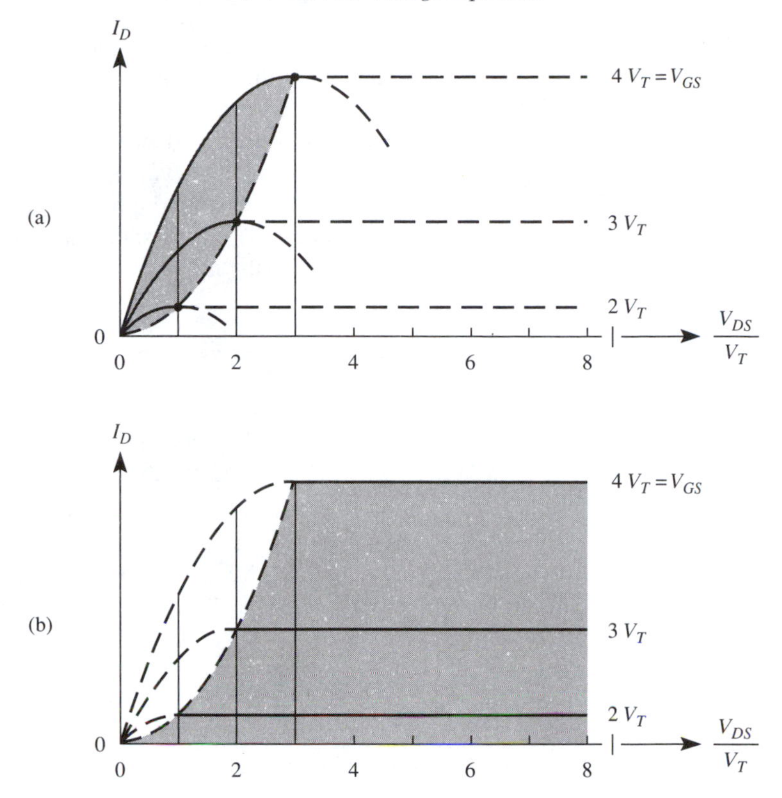

Figure 1–5 MOSFET output-plane characteristics predicted by rudimentary analysis. The input-voltage parameter V_{GS} and the output voltage are normalized using V_T. (a) The *curved* regime of operation (shaded). (b) The *saturation* regime of operation (shaded).

The physical quantities in the coefficient of Equation 1-14 deserve further description. The electron mobility μ_n in an inversion layer is typically two or three times smaller than the bulk mobility of a sample having the same doping as the substrate. At least a partial explanation for this fact is the diminished mean free path of the electron when it is confined in the very thin potential well. Its random motion is converted from free three-dimensional motion to a nearly two-dimensional confinement, and it is intuitively obvious that this kind of restriction will diminish average free-path length, with a consequent reduction of mobility. The oxide capacitance per unit area is given by

$$C_{\text{OX}} = \frac{\epsilon_{\text{OX}}}{X_{\text{OX}}}, \tag{1-15}$$

where

$$\epsilon_{\text{OX}} = \frac{\epsilon}{3}, \tag{1-16}$$

with ϵ being the permittivity of silicon. The oxide thickness X_{OX} is typically about 0.05 μm. Finally, as noted before, Z and L are the lateral dimensions of the channel, with L being in the current direction.

In much of the early literature, the symbol β was used to stand for the expression $\mu_n C_{OX} Z/L$, so that the coefficient of Equation 1-14 became $\beta/2$. But because today MOSFETs and BJTs are being combined in the same integrated circuit with increasing frequency, that choice is poor. Confusion with the BJT β, or static current gain, is likely. Therefore, let us use the symbol K to stand for the entire coefficient:

$$K \equiv \frac{\mu_n C_{OX}}{2} \frac{Z}{L}. \tag{1-17}$$

Thus Equation 1-14 becomes

$$I_D = K[2(V_{GS} - V_T)V_{DS} - V_{DS}^2]. \tag{1-18}$$

EXERCISE 1–7 Evaluate the coefficient K for a MOSFET in which $\mu_n = 580$ $\text{cm}^2/\text{V}\cdot\text{s}$, $X_{OX} = 500$ angstroms, and $(Z/L) = 1$.

■ Evidently,

$$C_{OX} = \frac{\epsilon}{3X_{OX}} = \frac{1.035 \times 10^{-12}\ \text{F/cm}}{3(500 \times 10^{-8}\ \text{cm})} = 6.9 \times 10^{-8}\ \text{F/cm}^2.$$

Thus

$$K = \frac{\mu_n C_{OX}}{2} \frac{Z}{L} = \frac{1}{2}(580\ \text{cm}^2/\text{V}\cdot\text{s})(6.9 \times 10^{-8}\ \text{F/cm}^2) = 20\ \mu\text{A/V}^2.$$

We shall identify the operating regime indicated in Figure 1-5(a) as the *curved* regime. Portions of the literature use the term "linear" to identify the same region of the output plane, but such a characterization is patently a poor way to describe a parabola. On the other hand, for operation extremely near the origin, linear approximations are meaningful and justified. Such operation is sometimes advantageous.

Physical reasoning was applied to the FET by Shockley (before any working devices had been fabricated), with the prediction that current would be essentially independent of V_{DS} after the value corresponding to the parabolic vertex was exceeded. This prediction was soon confirmed experimentally. The resulting operating regime for the MOSFET is shown in Figure 1-5(b), and is usually designated the *saturation* regime. The term is used here in the sense of *leveling off*, or becoming constant. It is both more important than the curved regime and easier to describe analytically. Given that each constant-current curve of the saturation regime corresponds to the vertex or maximum of a parabola, one can simply differentiate Equation 1-18 with respect to V_{DS} to obtain the key relation. The result is

$$\frac{\partial I_D}{\partial V_{DS}} = -K[2(V_{GS} - V_T) - 2V_{DS}]. \tag{1-19}$$

Setting the right-hand side equal to zero gives

$$V_{DS} = V_{GS} - V_T. \tag{1-20}$$

Thus the curve passing through the parabolic vertices in Figure 1-5 is the locus of points for which Equation 1-20 holds. This curve, plotted in Figure 1-6(a), is itself a parabola, and is in fact identical to the other parabolas, but with different orientation. It is rotated 180° in the plane of the paper with respect to the *I–V* curves.

EXERCISE 1–8 Why is the curve passing through the parabolic vertices of the output characteristics described as a "locus of points" for which Equation 1-20 is valid? Why do we not simply describe Equation 1-20 instead as "the equation for" the curve in Figure 1-6(a)?

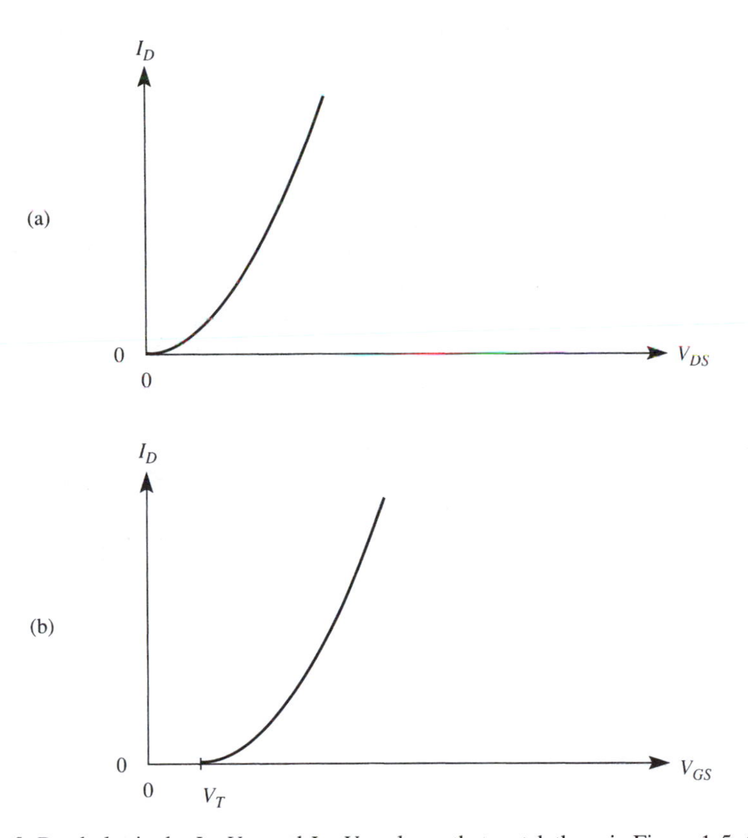

Figure 1–6 Parabolas in the I_D–V_{DS} and I_D–V_{GS} planes that match those in Figure 1-5. (a) Locus of the parabolic vertices in Figure 1-5, conforming to $V_{DS} = V_{GS} - V_T$, as Figure 1-5 confirms. (b) Current-to-voltage transfer characteristic of the MOSFET.

■ Equation 1-20 incorporates the variables V_{DS} and V_{GS}. But the equation for the curve in Figure 1-6(a) must involve the variables I_D and V_{DS}. (See Problem A6.)

Substituting Equation 1-20 into Equation 1-18 yields

$$I_D = K[2(V_{GS} - V_T)(V_{GS} - V_T) - (V_{GS} - V_T)^2], \tag{1-21}$$

or

$$I_D = K(V_{GS} - V_T)^2. \tag{1-22}$$

As noted earlier, operation of the MOSFET in the saturation regime is the most common and the most important situation. Equation 1-22 indicates that this operation is governed by an extremely simple relation, and also explains why the MOSFET is often described as a *square-law device*. Equation 1-22 is plotted in Figure 1-6(b), and is the current-to-voltage *transfer characteristic* for the MOSFET operating in the saturation regime. This transfer characteristic is identical in shape to the curve plotted in Figure 1-6(a) for equivalent calibration of the voltage axes.

1–5 Universal Transfer Characteristics

Treating the MOSFET simplistically as a parallel-plate capacitor with a lateral current and a resulting *IR* drop in one plate neglects important considerations, as we shall see. Nonetheless, Equation 1-22 is extremely useful. Figure 1-7 presents experimental data from long-channel MOSFETs of the early 1960s. Plotting $\sqrt{|I_D|}$ versus $|V_{GS}|$ should yield a straight line if Equation 1-22 is valid. This treatment has been accorded to two MOSFETs fabricated simultaneously, but having very different aspect ratios *Z/L*. Both curves have extensive linear regions. The curvature near the bottom, most evident in the higher-current device, exists because some conduction begins below the threshold voltage V_T, the phenomenon labeled *subthreshold conduction.* Nonetheless, because the two devices are identical in all respects except aspect ratio, extrapolating the linear portions of their *I–V* curves leads to a consistent and meaningful theoretical value of V_T. The value of 7 V observed for these early devices is extremely high by today's standards, and was the result of fabrication problems. A typical value today is of the order of 1 V.

The device described up to this point is known as the *enhancement-mode*, or *E-mode*, MOSFET. It is designated thus because it has no channel at equilibrium. A gate voltage is required to create the inversion layer that serves as the channel, and further increments of the gate voltage *enhance* the conductivity of the channel.

It is possible, however, to create a channel that exists even in the device at equilibrium. This is usually done today by conventional doping of the region just under the gate oxide, with ion implantation being an especially favored technique. The result is a structure somewhat like that shown in cross section in Figure 1-8. The channel is

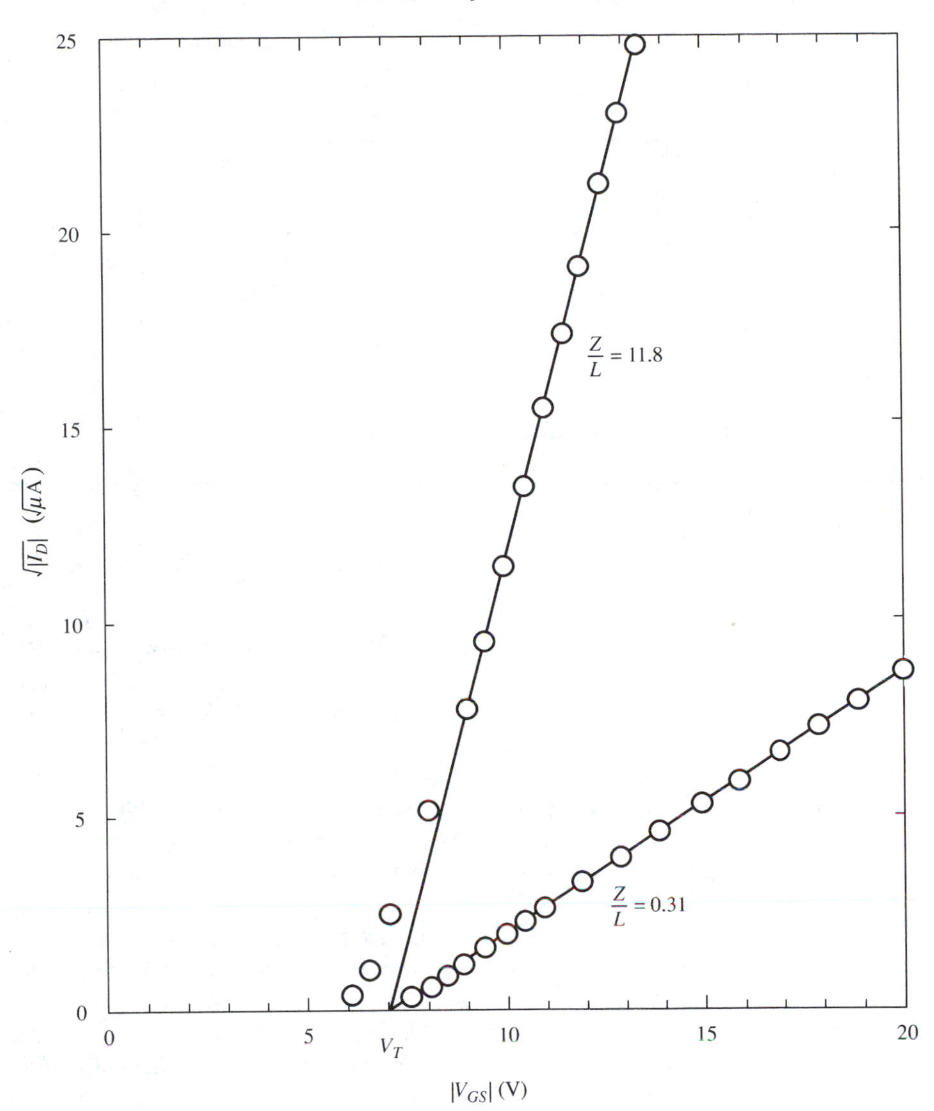

Figure 1–7 Experimental data from the long-channel MOSFETs of the early 1960s demonstrating the qualitative correctness of Equation 1-22.

bounded on the bottom by a *PN* junction, just as in a JFET. This time a *negative* voltage is applied to the gate, so that the MOS capacitor requires positive charge in the silicon. Donor-ion charge fulfills that requirement. In other words, a depletion layer is formed in the *N*-type channel region. Once again a positive V_{DS} gives rise to an *IR* drop in the channel. This time it results in a thicker depletion layer at the drain end of the channel than at the source end. For the same reason, the depletion layer of the *PN* junction below the channel is also thicker at the drain end of the channel. Because the

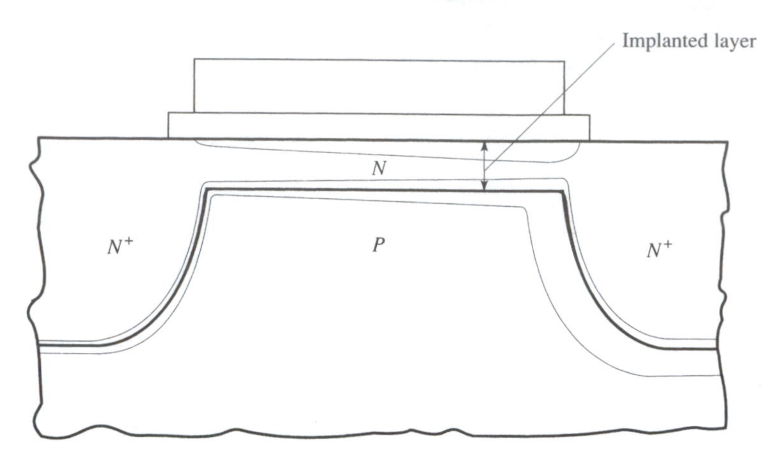

Figure 1-8 Schematic cross-sectional representation of a D-mode *N*-channel MOSFET, with a channel produced by ion implantation.

application of a negative voltage to the gate causes current reduction through channel depletion, this kind of device is termed a *depletion-mode,* or *D-mode*, MOSFET. Realized in the way just described, the D-mode MOSFET is a kind of hybrid device, with a MOSFET-like upper boundary for the channel and a JFET-like lower boundary.

Although the channel region in the D-mode MOSFET is a bit thicker than the inversion-layer channel of the E-mode device, the D-mode MOSFET channel still qualifies as "nearly two-dimensional," and as a result, the D-mode MOSFET is also very nearly a square-law device. Its essentially parabolic current-to-voltage transfer characteristic looks like that for the E-mode device, but is translated leftward along the V_{GS} axis, as can be seen in Figure 1-9(a). Thus the D-mode device has two structure-determined electrical constants. There is the characteristic current I_{DSS} at which the output current saturates when $V_{GS} = 0$. Then there is the value $V_{GS} = V_P$ that is required to "pinch off" the channel completely, so that no current can flow. Under these conditions, the depletion layer produced by gate voltage has moved down to touch that of the *PN* junction, all the way from source to drain; as a result, the channel is fully depleted of carriers. The output characteristics that accompany the transfer characteristic in Figure 1-9(a) are shown in Figure 1-9(b). Only the parameter labeling of the curves distinguishes them from the E-mode characteristics plotted in Figure 1-5.

There is an important aspect in which the D-mode MOSFET is distinguished from the JFET, with the JFET being also a D-mode device. The D-mode MOSFET possesses an enhancement regime of operation that augments its depletion regime. When positive gate bias V_{GS} is applied, additional electrons are brought into the channel from the source, and channel conductivity increases. This fact is represented in Figure 1-9(a) by the extension of the transfer characteristic to the right of the I_D axis, or into the regime of positive V_{GS}. The square-law behavior continues. Thus there are additional curves at higher current that can be drawn in Figure 1-9(b). In the JFET, by contrast, where the channel is bounded both top and bottom by *PN* junctions, positive bias on the *P*-type gates (the top and bottom gates are usually common) will forward bias

(a)

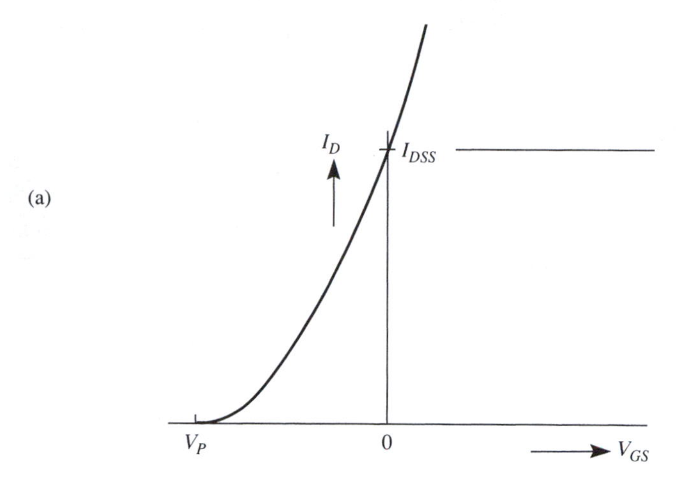

(b)

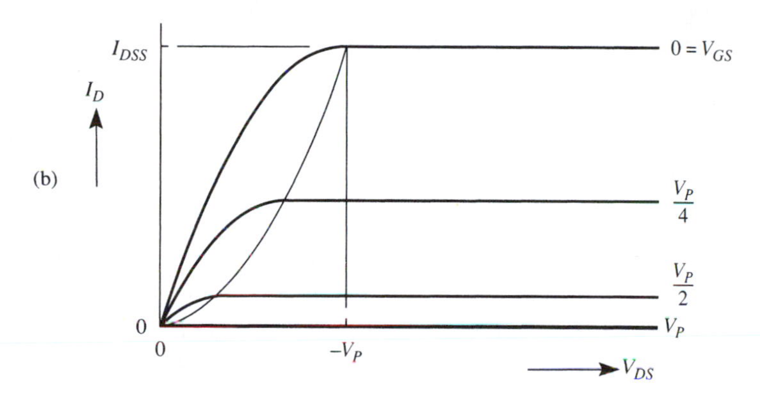

Figure 1–9 Current–voltage characteristics of a D-mode MOSFET. (a) Transfer characteristic, displaying the structure-determined constants V_P and I_{DSS}. (b) Output characteristics.

the gate junctions. Thus, for a gate bias of more than a few tenths of a volt, the gate junctions will conduct, and the high input resistance of the device is lost, normally an undesirable situation. As a practical matter, therefore, the transfer characteristic of the N-channel JFET can be extended only a short way into the positive-V_{GS} regime that defines enhancement-mode operation.

EXERCISE 1–9 Describe the output characteristics of the JFET with forward bias on the gate.

■ For small V_{GS} values, they would step upward just like those of the D-mode MOSFET, yielding a limited but sometimes useful E-mode regime. But for larger values, they would be closely spaced and would be accompanied by current from gate to source.

EXERCISE 1–10 Suppose that a large *negative* value of V_{GS} were applied to the D-mode MOSFET, causing the creation of an inversion layer of holes just under the oxide. Would the device start to conduct again, exhibiting a new regime of operation?

■ No. Extreme negative bias on the gate can indeed form an inversion layer of holes. But they will be unable to move to source, drain, or substrate, and will just "sit there," being confined in an isolating well, and not affecting terminal electrical properties.

Thus far we have considered only *N*-channel MOSFETs. But it is evident that by reversing all conductivity types, currents, and voltages, we can have E-mode and D-mode *P*-channel devices. In fact, for technological reasons *P*-channel devices were dominant early in the MOSFET era. The data in Figure 1-7, for example, were obtained from *P*-channel devices. It is possible to construct a transfer-characteristic diagram that summarizes all four possibilities—E and D, and *N* and *P*. Normalization once more is convenient. The D-mode device is best treated first because it possesses a characteristic current I_{DSS} as well as a characteristic voltage V_P. Using these quantities for normalization leads to the second-quadrant curve in Figure 1-10 for the D-mode *N*-channel MOSFET. It is simply a normalized version of the curve shown in Figure 1-9(a). Since the D-mode *P*-channel MOSFET involves reversing currents and voltages, we merely construct a second curve in the fourth quadrant, one that is symmetric in the origin to the second-quadrant curve.

While the E-mode device does not have an explicit characteristic current for normalization, it does have an implicit characteristic current, $I_D = KV_T^2$. Using this and the voltage $V_{GS} = V_T$ for normalization, we complete Figure 1-10, displaying four curves that represent all possible MOSFETs.

EXERCISE 1–11 Explain the implicit characteristic current used to construct Figure 1-10.

■ Refer back to the unnormalized transfer characteristic in Figure 1-6(b). Take the case of the E-mode *N*-channel device. Its parabolic transfer characteristic has a leftward branch that is without physical meaning, but that serves to define a characteristic current because it intersects the I_D axis. Its intercept on the current axis is the same as the current value at $V_{GS} = 2V_T$, by virtue of parabolic symmetry. Substituting $V_{GS} = 2V_T$ into $I_D = K(V_{GS} - V_T)^2$ yields $I_D = KV_T^2$.

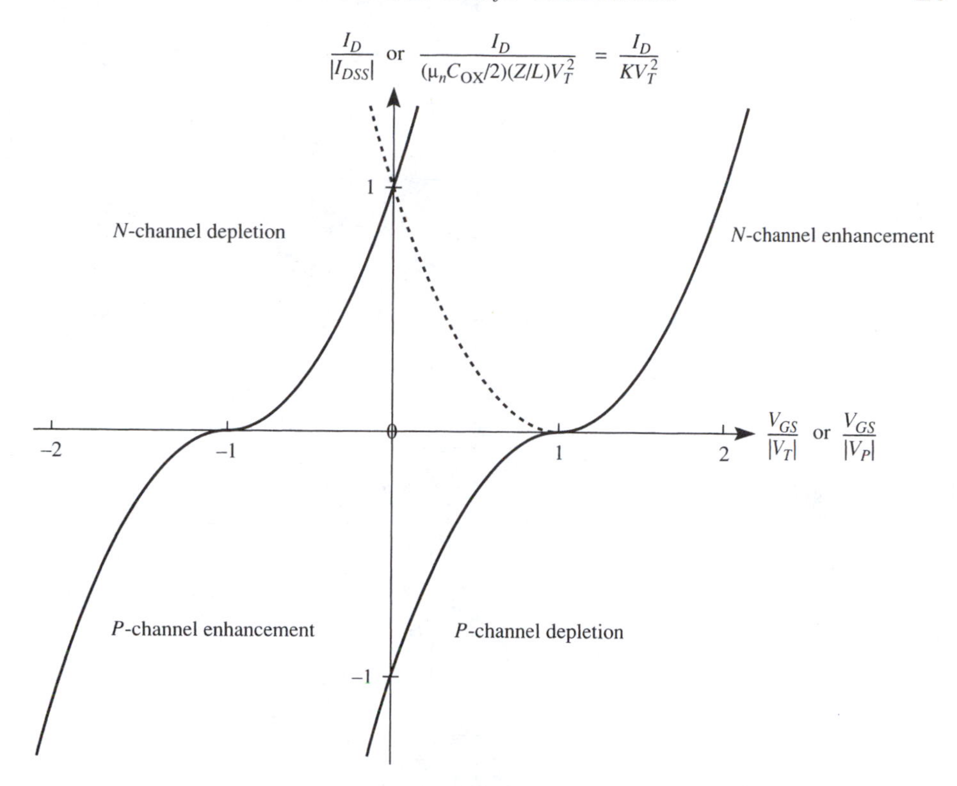

Figure 1–10 Universal transfer characteristics for the four possible kinds of MOSFET. Universality is achieved by normalization using V_P and I_{DSS} for the D-mode devices and V_T and KV_T^2 for the E-mode devices.

Performance advantages result from combining E-mode and D-mode devices in a circuit, and such *E–D* products are common today. However, the combination of *complementary* E-mode devices, or *N*-channel and *P*-channel E-mode devices, is even more advantageous. In fact, complementary MOSFET circuits, or *CMOS* circuits, constitute today's most rapidly growing technology. These matters are considered further in Section 1-7.

A number of MOSFET symbols are in common use. Three informative kinds are shown in Figure 1-11. A distinction between D-mode and E-mode devices is sometimes maintained by using a solid line to represent the channel in the former case and a broken line for the E-mode case, as in Figure 1-11(a). Another symbol, as in Figure 1-11(b), acknowledges that the substrate constitutes a fourth terminal of the MOSFET. Consistent with conditions in this section, we show the substrate connected (with a dotted line that is not part of the symbol) to the source. This symbol also indicates an *N*-channel device, with the arrow direction, as usual, from *P* to *N*, or from the *P*-type substrate to the channel. Finally, one sometimes wishes to denote a "preferred source" in spite of the intrinsic symmetry of the MOSFET, and this can be done by offsetting the gate terminal in the source direction.

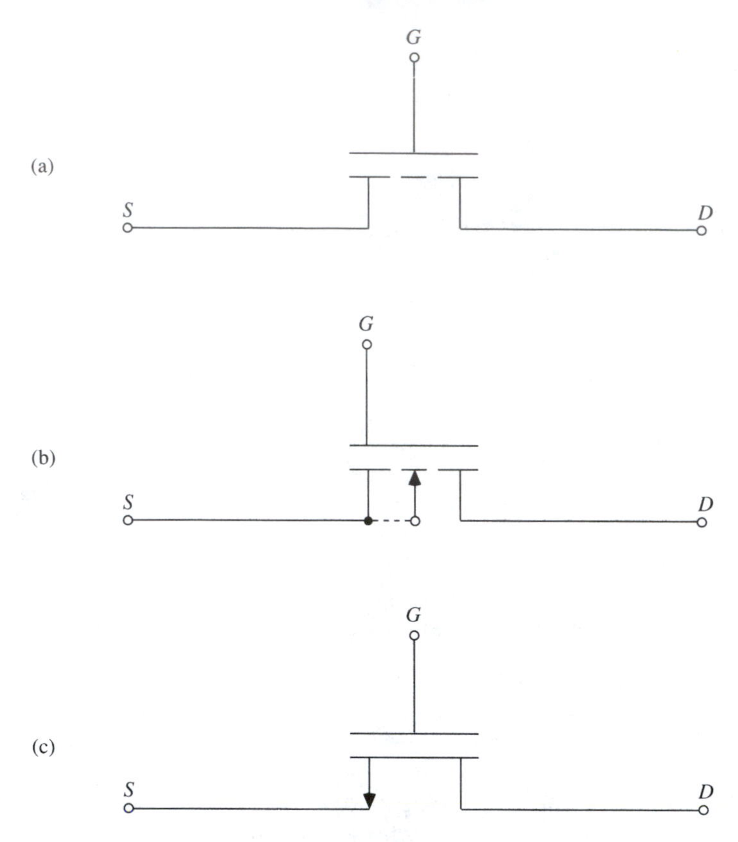

Figure 1–11 MOSFET symbols, with orientation corresponding to the diagrams in Figures 1-2 and 1-4. (a) E-mode device, indicated by the broken line representing the channel. D-mode device can be indicated by substituting a solid line. (b) More detailed symbol, indicating an *N*-channel device (arrow points from *P*-type substrate to channel), the substrate as a potential fourth terminal, and the "preferred source." (c) MOSFET analog of BJT symbols using arrow to designate source (emitter) terminal, and current direction.

A third symbol in common use is shown in Figure 1-11(c). Its virtues are simplicity and parallelism with the BJT symbol. That is, an arrow is used to designate the source terminal and also to indicate normal current direction, just as an arrow designates the BJT emitter terminal and its normal current direction. The line connecting source and drain represents the channel, and is sometimes made thicker to distinguish a D-mode from an E-mode device.

1–6 Transconductance

The rate of change of output current from a device with respect to its input voltage is an important characteristic property and is termed *transconductance*, g_m. For the MOSFET,

$$g_m \equiv \frac{\partial I_D}{\partial V_{GS}}. \tag{1-23}$$

The partial derivative has conductance dimensions and has "across" or *trans* properties, to repeat, because it describes a result at the output port produced by a change at the input port. Transconductance exhibited in the saturation regime is of particular importance, and is of course simpler than for the curved regime. Differentiating Equation 1-22 yields

$$g_m = 2K(V_{GS} - V_T). \tag{1-24}$$

Experimental data confirming this simple linear relationship are shown in Figure 1-12, once again drawing data from *P*-channel devices of the early 1960s. The symbol S in Figure 1-12 stands for *siemens*, the unit of conductance, identified formerly as the *mho*.

Transconductance has great importance because it plays a large part in determining the switching speed of an active device, topics discussed in Section 1-7 and Chapter 6. High-transconductance devices yield circuits capable of high-speed operation. Equations 1-17 and 1-24 make it evident that MOSFET transconductance is fixed by structure and technology. That is, it depends on properties more or less under the designer's control. This explains the preference for "*N*-MOS" over "*P*-MOS," to take ad-

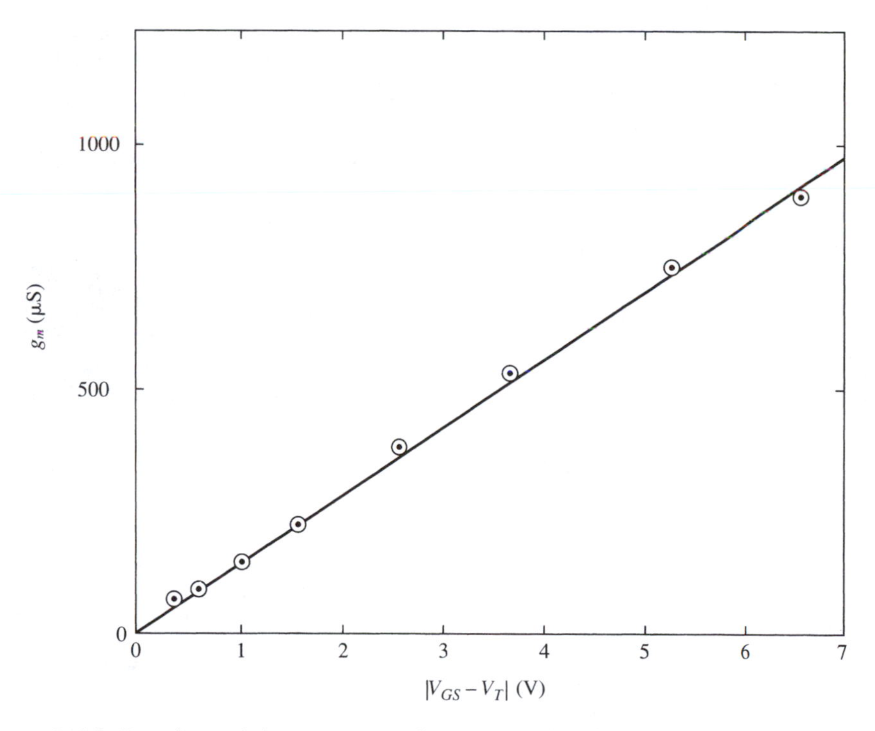

Figure 1-12 Experimental data on transconductance as a function of gate voltage, showing the linear relationship predicted by simple theory.

vantage of the higher electron mobility. It is partly responsible for the constantly declining values of X_{OX} in recent decades (thus increasing C_{OX}). Finally, the presence of the factor Z/L in g_m means that the designer makes a g_m adjustment at the graphics console (or in an earlier era, at the drawing board). The designer's aim, however, is to make g_m just large enough and no larger, because overdesign would involve an oversized MOSFET, requiring more silicon area. Circuit costs are directly tied to circuit area.

This brings us to another fact that helped the MOSFET toward its present dominance. Equations of simple form, and coming from the simplistic analysis just presented, give an excellent description of MOSFET properties after empirical adjustment. Equation 1-22 is a prime example. As a result, computer design and optimization of MOS integrated circuits has been practical since the mid-1960s. Hence the necessary fine tuning of device sizes needed to give just enough and not too much transconductance was possible, if not easy. (It is a curious fact that the BJT is more difficult to describe with equal accuracy, but has transconductance reserves that make such description less necessary.)

Transconductance in the BJT is remarkably independent of structure and technology. It is independent of emitter area and even of the choice of semiconductor material! This is because BJT operation involves modulation of a naturally occurring potential barrier with invariant incremental properties. But in the MOSFET case, one is designing a variable resistor. Many BJT–MOSFET transconductance comparisons have been made over the years. These are not straightforward, because the two devices are so different, a subject addressed more fully in Chapter 6, but they typically indicate that the BJT is superior in transconductance by a factor ranging from 10 to 100. The fact that the MOSFET is today's dominant device is the result of a large collection of other MOSFET attributes and advantages that will be treated later. The designer, seeking always to have the best of both worlds, is today introducing the BJT into MOSFET circuits—especially CMOS circuits—in places where high transconductance is crucial, and the resulting *BiCMOS* products have assumed great importance.

1–7 Inverter Options

Most of the world's transistors, whether BJT or MOSFET, are used as switches, and in most such applications, fast switching is earnestly desired. A qualitative discussion of factors affecting the switching speeds of simple MOSFET voltage-inverter circuits is in order here. The first circuit is shown in Figure 1-13(a), consisting of the E-mode MOSFET M_1 and a resistor as a load device. The device M_1 in the indicated role is often termed a *driver*. Examine first the *turn-on* properties of the inverter. Initially the MOSFET is OFF, with $v_{GS} < V_T$, and input voltage LOW. As a result, the output voltage is HIGH, with v_{DS} essentially equal to V_{DD}. Inevitably, there is a parasitic capacitance C shunting the output port, and this capacitor is fully charged to V_{DD}.

EXERCISE 1–12 What accounts physically for the capacitance C in Figure 1-13(a)?

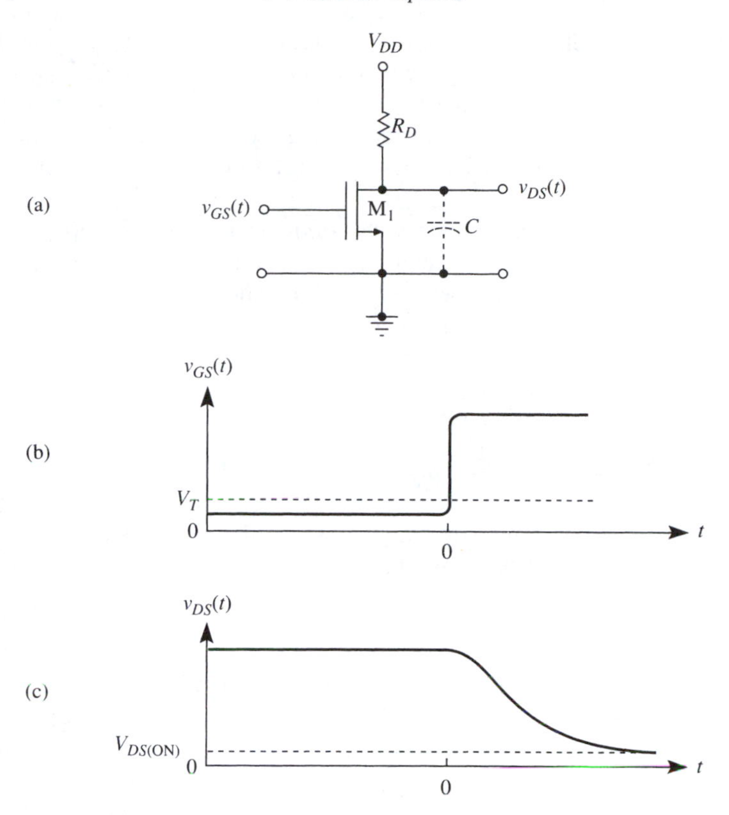

Figure 1–13 MOSFET inverter with resistive load. (a) Schematic circuit, showing parasitic capacitance C shunting output port. (b) Idealized input *turn-on* waveform. (c) Output *turn-off* waveform, showing time delay connected with discharging C.

■ The capacitance C is a parallel combination of three parasitic capacitances. First is the output capacitance of M_1, mainly depletion-layer capacitance associated with the drain junction that can be seen in Figure 1-1. Second is capacitance between the conductor that extends from the output terminal of M_1 to the input terminal of the next stage and the material that underlies the conductor. The underlying material is typically at ground potential and separated from the conductor by a layer of SiO_2. The third and last contribution comes from the capacitance shunting the input port of the next stage. In the likely event that the next-stage driver device is another MOSFET, this element is essentially its MOS input capacitance.

From this description it is clear that the parasitic capacitance is voltage-dependent, but that fact is not overly important in the present qualitative discussion. Now let the input signal shown in Figure 1-13(b) be applied, a positive-going voltage step that takes the device from below threshold to well above threshold in a very short time. The speed at which the output voltage $v_{DS}(t)$, shown in Figure 1-13(c), can fall from the HIGH of V_{DD} to its LOW value is fixed by the time required to discharge C. The only agency for discharging C is M_1. The greater the increase of drain current for an increment of input voltage, which is to say, the greater the transconductance of M_1, the faster M_1 can discharge C. In sum, $i_D(t)$ determines discharge time, and $i_D(t)$ is related to $v_{GS}(t)$ by g_m.

To complete the switching process, let a downward voltage step, equally ideal, turn off M_1. At this point M_1 is essentially an open circuit. The only agency for charging C back up to V_{DD} is the load device R_D, often called a "pull-up" device. Treating C as a fixed or voltage-independent capacitor for purposes of approximation, we clearly see that the *turn-off* time at the output port of the inverter is characterized by R_DC, with an exponential rise resulting.

EXERCISE 1–13 Why not simply reduce R_D to achieve fast *turn-off*?

■ The total voltage change, or *logic swing*, at the output port of the inverter is given by $R_DI_{D(\mathrm{ON})}$ and should not be reduced arbitrarily.

Resistive loads have rarely been used in MOS circuits, especially in integrated circuits. The diode-connected E-mode MOSFET, however, was used extensively in the earliest MOS integrated-circuit products. (See Problem A7.) Connecting gate to drain yields a nonlinear resistive diode that begins to conduct at approximately the voltage V_T. This *saturated load* is shown as the upper device M_2 in Figure 1-14(a). With this arrangement, *turn-on* of the inverter is virtually unaltered, since we have seen that the driver M_1 dominates that case. But *turn-off* by contrast is now over an order of magnitude slower! Figure 1-14(b) shows the output plane for M_1, with the characteristic of M_2 entered as a load line. For comparison, the R_D characteristic is also shown, with a dashed line. The voltage difference $V_{DD} - V_{DS(\mathrm{ON})}$ constitutes the logic swing discussed in Exercise 1-13, so Figure 1-14(b) constitutes a graphical presentation of the point made there verbally.

EXERCISE 1–14 Explain qualitatively why the saturated load is so inferior to a resistive load in the *turn-off* phase of switching.

■ With a resistive load, the charging of C is accomplished by a current that declines linearly with capacitor voltage. With the saturated load, the capacitor-charging current declines *faster* than linearly, and essentially vanishes at $v_{DS} = V_{DD} - V_T$.

It might be surprising that an option as inferior in performance as the saturated E-mode load was used so extensively. But it is small in area, and hence cost, and its

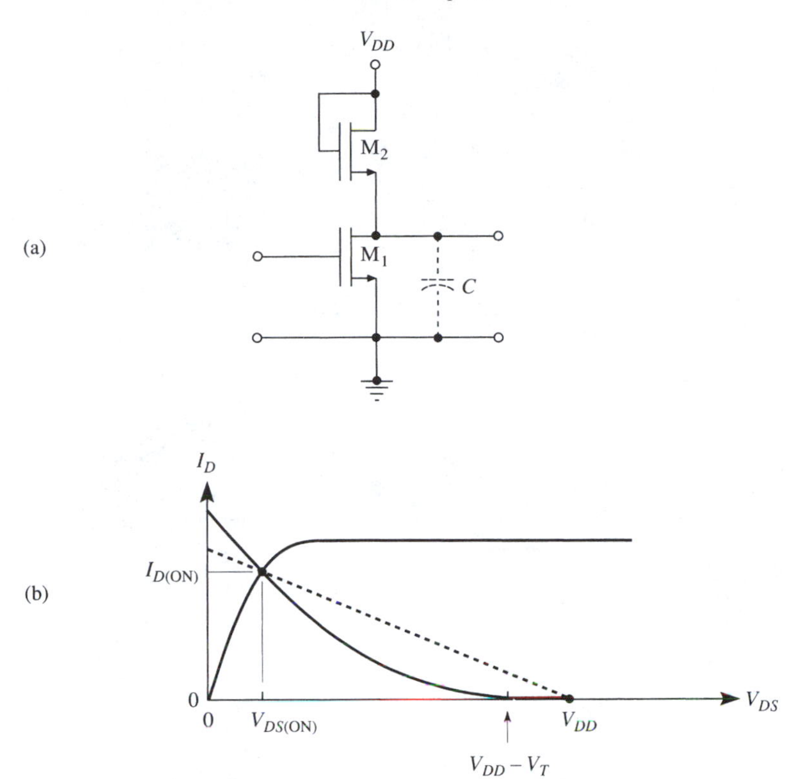

Figure 1–14 MOSFET inverter with saturated load. (a) Schematic circuit. (b) Output plane for driver device M_1, showing nonlinear loadline (solid line) of saturated-load device M_2 and, for comparison, the linear loadline (dashed line) of a resistive load as in Figure 1-13.

fabrication is totally compatible with that of an active device such as M_1. Also, there are some applications (inexpensive calculators, for example) where the importance of cost far outweighs that of speed.

The advent of ion implantation made it feasible, as noted earlier, to create D-mode devices as companions to E-mode devices in the same integrated circuit. The resulting *D-mode load* outperforms the saturated load by a wide margin [13]. This *enhancement–depletion*, or E–D, combination is extensively used today. An E–D inverter is shown in Figure 1-15(a). The substantial benefit here, of course, is that the parasitic capacitance C is now in principle charged at *constant current*, as can be seen in Figure 1-15(b), and hence the charging curve is approximately a steep ramp rather than a saturating curve.

It is actually possible to achieve a relationship as favorable as that depicted in Figure 1-15(b) by using a pair of separate, or *discrete*, devices in the circuit of Figure 1-15(a). However, for compelling reasons of technology and economy this is rarely done. When instead the two devices are fabricated on the same substrate, as is usual (in an integrated circuit, in other words), a phenomenon known as *body effect* (see

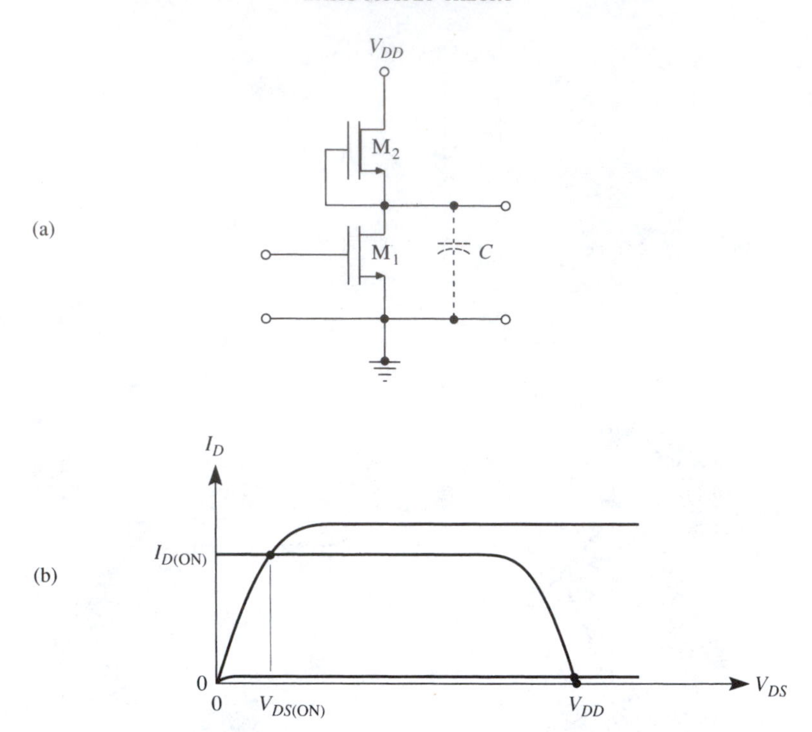

Figure 1–15 MOSFET inverter with D-mode load. (a) Schematic circuit. (b) Output plane for driver device M_1, showing nonlinear loadline of D-mode device M_2.

Section 4-3) causes the *I*–*V* characteristic of the D-mode load to be significantly degraded, roughly approaching linearity. This real-world outcome is treated numerically in Section 5-4.

The final inverter option incorporates an *active load*. It is a *P*-channel E-mode device that is under control of the input signal along with the *N*-channel driver. These two devices are homologous, or *complementary*, and in an inverter they are the basis of *CMOS* technology. Such an inverter is shown in Figure 1-16(a). Let us start with input voltage LOW again. This causes the *N*-channel device M_1 to be OFF, while the *P*-channel device M_2 is in its conducting state, as shown by solid lines in Figure 1-16(b). As a result, the output voltage is HIGH, approximately V_{DD}, as indicated by the intersection of the solid lines. The up-transition in input voltage then discharges *C* as before, with growing current as switching proceeds.

When M_1 is turned ON, the output voltage $V_{DS(ON)}$ approximates zero, as shown by the intersection of the dashed lines in Figure 1-16(b). On an integrated circuit, the V_{DD}-bus and ground-bus lines are often parallel, resembling "rails," and this very desirable switching property is therefore described as a "rail-to-rail logic swing." In contrast, note that $V_{DS(ON)}$ in Figures 1-13(b), 1-14(b), and 1-15(b) is well above zero. And as a further shortcoming, in Figure 1-14(b) the OFF voltage is lower than V_{DD} as well.

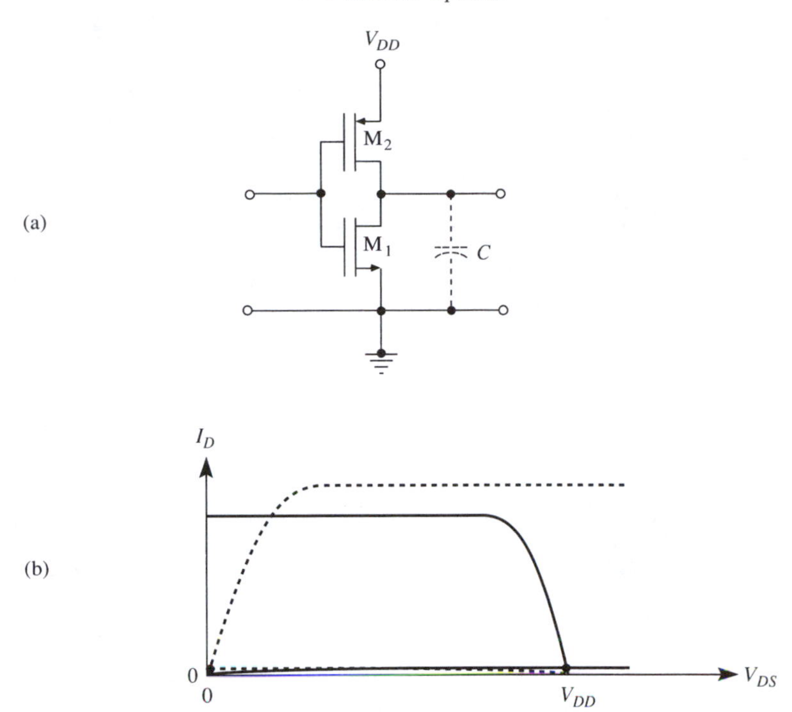

Figure 1–16 CMOS inverter. (a) Schematic circuit, showing *N*-channel driver device M_1 and *P*-channel pull-up device M_2. (b) Output plane for driver device M_1, showing active-load characteristic, and an intersection point for each state of the inverter.

An additional important benefit of the CMOS configuration enters during the *turn-off* phase of operation. This time M_2 is able to *charge C* with growing current as switching proceeds, just as *C* had been discharged with growing current. As a result, *turn-off* speed approximates *turn-on* speed.

Yet another advantage of CMOS circuitry is near-zero *standby current.* This is the static current through the inverter when the driver device is ON. In Figure 1-16(b), both currents (curve intersections) are actually much smaller than the diagram suggests, because they have been exaggerated for clarity. Note also that in both Figures 1-14(b) and 1-15(b), $I_{D(\mathrm{ON})}$ is large. There is, however, a brief time during the switching of a CMOS inverter when both devices conduct. Hence, switching is accompanied by a pulse of current, known as *totempole current*, that is inevitable. (This term arose because one device sits on top of the other in a CMOS inverter.) As a result of totempole current, power dissipation in a CMOS inverter increases rapidly as switching frequency is increased. But for low-frequency applications, power dissipation can be very low.

MOS-Capacitor Phenomena

The rudimentary model of Chapter 1 treats the MOS capacitor as a simple parallel-plate capacitor. In fact, the situation is appreciably more complicated than that. A more accurate model of the MOS-capacitor system incorporates a parallel-plate capacitor (the oxide portion) that is in series with the parallel combination of two voltage-dependent capacitors (residing in the silicon portion). Chapter 3 examines the MOS capacitor in these terms, while the present chapter describes real-life complications, such as parasitic charges, which intrude into otherwise straightforward capacitive phenomena. The relative complexity of the true picture makes it surprising that the simple model of Chapter 1 is able, with a measure of empirical adjustment, to describe the MOSFET so accurately.

From the analytic point of view, the problem posed by the silicon portion of the MOS capacitor is *identical* to the problem posed by one side of a step junction at equilibrium. Consequently, we shall start with an approximate analysis using the depletion approximation. It is important to note that still other physical mechanisms (such as foreign charged species residing on a semiconductor surface) can cause conditions in the silicon like those found in the MOS capacitor. For many years this was labeled the *semiconductor-surface problem.* It was study of the MOS capacitor that ultimately clarified these problems and caused most workers to appreciate that the three problems are identical.

2–1 Oxide-Silicon Boundary Conditions

Both conventional and MOS capacitors are well described by a lucid and quantitative technique that is worth reviewing here, starting with a capacitor having air as its dielectric material. For more general kinds of capacitors, see Appendix C. Problems involving electric field can be made graphic and palpable by employing a fictional entity—the line of force. This is a directed line that is "launched" by a positive charge and terminates on a negative charge of equal magnitude; the charge magnitude involved is a matter of arbitrary definition. That is, the line is seen to originate on a positive charge of a magnitude one may specify at will, and to end on an equal and opposite charge. In the general case, these lines are curved. The electric-field vector at any point along a line of force is tangent to the line of force at that point and has a direction (sense) that is the same as that of the line of force.

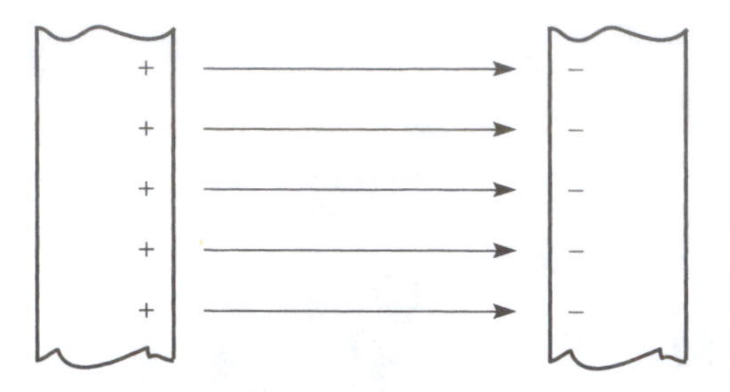

Figure 2-1 The uniform electric field deep inside a parallel-plate capacitor involves parallel, linear *lines of force* of fixed spacing, represented here in two dimensions.

The magnitude of the electric field at a particular position is proportional to the *density of lines of force* at that position (that is, to their degree of "packing"). Since the amount of charge required to "create" a line of force is arbitrary, it follows that the relationship between line density and field magnitude is also a matter of arbitrary definition.

In the uniform-field concept we can claim that it is obvious that the simultaneous requirements of field-vector parallelism everywhere and field-magnitude equality (line-density equality) everywhere can be met only by parallel and equally spaced lines of force. These conditions in turn require uniform (areal) charge densities on the plates of a parallel-plate capacitor when this means is used for creating the region of uniform field, a situation illustrated in two-dimensional fashion in Figure 2-1.

EXERCISE 2-1 For an air-dielectric parallel-plate capacitor, let us make the arbitrary definition (3 lines of force/cm^2) $\equiv$ (1 V/cm) $= E$, as illustrated in Figure 2-2. Calculate the amount of charge necessary to launch one line of force.

■ The elementary expression for parallel-plate capacitance is

$$C = \frac{A\epsilon_0}{d},$$

where A is the area under consideration on one plate, d is the plate spacing, and ϵ_0 is the *permittivity of free space* $= 8.85 \times 10^{-14}$ F/cm. Thus capacitance per unit area is given by

$$\frac{C}{A} = \frac{\epsilon_0}{d}.$$

From the charge-voltage capacitor law, $Q = CV$, we can obtain an expression in C/A by dividing both sides by A:

$$\frac{Q}{A} = \frac{C}{A}V.$$

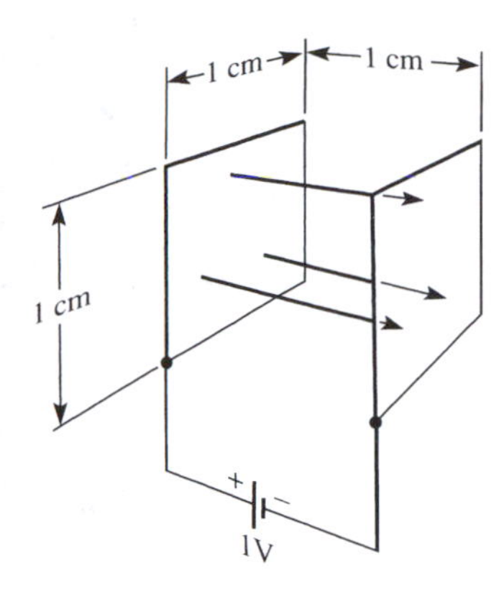

Figure 2–2 Line-of-force *density* (number per cm^2) is proportional to electric-field magnitude, with the proportionality factor being a matter of arbitrary definition.

Hence the desired quantity, Q/A, is given by

$$\frac{Q}{A} = \frac{\epsilon_0 V}{d}.$$

Let us take $V = 1$ V and $d = 1$ cm, which gives us a field of 1 V/cm. Then

$$\frac{Q}{A} = \frac{(8.85 \times 10^{-14}\ \mathrm{F/cm})(1\ \mathrm{V})}{1\ \mathrm{cm}} = 8.85 \times 10^{-14}\ \mathrm{C/cm^2}.$$

(If V and d are altered by the same factor, Q/A is unaffected, as the last expression shows.) Dividing the last result by 3 lines/cm^2 yields 2.95×10^{-14} C/line, which is the answer sought.

As an extension, determine the number of electrons necessary to terminate a line of force so defined:

$$\frac{2.95 \times 10^{-14}\ \mathrm{C/line}}{1.602 \times 10^{-19}\ \mathrm{C/electron}} = 1.84 \times 10^{5}\ \mathrm{electrons/line}.$$

As already noted, uniform-field conditions do not exist near the edges of a capacitor, where the lines of force curve outward, an effect often described as "fringing." But this further illustrates the utility of the line-of-force picture. In the representation of Figure 2-3 several qualitative features of the field distribution become immediately apparent. It is evident that "crowding" of the lines occurs at the inner corners of the plate edges, corresponding to a local increase in field intensity, or magnitude. But in the median plane of the capacitor, field magnitude diminishes as the plate edges are approached and passed. Furthermore, a question as to field direction at a point such as P is answered by the vector $\mathbf{E}$. Thus the line-of-force representation conveys electric-field information with respect to both magnitude and direction.

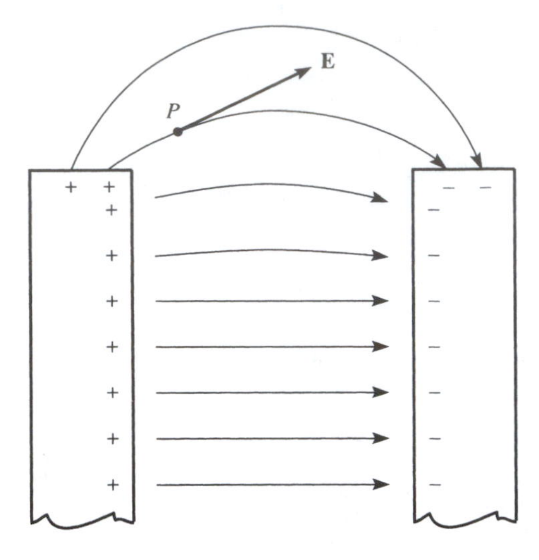

Figure 2–3 The "fringe" effects at the edges of capacitor plates, illustrating line-of-force nonlinearity, electric-field concentration (closely spaced lines of force) at a corner, and gradual electric-field decrease as one departs from the interior of the capacitor along the median plane.

The primary distinction between a conventional parallel-plate capacitor and the MOS capacitor is of course the substitution of a semiconducting region for a conducting region as one of the plates. An important result of this substitution is a region of charge in the semiconductor (which we take to be silicon in what follows) that is appreciably thicker than the corresponding layer in the conductor. Figure 2-4(a) presents this difference pictorially, showing a pair of metallic plates with field penetration, and hence space-charge layers, amounting to a few atomic layers at most. In Figure 2-4(b), on the other hand, is the MOS capacitor. Let us assume that the insulator here is silicon oxide. In typical situations, the layer of significant charge in the silicon is several times

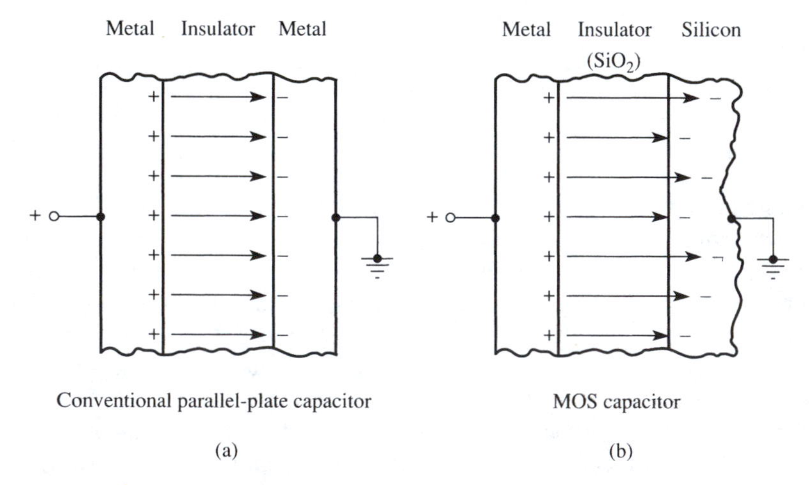

Figure 2–4 Elementary comparison of capacitors. (a) Conventional parallel-plate capacitor. (b) MOS capacitor, emphasizing a charge layer of appreciable thickness in the silicon.

the oxide thickness, which certainly qualifies as a nonnegligible penetration by the associated electric field. Now let us examine conditions at the oxide–silicon boundary in Figure 2-4(b). We have here a problem of extreme practical importance that is nonetheless very nearly one-dimensional. Hence, combining Equations C3 and C5 of Appendix C gives

$$\frac{Q}{A} = \epsilon_{OX} E_{OX} = D_{OX}, \tag{2-1}$$

where ϵ_{OX}, the absolute permittivity of the oxide, has been substituted for ϵ_0, the permittivity of the free-space dielectric in the previous capacitor, and where E_{OX} and D_{OX} are the uniform values of electric field and electric displacement existing in the oxide. Recalling from Appendix C that it is the normal component of electric displacement D that is continuous through an interface such as the oxide–silicon boundary, we can write

$$D_{OX} = D, \tag{2-2a}$$

or

$$\epsilon_{OX} E_{OX} = \epsilon E, \tag{2-2b}$$

or

$$\kappa_{OX} \epsilon_0 E_{OX} = \kappa \epsilon_0 E, \tag{2-2c}$$

where all symbols without subscripts are taken to apply to the silicon. The subscript "OX" of course refers to the oxide, and ϵ_0 is the permittivity of free space. Hence we have

$$E = \frac{\kappa_{OX}}{\kappa} E_{OX} \approx \frac{3.9}{11.7} E_{OX} = \frac{1}{3} E_{OX} \tag{2-3}$$

on opposite sides of the oxide–silicon interface, where the values of κ_{OX} and κ are taken from the front-matter table Properties of Silicon. The meaning of Equation 2-3 is that there exists a discontinuity in E at the interface. This situation can best be illustrated by adopting an arbitrary but specific MOS capacitor and bias condition. Computation of its field and potential profiles will then show the practical consequences of the boundary conditions just determined. Of particular importance is the computation of areal charge density in the silicon that follows from the knowledge of boundary field values and Equation 2-1.

EXERCISE 2–2 Equations 2-1 through 2-3 are all scalar by virtue of the convenient one-dimensional simplicity of the MOS capacitor. Generalize Equation 2-1 into three-dimensional form and identify the resulting expression.

■ A Gaussian pillbox placed with one face in the oxide and the other face in the neutral bulk region of the silicon would have flux through only one face—the face in the oxide. No flux will pass through the sides of the box or through the face in the neutral silicon, and these facts account for the form of Equation 2-1. In the general case we can assert that the integral over a closed surface of the electric displacement **D** can be related to the total charge Q_T within the surface by

$$Q_T = \oint_s \mathbf{D} \cdot d\mathbf{s},$$

where $d\mathbf{s}$ is a unit vector normal to the surface. This is a statement of Gauss's law.

2-2 Approximate Field and Potential Profiles

Assuming a specific example facilitates a quantitative but approximate examination of the MOS capacitor. Let us assume that it consists of an oxide layer 0.1 μm (1000 Å) thick on *P*-type silicon having a normalized bulk potential of $U_B = 12$ (where $U_B \equiv q\psi_B/kT$), corresponding to a net doping of $N_A - N_D \approx N_A = 1.6 \times 10^{15}/\text{cm}^3$, and a resistivity of about 8 ohm·cm. Assume that a positive voltage is applied to the gate sufficient to produce a space-charge layer 0.5 μm thick, as shown in Figure 2-5. Let this be an ideal MOS capacitor. Its other plate is metallic, as is also illustrated in Figure 2-5, and of course will become the gate of a MOSFET, as Figure 1-1 has indicated.

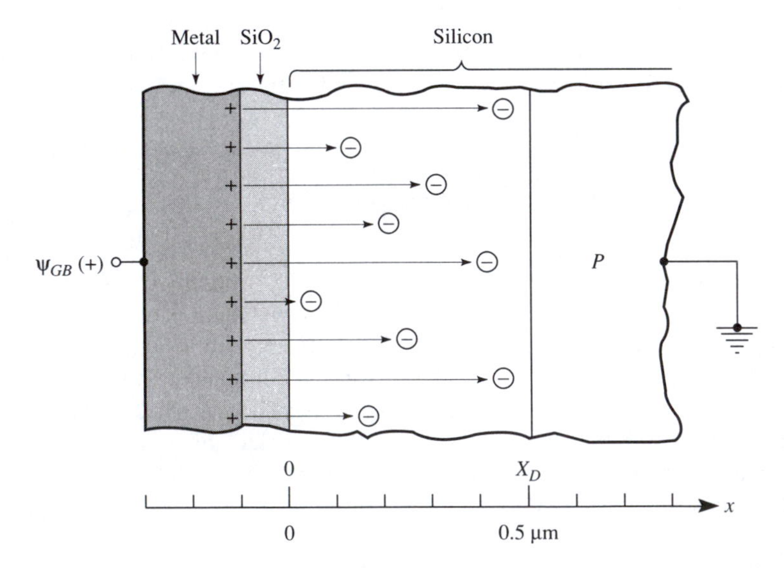

Figure 2-5 A specific arbitrary MOS capacitor and applied voltage. The silicon net doping is $1.6 \times 10^{15}/\text{cm}^3$, and the oxide-layer thickness is 0.1 μm. Depletion-approximation space-charge-layer thickness is chosen to be 0.5 μm.

Consider electrostatic potential first. Let the subscript S (passed over so many times before on grounds of ambiguity) designate the silicon surface that we have referred to as the oxide–silicon interface. We have already used B (bulk) for the silicon interior and will of course use G for the gate electrode or terminal. In these terms, the total electrostatic potential difference between the MOS-capacitor terminals becomes in double-subscript notation parallel to that described in Appendix A, $\psi_G - \psi_B \equiv \psi_{GB}$. Then, according to the rules of double-subscript notation, this potential difference can also be written $\psi_{GB} = \psi_{GS} + \psi_{SB}$. The first term on the right is potential drop through the oxide, the conventional-capacitor portion of the MOS device, and the second term represents potential drop in the silicon.

Now assume that the charge in the silicon is purely ionic and that it can be modeled using the depletion approximation. The voltage drop in the silicon can be computed by simply integrating the electric field through the space-charge layer:

$$\psi_{SB} = +\int_0^{X_D} E\, dx, \tag{2-4}$$

where the plus sign on the integral is consistent with our use of the Ohm's-law and double-subscript sign conventions described in Appendix A, and where X_D is the well-defined depletion-layer thickness associated with the depletion approximation. Assuming that no inversion layer is present, an assumption that can be checked at the end of the calculation, we may apply the relation between depletion-layer thickness and voltage drop that applies to the one-sided step junction. Recall that in such a case, the total voltage drop is approximately that on the lightly doped side. Making obvious substitutions to adapt Equation 3 of Table 1 to the present problem gives

$$\psi_{SB} = \frac{qN_AX_D^2}{2\epsilon} = \frac{(1.6\times10^{-19}\text{ C})(1.6\times10^{15}/\text{cm}^3)(5\times10^{-5}\text{ cm})^2}{2(1.035\times10^{-12}\text{ F/cm})} \tag{2-5}$$
$$= 0.31\text{ V}.$$

The next step is to realize that the ionic charge in the silicon constitutes the charge on the negative plate of the MOS capacitor. The fact that this charge is distributed through a volume of significant thickness is unimportant with respect to computing the potential drop *through the oxide*. That is, we can apply the law of the parallel-plate capacitor here because the problem is one-dimensional. We will get the same result for ψ_{GS} whether the charge on the negative plate lies in a mathematical plane at the interface or is spread through a thick region. Letting Q_S be the charge in the silicon per unit area of the MOS capacitor, and letting C_{OX} be the capacitance per unit area of the oxide region,

$$\psi_{GS} = -\frac{Q_S}{C_{\text{OX}}} = -\frac{X_{\text{OX}}}{\epsilon_{\text{OX}}}Q_S = -\frac{X_{\text{OX}}}{\epsilon_{\text{OX}}}(-qN_AX_D) = \frac{X_{\text{OX}}qN_AX_D}{\epsilon_{\text{OX}}}, \tag{2-6}$$

where X_{OX} is the oxide thickness. The first minus sign enters Equation 2-6 because ψ_{GS} is applied at the gate side, whereas the charge per unit area in the silicon is on the opposite side. The second minus sign enters because the silicon charge is negative. Substituting presently applicable values into Equation 2-6 gives

$$\psi_{GS} = \frac{(10^{-5}\ \text{cm})(1.6 \times 10^{-19}\ \text{C})(1.6 \times 10^{15}/\text{cm}^3)(5 \times 10^{-5}\ \text{cm})}{0.345 \times 10^{-12}\ \text{F/cm}} \tag{2-7}$$

$$= 0.37\ \text{V}.$$

Thus, we now have the information necessary for a quantitative presentation of the potential profiles. The potential distribution, obtained by integrating the linear field profile of the depletion approximation, is parabolic.

Figure 2-6 summarizes the findings. In Figure 2-6(a) the MOS capacitor is shown with acceptor ions represented as circled symbols. Figure 2-6(b) shows the triangular

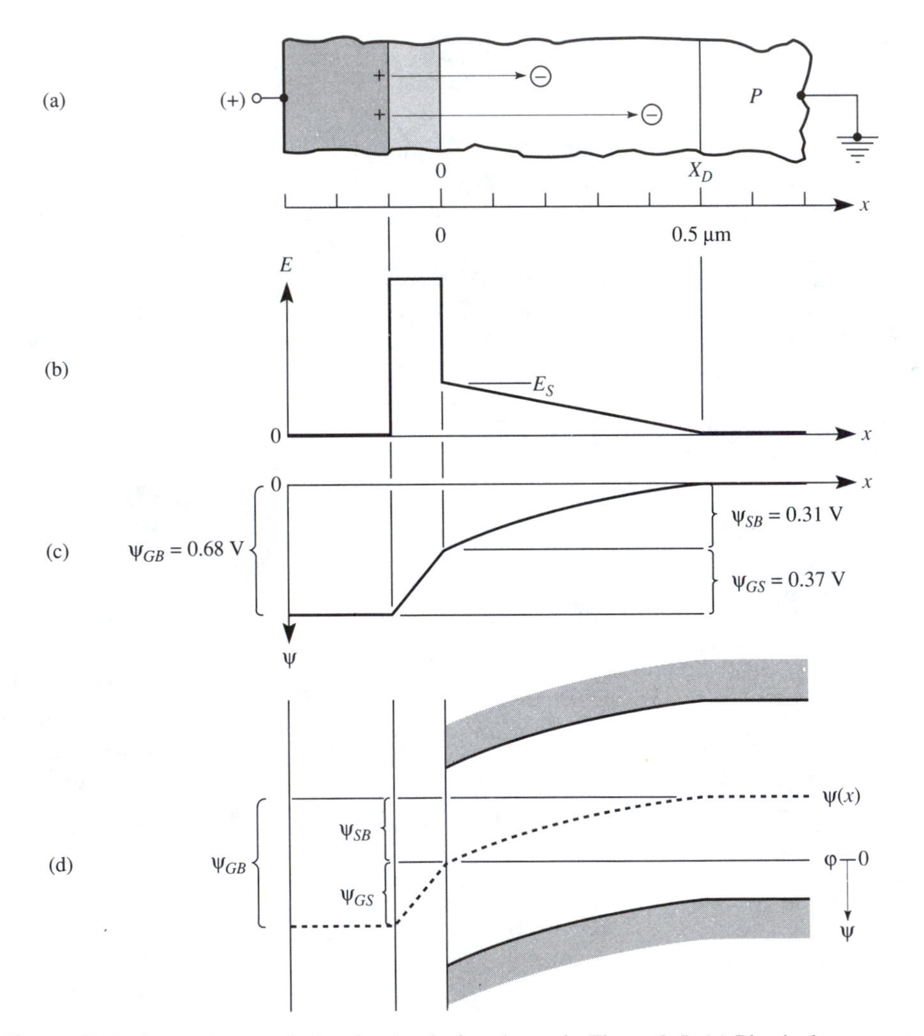

Figure 2–6 Approximate solution for the device shown in Figure 2-5. (a) Physical representation. (b) Electric-field profile. (c) Electrostatic-potential profile. (d) Band diagram.

field profile in the silicon that is consistent with the depletion approximation. The field discontinuity at the oxide–silicon interface is a consequence of Equation 2-3. We assume that no charges exist in the oxide, and hence all lines of force must pass entirely through it. Their parallelism, enforced by the one-dimensional geometry, translates into a constant electric field in the oxide. At the oxide–metal interface, the electric field drops abruptly to zero because lines of force cannot penetrate far into a metal. Turning to Figure 2-6(c), we observe a parabolic potential profile in the silicon (depletion approximation again), a slope discontinuity by a factor of three at the oxide–silicon interface, consistent with Figure 2-6(b), and a linearly rising potential through the oxide. Note that the downward direction has been chosen as positive for ψ, for consistency with the band-diagram convention of Appendix D. The neutral silicon bulk and the field plate are equipotential regions, as $\psi(x)$ clearly shows. Since a band diagram is a potential plot, we can go directly from Figure 2-6(c) to a band diagram for the MOS capacitor under bias. The only additional information that is needed is the bulk potential ψ_B. Let us use the majority-carrier Fermi level in the bulk ϕ_B as the reference for all potentials. The bulk potential ψ_B is negative for this P-type case, and can be obtained from Equation E4 of Appendix E as

$$\psi_B = -\frac{kT}{q}\ln\frac{p_n}{n_i} = -(0.0257\text{ V})\ln\left(\frac{1.6\times10^{15}/\text{cm}^3}{10^{10}\text{cm}^3}\right) = -0.31\text{ V}. \quad (2\text{-}8)$$

Combining this result with those in Figure 2-6(a) through (c) leads to the band diagram in Figure 2-6(d) for the ideal MOS capacitor under bias, where the bias amounts to

$$\psi_{GB} = \psi_{GS} + \psi_{SB} = 0.68\text{ V}. \quad (2\text{-}9)$$

For reasons that are evident in Figure 2-6(d), the quantity ψ_{SB}, the amount of the potential drop in the silicon, is often referred to as the *band bending* in the MOS-capacitor problem, or in the equivalent surface problem. Notice that the dashed "intrinsic-line" profile is identical to that in Figure 2-6(c), because the bulk Fermi level has been selected as the potential reference.

EXERCISE 2–3 How can an electric-field discontinuity exist at the oxide–silicon interface?

■ The oxide and the silicon both possess significantly large dielectric constants, and hence a sheet of charge develops at the interface that is analogous to the charge sheets shown in Figures C1(b) and C2(b) of Appendix C.

EXERCISE 2–4 What is the sign of the charge sheet present (but not shown) in Figure 2-6(a), and why is it there?

■ The charge sheet is negative and is completely analogous to the central charge sheet in Figure C3(a). In both cases, the negative charge is present because positive bias is applied to the left-hand capacitor plate, and the dielectric material of greater permittivity is on the right.

Now we are in a position to confirm that there is no inversion layer at the silicon surface. Note in Figure 2-6(d) that the surface potential ψ_S is near zero. When the surface potential is precisely zero, very nearly the case in Figure 2-6 (by coincidence, since the initial values were chosen arbitrarily), the surface is said to be at the *threshold of weak inversion.* It was explained in Section 1-2 that for the presence of an inversion layer, it is necessary for the gate-to-bulk voltage (and hence ψ_{GB}) to have a value beyond the threshold voltage V_T. And the threshold voltage in turn is defined as the voltage that causes electron density (volumetric density!) at the silicon surface (at the oxide–silicon interface) to equal hole density in the bulk region. With our assumption of band symmetry, this tells us that the surface potential in Figure 2-6(a) must be as far below the Fermi level as the bulk potential is above it, or $\psi_S = -\psi_B$, a succinct statement of the threshold condition. But in the arbitrary situation of Figure 2-6, the band bending is too small by about a factor of two to meet this condition. Hence, the case represented is far below threshold.

2-3 Accurate Band Diagram

Now let us employ the tools of Appendix F, methods for approximate-analytic modeling of semiconductor regions at equilibrium. We shall carry out an accurate examination of the problem that has just been analyzed by means of the depletion approximation, and will compare the results. The only differences that will exist between the two are in the silicon region of the device because the rest remains idealized. Let us continue with the case of $U_B = -12$, or about $1.6 \times 10^{15}/\text{cm}^3$ of net *P*-type doping. To compare an approximate analysis with an accurate one, the first thing to decide is what should be held constant from one to the other. When the depletion-approximation descriptions are to be compared to exact solutions for step junctions, we usually choose to hold contact potential, or the total voltage drop in the equilibrium junction, constant. This choice produced the greatest congruence of the band diagrams. In the present case, however, it is reasonable to hold the field in the oxide constant, because we have not yet altered that portion of the model. Given this different choice, it follows that the peak field in the silicon, the value just inside the oxide–silicon interface, will be unchanged, because Equation 2-3 still applies. The result will be a small change in the charge, field, and potential profiles in the silicon. Let us start at the oxide–silicon interface and work toward the silicon interior.

Given the knowledge that the area of the triangle in Figure 2-6(b) is equal to ψ_{SB}, the voltage drop in the silicon, it is easy to find the peak electric field in the silicon for this depletion-approximation case. It becomes

$$E_S = \frac{2\psi_{SB}}{X_D} = \frac{2(0.31\ \text{V})}{0.5 \times 10^{-4}\ \text{cm}} = 1.24 \times 10^4\ \text{V/cm}. \tag{2-10}$$

Next, it is necessary to convert this field value to the fully normalized form plotted on the ordinate of Figure F3 of Appendix F in order to determine which regime of this diagram applies to the present problem. Since the normalizing quantity for potential is kT/q and that for distance is L_D, it follows that fully normalized electric field at the silicon surface is

$$\left.\frac{dW}{d(x/L_D)}\right|_S = \frac{E_S L_D}{kT/q}. \tag{2-11}$$

Since $L_D \approx L_{De}$ in our extrinsic sample, and using Equation F15 of Appendix F, we have

$$E_S\left(\frac{q}{kT}\right)\left[\frac{\epsilon}{qN}\frac{kT}{q}\right]^{1/2} = E_S\left[\frac{\epsilon}{qN}\frac{q}{kT}\right]^{1/2}, \tag{2-12}$$

where N is net doping. Evaluating the right-hand side of this expression gives

$$(1.24 \times 10^4 \text{ V/cm}) \times \left[\frac{1.035 \times 10^{-12} \text{ F/cm}}{(1.602 \times 10^{-19} \text{ C})(1.6 \times 10^{15}/\text{cm}^3)(0.02566 \text{ V})}\right]^{1/2} = 4.92. \tag{2-13}$$

Figure F3 shows the near-congruence of the exact solution and the linear electric-field asymptote at this value, so it follows from Equation F26 that the oxide–silicon interface falls at $-x/L_D = 4.92$. Figure F4 shows that the parabolic potential asymptote also applies at this position (as it must, since the field and potential asymptotes correlate). So from Equation F34,

$$W = \frac{1}{2}\left(-\frac{x}{L_D}\right)^2 + 1 = 13.1. \tag{2-14}$$

Thus, using $kT/q = 0.02566$ V, we have

$$\psi_{SB} = \frac{kT}{q} W = 0.336 \text{ V}, \tag{2-15}$$

a value about 8% higher than the value obtained from the depletion approximation. The increment is of course associated with the extra area under the field profile near the depletion-layer boundary, shown in Figure 2-7(a), that was also emphasized in Figure F3. The corresponding charge-density and potential profiles are shown in Figure 2-7(b) and (c). Dotted lines are used to indicate the depletion-approximation results in these three diagrams for comparison with the more exact profiles. Finally, an adjusted band diagram is presented in Figure 2-7(d), one that differs only slightly from that of Figure 2-6(d).

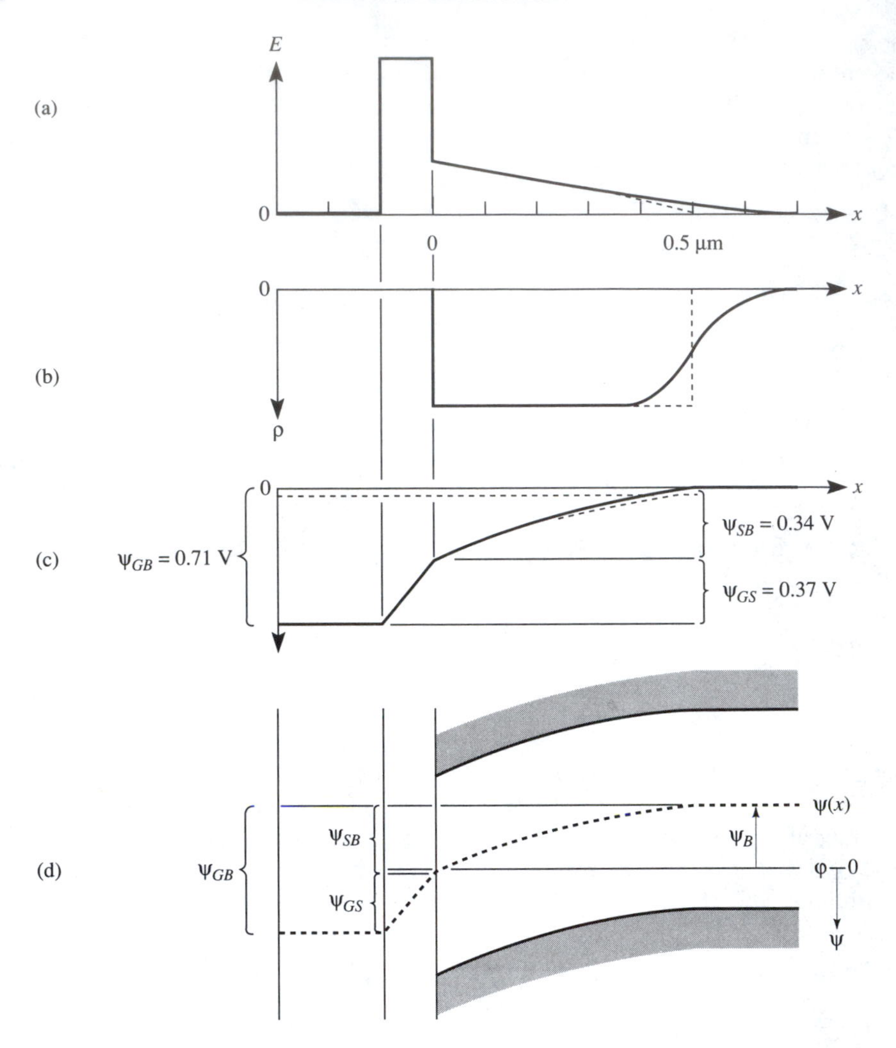

Figure 2–7 Accurate solution (solid lines) for the device shown in Figure 2-5, keeping field in the oxide the same as in the approximate solution (dotted lines) of Figure 2-6. (a) Electric-field profile. (b) Space-charge profile. (c) Electrostatic-potential profile. (d) Band diagram.

EXERCISE 2–5 Account quantitatively for the 8% greater band bending in the accurate analysis than in the approximate analysis.

■ One normalized unit, or 0.02566 V (one thermal voltage), is about 8% of the band bending calculated in either case.

EXERCISE 2–6 Account for the difference also by comparing Equation 2-14 with its approximate analog.

■ The unity term in the normalized Equation 2-14 contributes one normalized unit to the depletion–approximation–replacement (DAR) solution. Dropping the unity term yields the depletion–approximation (DA) equation.

In Figure 2-7(d), examine the region of the silicon near the oxide–silicon interface. Band bending causes the conduction-band edge to approach the Fermi level, going a bit beyond the threshold of weak inversion. This guarantees a few surplus electrons there, an incipient inversion layer, but at extremely low densities. Band bending causes the valence-band edge to move away from the Fermi level, so that acceptor ionization is even more complete than in the neutral sample. Thus, negative acceptor ions and negative mobile carriers (the beginning of an inversion layer) coexist in this region. They do not in any way "interfere" with one another and are in fact additive. At threshold, right at the surface, the volumetric space-charge density has twice the value it has in the fully depleted region close to (but outside) the inversion layer. Half the charge-density contribution right at the silicon surface is therefore ionic, negatively charged acceptors; the remainder consists of electrons.

2–4 Barrier-Height Difference

By the 1950s there was a good understanding of the principles governing ideal semiconductor surfaces, and by extension, the ideal MOS capacitor that has occupied us up to this point. The MOS capacitor made its appearance at the end of the decade. It was in the 1960s that a comparable grasp of the nonideal MOS capacitor was acquired, a step that was essential to practical MOS integrated-circuit technology. The fact that the MOS market segment swept past the older bipolar integrated-circuit technology in overall importance is convincing evidence that these practical problems were solved. Now we examine a feature that varies with the identity of the semiconductor, the insulator, and the gate material in the MOS capacitor.

When an electron is withdrawn from the surface of a conductive solid, an "image" charge appears in the solid and the resulting pair of opposite-sign charges are mutually attractive. Consequently, to extract the electron we must do work on it. If it is removed to a distance such that the image force is essentially zero, then, at rest, it resides at an energy higher than the Fermi level in the solid. (The necessary distance, to be sure, is small.) This higher energy is termed the *vacuum level.* The energy interval from the vacuum level to the Fermi level is termed the *work function* Φ. Thus there exists a barrier at a silicon surface that prevents electrons from escaping, and the work function is a measure of the height of that barrier. For silicon the work function is about 4.8 eV and appears not to be a function of doping [25].

Figure 2-8 shows a band diagram for silicon in the ideal *flatband* condition, or the

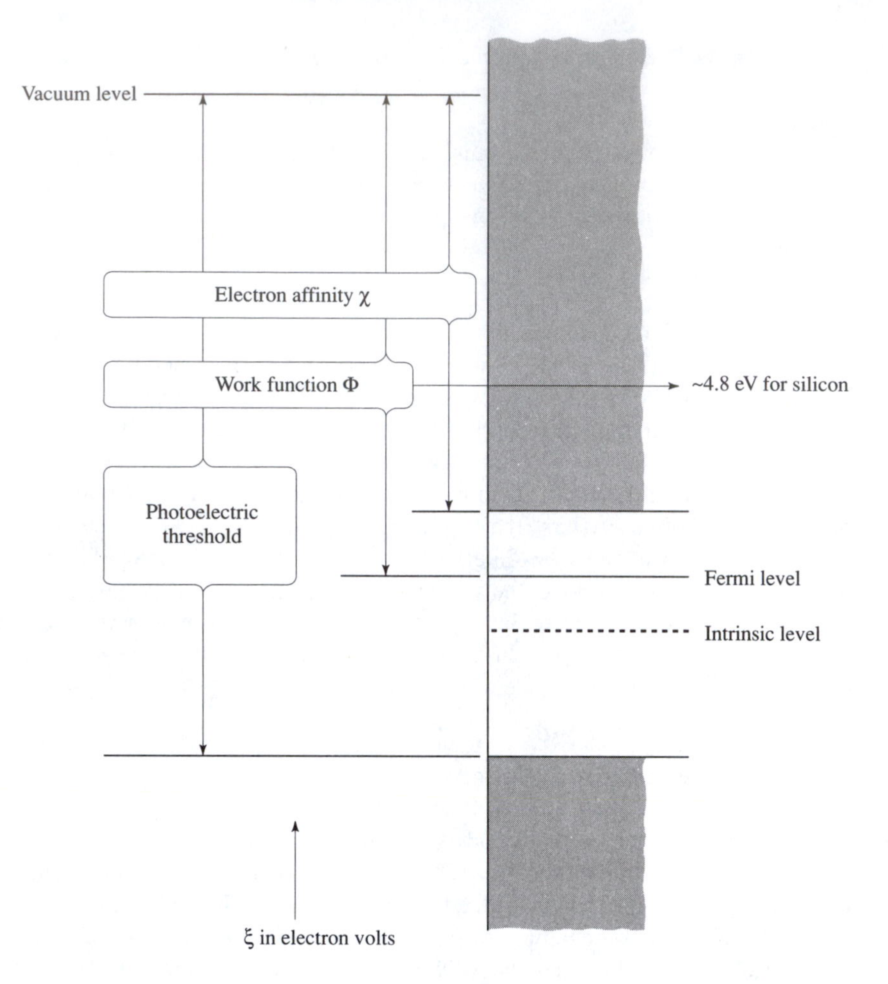

Figure 2-8 Qualitative band diagram for the surface region of a semiconductor sample, defining electron affinity, work function, and photoelectric threshold.

condition where $\psi_{SB} = 0$. Indicated there, though not to scale, is the interval Φ, along with two other intervals often cited in surface work; these are the *photoelectric threshold* and the *electron affinity* χ. The former can be appreciated as follows. Let photons of gradually increasing energy be incident upon the surface of the sample shown in Figure 2-8. Since electrons are most abundantly available in the valence band of the semiconductor sample, we will see a sharp rise in the rate of electron ejection from the sample (that is, in the rate of photoelectron production) when the photon energy corresponds to the energy required to raise an electron from the top edge of the valence band to the vacuum level, at which energy the departing electrons have been liberated from the solid. This is the photoelectric threshold. The electron affinity is defined as the energy that binds a conduction electron at the edge of the conduction band to the solid—or the energy a conduction electron must acquire to reach the vacuum level and hence to escape.

Let us concentrate on the work function Φ. As in our previous band diagrams, we can convert from electron energy in electron volts to electrostatic potential in volts. The work function can be cited in either, but the one intended must be clearly stated to avoid the ever-present sign ambiguity. Since the literature is inconsistent on this score, let us adopt the symbol H to stand exclusively for barrier height in *volts.*

In the MOS capacitor we are concerned not with free surfaces, but with two interfaces, namely, the oxide–silicon interface and the oxide–metal (or other gate material) interface. A barrier to electron passage into the oxide exists at each interface. The oxide is an insulator with a gap on the order of 8 V. One barrier height of interest to us is that from the Fermi level in the silicon bulk to the conduction-band edge in the oxide. An electron in the silicon acquiring sufficient energy can enter the conduction band of the oxide and can then penetrate the normally insulating layer, arriving at the gate electrode. Such an energetic electron is sometimes termed a "hot" electron.

When an oxide is grown on silicon, its conduction-band edge is separated from the valence-band edge of the silicon by a potential interval of 4.35 V [26]. Since the energy gap is 1.12 V in silicon, it follows that the conduction bands of the insulator and the semiconductor are separated by 3.23 V on the potential scale. These values are apparently independent of doping and crystalline orientation [26]. Thus, the band structures of the oxide and the silicon have a fixed relationship. When the surface potential (the midgap potential on the silicon side of the interface) is manipulated in the MOS capacitor, all four band edges move up and down together along with the surface potential.

Now recall that at equilibrium, the Fermi level in the silicon remains constant or "flat" irrespective of band bending in the silicon. It follows, of course, that the spacing (in terms of potential) of the conduction-band edge in the oxide from the Fermi level in the silicon will change as band bending is altered by bias adjustments on the MOS capacitor. So that the barrier height H_B will be unequivocally defined, let us therefore specify that the silicon must be in the *flatband* condition, $\psi_{SB} = 0$, when we observe the interval between the Fermi level in the silicon and the conduction band in the oxide. For the MOS capacitor shown in Figure 2-9(a), the desired flatband situation is depicted in Figure 2-9(b). It is evident on the diagram that the desired barrier height H_B can be calculated by adding the values of three quantities: ψ_B from Equation 2-8, half the gap, and the spacing of the conduction-band edges, $\psi_{CO} - \psi_{CS}$. The last item is written this way because we wish to refer all elements of the MOS capacitor *to* the silicon. Thus,

$$\begin{aligned} H_B = \psi_{CO} - \phi_B &= \psi_B - \frac{\psi_G}{2} + (\psi_{CO} - \psi_{CS}) \\ &= -0.31 \text{ V} - 0.56 \text{ V} - 3.23 \text{ V} = -4.10 \text{ V}. \end{aligned} \tag{2-16}$$

(As usual, we have let $\phi_B \equiv 0$.) Suppose we assume an aluminum field plate, or gate electrode. Its barrier height has been given in the literature [26] as

$$H_G = \psi_{CO} - \phi_G = -3.20 \text{ V}. \tag{2-17}$$

This is the final value needed for the construction of Figure 2-9(b).

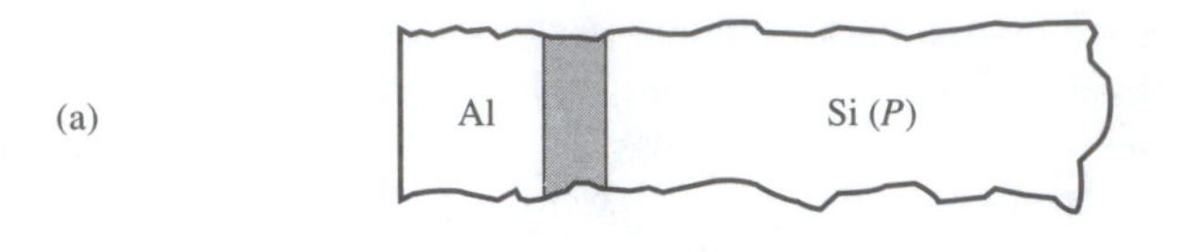

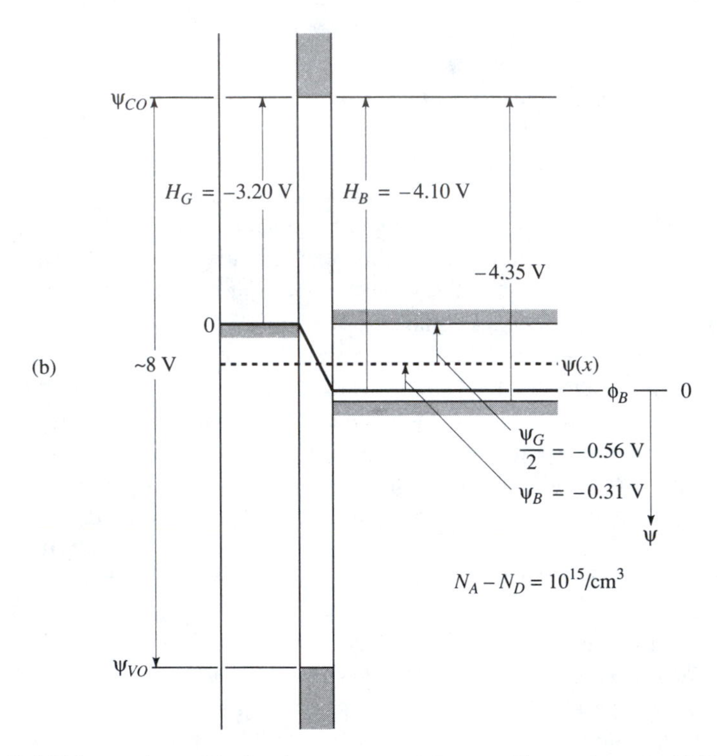

Figure 2–9 MOS capacitor with aluminum gate and P-type silicon substrate. (a) Physical representation. (b) Band diagram for flatband condition, assuming that barrier-height difference is the only nonideal feature present.

Normally we will not work with the individual barrier heights, but rather with the barrier-height *difference* H_D. Let us define this quantity as the position of the gate Fermi level ϕ_G, referred under flatband conditions to the Fermi level in the silicon, $\phi_B = 0$. That is,

$$H_D \equiv \phi_{G|FB} = \phi_G - \phi_B, \tag{2-18}$$

a quantity that obviously can be positive *or* negative. Adding and subtracting the oxide conduction-band-edge position ψ_{CO} on the right-hand side of Equation 2-18 gives us

$$H_D = (\psi_{CO} - \phi_B) - (\psi_{CO} - \phi_G) = H_B - H_G$$
$$= -4.10 \text{ V} - (-3.20 \text{ V}) = -0.90 \text{ V}. \tag{2-19}$$

The quantity H_D will be used in what follows, and no further use will be made of the work function Φ. Because H_D depends on H_B, which in turn depends on doping [refer again to Figure 2-9(b)], it follows that H_D is doping-dependent. Figure 2-10(a) shows how H_D varies with doping for the case of silicon, with gold and aluminum field plates. From Figure 2-9(b), the equations for these curves can be written in terms of equilibrium electron density n_0 (instead of net doping) as

$$H_D = H_B - H_G = \frac{kT}{q} \ln \frac{n_0}{n_i} - 3.79 \text{ V} - H_G. \tag{2-20}$$

For our P-type sample, $n_0 = 6.25 \times 10^4/\text{cm}^3$, so that the logarithmic term is negative. The value of H_G for an aluminum field plate is -3.20 V, as noted above, and it is -4.10 V for gold [26].

EXERCISE 2–7 Explain the relationship of Equation 2-20 and Figure 2-9(b).

■ The first term on the right-hand side of Equation 2-20 is a way of writing the bulk potential ψ_B, in view of Equation E3. This is the distance from ϕ_B to $\psi(x)$ in Figure 2-9(b). The second term is the distance from $\psi(x)$ to ψ_{CO}, or

$$-0.435 \text{ V} - \psi_G/2 = -0.435 \text{ V} - 0.056 \text{ V} = -3.79 \text{ V}.$$

In modern integrated-circuit practice (since approximately 1970), aluminum and gold field plates have been replaced by polycrystalline silicon, or "polysil." The polysil is usually heavily doped to simulate the metallic behavior that is desired in a field plate and also in the interconnection pattern that is fabricated using the same deposited-polysil layer. The dependence of H_D upon doping for P^+ and N^+ polysil field plates is shown by the solid lines in Figure 2-10(b). These curves are based upon experimental determinations made by Werner [27] for P^+ polysil boron-doped to $1 \times 10^{20}/\text{cm}^3$, and for N^+ polysil phosphorus-doped to $3 \times 10^{20}/\text{cm}^3$. These curves too can be represented by Equation 2-20 by employing an appropriate value for H_G. The Fermi level is near the conduction-band edge in N^+ silicon and near the valence-band edge in P^+ silicon; we can in fact take the values of H_G as approximately -3.23 V and -4.35 V for the N^+ and P^+ cases, wherein net doping is $N \approx 2 \times 10^{19}/\text{cm}^3$. The dotted curves in Figure 2-10(b) show the corresponding approximate values used in SPICE modeling of the MOSFET, as explained in Section 5-1.

EXERCISE 2–8 Determine the barrier-height difference H_D for an MOS capacitor with a polysil gate. Assume the gate is heavily doped with acceptors, and that $H_G = -4.35$ V. Also assume that the device has the same oxide and silicon struc-

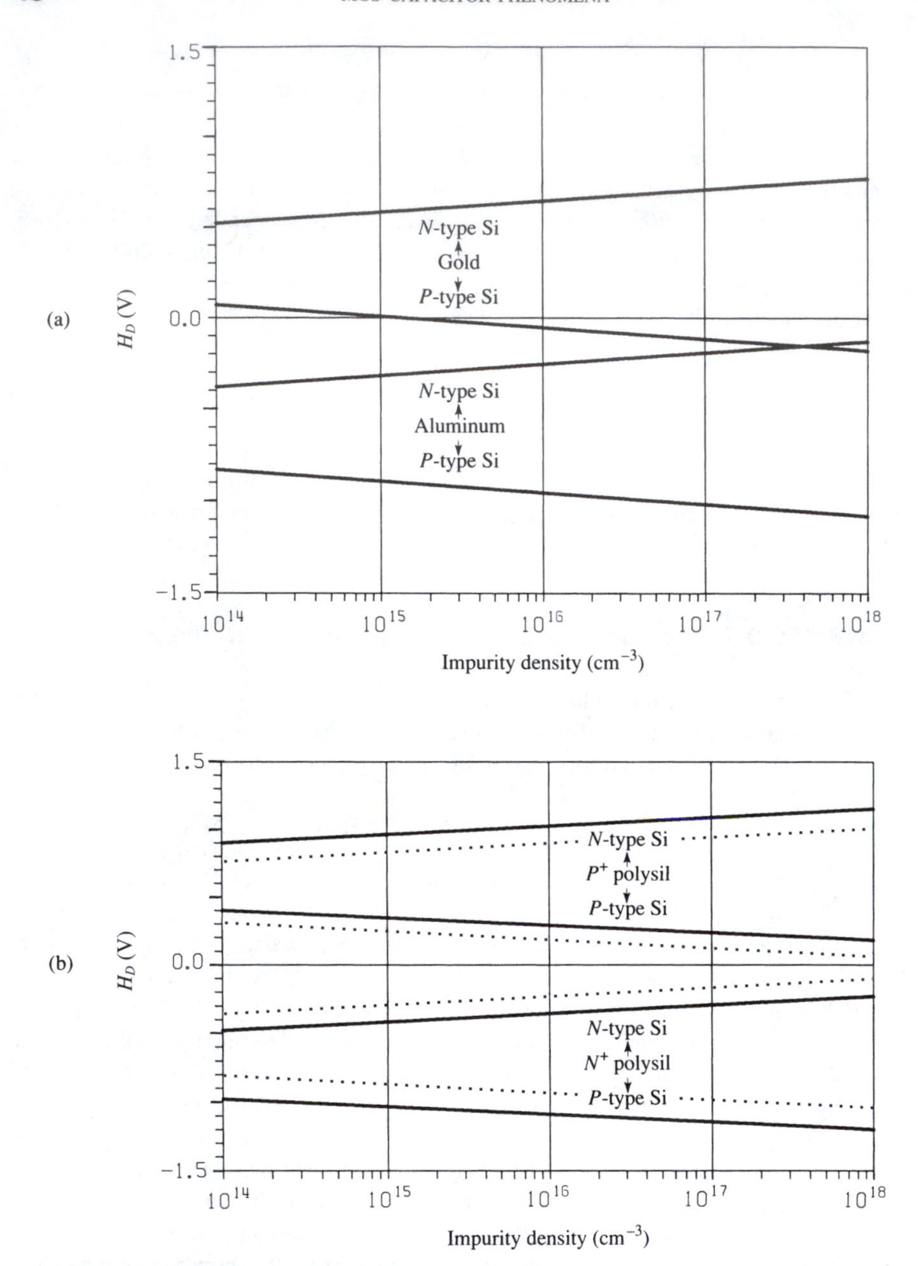

Figure 2–10 Barrier-height difference H_D as a function of doping type and density. (a) Data for gold and aluminum field plates. (After Deal and Snow [26] with permission, copyright 1966, Pergamon Press Ltd.) (b) Data for P^+ and N^+ polysil field plates. (Top and bottom solid lines are from Werner [27] with permission, copyright 1974, Pergamon Press Ltd.) Center two solid lines are calculated from top and bottom lines by adding and subtracting, respectively, the value 1.38 V (also obtained from the data of Werner [27]). Dotted lines show approximate values used in SPICE modeling of the MOS system.

ture that was used in the preceding calculations. Comment on the properties of this MOS capacitor.

■ Using the value of $H_B = -4.10$ V given in Equation 2-16, then Equation 2-20 gives us

$$H_D = H_B - H_G = -4.10 \text{ V} - (-4.35 \text{ V}) = 0.25 \text{ V}.$$

Thus the barrier-height difference is positive. If this were the only factor present, it would be necessary to apply a quarter-volt *positive* bias to bring the device to the flatband voltage, and the threshold condition would occur at a still more positive bias.

Let us now remind ourselves that a voltmeter measures Fermi-level difference. The MOS capacitor depicted in Figure 2-9(a) was brought into the flatband condition by applying an external bias of −0.9 V to the gate with respect to the silicon, a voltage equal to the Fermi-level difference. When the capacitor is in this bias condition, the difference in *electrostatic potential* between gate and silicon (or bulk), the quantity ψ_{GB}, is zero. This fact is exhibited in Figure 2-9(b) by the horizontal dotted line representing $\psi(x)$. The coexistence of an externally measurable voltage difference and no electrostatic-potential difference is attributable to barrier-height difference H_D.

Now suppose we arbitrarily change the bias on the same MOS capacitor to +0.2 V. Because a positive voltage on the field plate will drive majority holes away from the silicon surface, this bias is clearly in the depletion-inversion direction. The result of applying such a bias is shown in the partial band diagram of Figure 2-11. Once again we will defer until the end of the calculation the question of whether an inversion layer

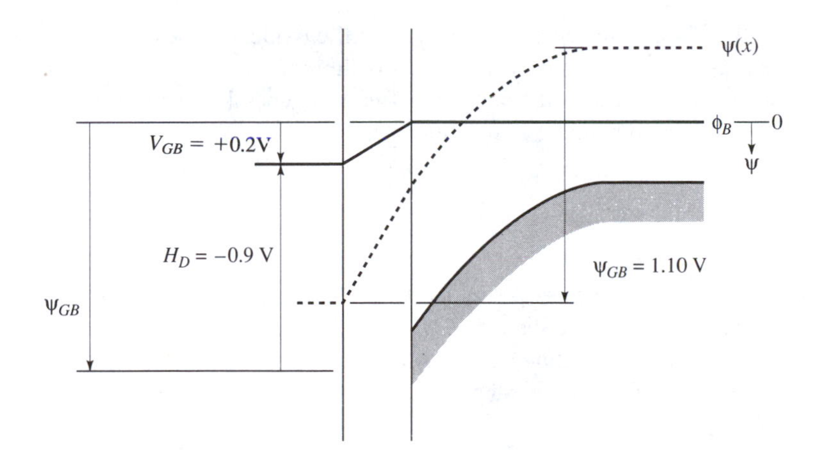

Figure 2–11 Band diagram (lower half) for the structure of Figure 2-5 with $V_{GB} = H_D + \psi_{GB} = +0.2$ V applied arbitrarily.

is present. However, comparing Figure 2-11 with Figure 2-7(d) makes it clear that a well-depleted condition exists near the silicon surface, since that was true in the earlier case and the present case has even more band bending.

It is evident in Figure 2-11 that the external bias voltage V_{GB} can be written as the algebraic sum of H_D and the total electrostatic potential difference ψ_{GB}, an important result:

$$V_{GB} = H_D + \psi_{GB}. \tag{2-21}$$

In the equipotential or flatband condition, by definition $\psi_{GB} = 0$. Thus it follows that when a nonzero barrier-height difference is the only nonideal factor, the bias voltage that must be applied to produce the flatband condition, a bias value called the *flatband voltage* V_{FB}, is equal to the barrier-height difference. In the present case, as illustrated in Figure 2-9,

$$V_{FB} = H_D = -0.9 \text{ V}. \tag{2-22}$$

To demonstrate the utility of Equation 2-21, let us ask this question: What amount of band bending ψ_{SB} exists under the conditions shown in Figure 2-11? Breaking ψ_{GB} into its two components in the manner of Equation 2-9, causes Equation 2-21 to become

$$V_{GB} = H_D + \psi_{GS} + \psi_{SB}. \tag{2-23}$$

From Equation 2-6, $\psi_{GS} = -Q_S/C_{OX}$, where

$$C_{OX} = \frac{\kappa_{OX}\epsilon_0}{X_{OX}} = \frac{\epsilon_{OX}}{X_{OX}} = \frac{3.9(8.85 \times 10^{-14}\text{ F/cm})}{10^{-5}\text{ cm}} = 3.45 \times 10^{-8}\text{ F/cm}^2, \tag{2-24}$$

and ϵ_{OX} is the absolute dielectric permittivity of the oxide. On the other hand, the band bending ψ_{SB} is directly related to W_S, which is the normalized potential at the surface of the silicon. Normalized potential W is defined in general terms in Equation F23. In terms of W_S, the band bending in the silicon is

$$\psi_{SB} = -\frac{kT}{q}W_S. \tag{2-25}$$

The minus sign enters because W increases upward on the band diagram and ψ increases downward, as can be seen in Equation F23. Given the condition of strong depletion that exists here, Equation F34—or the left-hand part of Equation 2-14—is valid, and is repeated here for convenience:

$$W = \frac{1}{2}\left(-\frac{x}{L_D}\right)^2 + 1. \tag{2-26}$$

Equation F26 is also valid, and can be rewritten here as

$$\left|\frac{dW}{d(x/L_D)}\right| = \left|\frac{dU}{d(x/L_D)}\right| = -\frac{x}{L_D}. \qquad (2\text{-}27)$$

The magnitudes of both normalized-field expressions are used because W and U have opposite signs. (For our conventions, $-x/L_D$ is positive.) Placing the first normalized-field expression in Equation 2-26 gives us

$$W = \frac{1}{2}\left[\frac{dW}{d(x/L_D)}\right]^2 + 1. \qquad (2\text{-}28)$$

Next, recall that

$$|Q_S| = \epsilon|E_S| = |D_S|, \qquad (2\text{-}29)$$

where Q_S is the areal charge density in the silicon, previously used in Equation 2-6, while E_S and D_S are the electric field and displacement, respectively, just inside the silicon at the oxide–silicon interface. Absolute value has been indicated because the sign must be dropped in what follows. From Equation 2-12 we may write

$$\left.\frac{dW}{d(x/L_D)}\right|_S = \frac{|Q_S|}{\epsilon}\left[\frac{\epsilon}{qN}\frac{q}{kT}\right]^{1/2} = \frac{|Q_S|}{K}, \qquad (2\text{-}30)$$

where evidently

$$K = \sqrt{\epsilon qN(kT/q)} = 2.61 \times 10^{-9}\ \text{C/cm}^3. \qquad (2\text{-}31)$$

Substituting Equation 2-30 into Equation 2-28 yields

$$|W_S| = \frac{1}{2}\left(\frac{Q_S}{K}\right)^2 + 1, \qquad (2\text{-}32)$$

where W_S is of course the normalized potential at the silicon surface (normalized surface potential). Placing Equations 2-6 and 2-32 in Equation 2-23 gives

$$V_{GB} = H_D - \frac{Q_S}{C_{OX}} + \frac{kT}{q}\left[\frac{1}{2}\left(\frac{Q_S}{K}\right)^2 + 1\right], \qquad (2\text{-}33)$$

or

$$Q_S^2 - \frac{q}{kT}\frac{2K^2}{C_{OX}}Q_S + \frac{q}{kT}2K^2\left(\frac{kT}{q} + H_D - V_{GB}\right) = 0. \qquad (2\text{-}34)$$

Applying the quadratic formula yields

$$Q_S = \frac{q}{kT}\frac{K^2}{C_{OX}} \pm \sqrt{\left(\frac{q}{kT}\frac{K^2}{C_{OX}}\right)^2 - \frac{q}{kT}2K^2\left(\frac{kT}{q} + H_D - V_{GB}\right)} \qquad (2\text{-}35)$$

$$= -1.74 \times 10^{-8}\ \text{C/cm}^2.$$

The negative sign on the radical must be taken because the charge in the silicon is necessarily negative. Using Equations 2-6, 2-24, and 2-35 yields

$$\psi_{GS} = -\frac{Q_S}{C_{OX}} = 0.50\ \text{V}. \qquad (2\text{-}36)$$

Hence, from Equation 2-23,

$$\psi_{SB} = V_{GB} - H_D - \psi_{GS} = 0.2\ \text{V} - (-0.9\ \text{V}) - 0.50\ \text{V} = 0.60\ \text{V}. \qquad (2\text{-}37)$$

This value is just below the band-bending value that produces the threshold condition, or $|2\psi_B| = 0.62$ V in the present case, so the inversion-layer charge is very small and may be neglected.

As a check on the solution, let us calculate the electric field in the oxide:

$$E_{OX} = \frac{\psi_{GS}}{X_{OX}} = \frac{0.5\ \text{V}}{10^{-5}\ \text{cm}} = 5.0 \times 10^4\ \text{V/cm}. \qquad (2\text{-}38)$$

As shown in Equation 2-3, the field at the silicon surface should be about three times smaller. Using Equation 2-29 and introducing appropriate signs,

$$E_S = \frac{-Q_S}{\epsilon} = \frac{1.74 \times 10^{-8}\ \text{C/cm}^2}{1.04 \times 10^{-12}\ \text{F/cm}} = 1.67 \times 10^4\ \text{V/cm}, \qquad (2\text{-}39)$$

confirming the expectation.

Examination of Figure 2-11, which has just been treated quantitatively, shows an important point. The effect of a negative barrier-height difference, such as that in the aluminum–silicon system, is indeed to move the MOS capacitor's condition in the direction of depletion inversion. This is one of several factors that accounted for the early exclusive use of *P*-channel MOS technology—involving devices made on *N*-type material and having hole inversion layers. Efforts to use *P*-type material, as in Figure 2-11, often resulted in devices with thresholds much more negative than the desired value, in many cases producing a D-mode FET when an E-mode FET was intended.

2-5 Interfacial Charge

A departure from crystalline perfection in a silicon sample introduces electronic states that reside within the forbidden band. An example that is familiar at this point is an impurity atom placed substitutionally in the crystal. The surface of a crystal represents a gross interruption of crystalline perfection. In a primitive model of the surface we can visualize unsatisfied, or "dangling," bonds associated with the atoms at the surface. Significant thought along these lines goes all the way back to the work of J. Willard Gibbs in the last century, although these ideas were not published until the first decade of this century [28]. In the early 1930s Igor E. Tamm put forward a quantitative theoretical model for the states associated with dangling bonds, leading to the expression "Tamm states," which is still in use [29]. In the late 1930s William Shockley offered a related but different model [30]. Increasing numbers of gifted theoreticians and experimentalists were attracted to the challenging surface problems. As it turned out, such studies lay directly on the path to the transistor. A useful survey of early research on surface states has been given by Many et al. [31].

Figure 2-12 offers a heuristic picture of dangling bonds at a silicon surface. Letting the short lines be interpreted as electrons, we can represent in the manner shown there the case where the silicon surface atoms (as well as the underlying atoms) all have the right number of electrons to ensure neutrality. If we employed some unspecified means to remove all of the "dangling" electrons at the surface, there would then be uncompensated positive charge associated with each surface atom. The diagram suggests that the charge would be $+2qN_S$, where N_S is the density of the surface atoms, but we must remind ourselves at this point that the picture used here is but a two-dimensional representation of a three-dimensional monocrystal.

The states portrayed in this simplistic picture can be described as *donorlike*. That is, they carry a positive charge when they are empty of electrons and are neutral when filled by electrons. However, with the numerous differing possibilities for crystalline imperfection at a surface, we should not be surprised to learn that surface states may

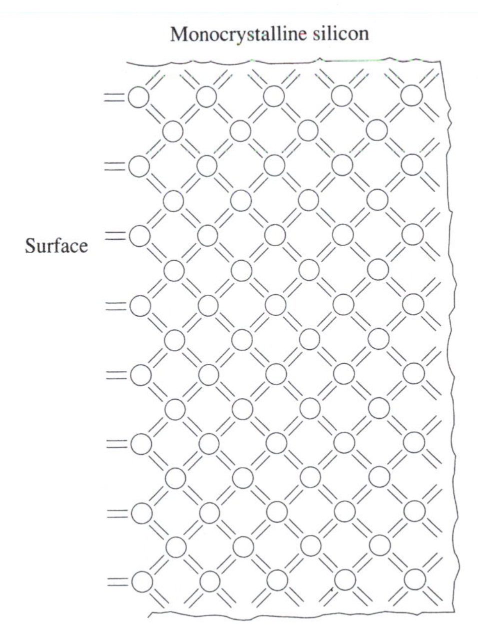

Figure 2–12 Pictorial representation of single-crystal surface. Each silicon atom is assigned the number of electrons necessary for neutrality, and the result is "dangling" or unsatisfied bonds at the surface.

be either donorlike or acceptorlike. The latter states are neutral when empty and negatively charged when filled by an electron. Likewise we would expect these varied kinds of surface states to be distributed through a range of energies within the bandgap at the surface. Continuing with simplistic descriptions, we can represent the situation in the manner of Figure 2-13, where short lines now represent the surface states themselves, which are also assumed to be uniformly distributed in energy.

Next, assume that these states are in good communication with the "bulk" of the crystal. That is, assume that electrons can move freely from the surface states to the interior of the crystal and back again, when it is energetically favorable to do so. If the surface potential ψ_S is then modified by some means, charge will flow into or out of the surface states until virtually all of those lying above the Fermi level will be empty of electrons, and those below, filled. Descriptive terminology evolved concerning these states. Although the terms are not now widely accepted, they are worth invoking for their graphic qualities. When the exchange of charge was rapid—requiring less than a millisecond—the states responsible were termed "fast" surface states. If the exchange required seconds, minutes, or more, the states involved were called "slow" surface states.

Assuming that "fast" surface states like those represented in Figure 2-13 are also donorlike, we can see that the effect of having such states will be to produce band bending. Those states lying above the Fermi level will primarily manifest the positive charge of an empty donor state. Lines of force will extend into the silicon to terminate

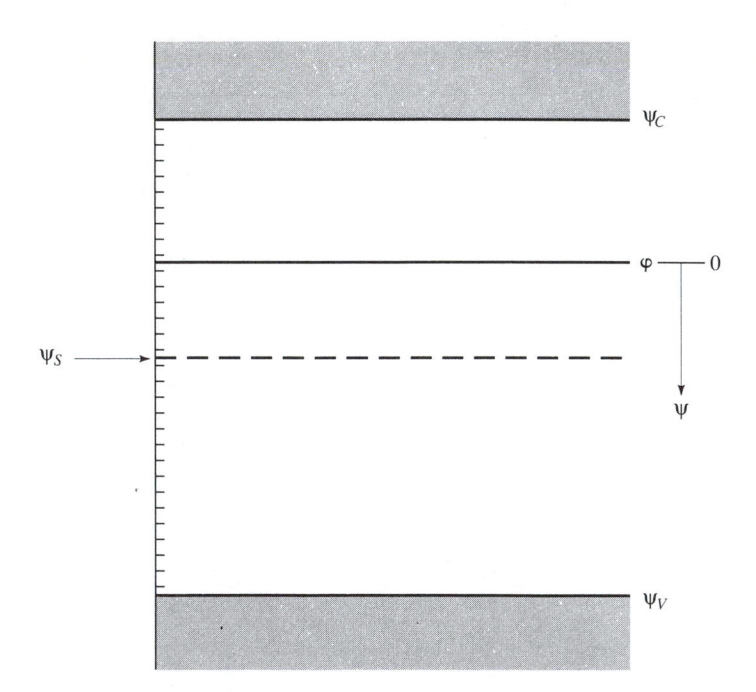

Figure 2-13 Using short lines to represent electronic states available at the surface of an N-type sample in the flatband condition. Electrons residing in these states constitute a form of interface trapped charge.

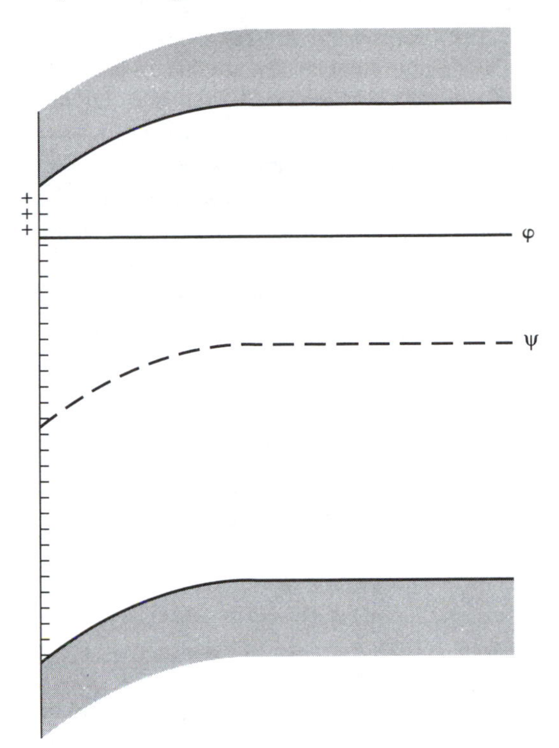

Figure 2–14 Donorlike electronic surface states on an *N*-type sample. Filled states (primarily below the Fermi level) are neutral, while empty states account for positive charge at the surface and a resulting positive (rightward) field as well as the band bending shown.

on the most available negative charges—conduction electrons. That is, the silicon near the surface will experience *accumulation.* The bands will bend down, as shown in Figure 2-14. An energy well is thus formed by the insulating oxide on the left-hand side and the conduction-band edge on the other side, confining the *surplus* electrons. (See Appendix F.)

The density of "fast" surface states on a silicon surface is subject to wide variations, depending on the details of surface treatment. A silicon sample cleaved in a vacuum exhibits a density of such states amounting to about $10^{15}/cm^2$, which approximates the density of surface atoms [25]. This semiquantitative agreement calls forth the dangling-bond picture of Figure 2-12. When the sample is removed from the vacuum, a layer of silicon oxide forms spontaneously because of the great affinity of silicon for oxygen. Other atoms present in the atmosphere can also react or be adsorbed. At this point the density of "fast" surface states drops into the range of 10^{11} to $10^{12}/cm^2$ [32]. These are accompanied, however, by some 10^{12} to $10^{13}/cm^2$ of "slow" surface states. The picture suggested by this observation is that the spontaneous oxide has satisfied some of the surface-atom bonds that were unsatisfied in the cleaved-surface condition, but that there are new states formed (perhaps in the oxide) that electrons can reach with more difficulty, and hence more slowly. The spectrum of time constants is a continuum, and the areal density of such states is proportional to characteristic time. This kind of charge exchange can play a part in a form of electrical noise known as "flicker noise," which is characterized by a $1/f$ spectrum [33].

The problem of interest to us, of course, is the problem of such states at an oxide–silicon interface. Hence the term *interfacial states* becomes more appropriate than *surface states* and will be used henceforth. Modern parlance lumps together the charge in "fast" and "slow" interfacial states and refers to it collectively as *interface trapped charge*. This agreed-upon terminology [34] is accompanied by the now-standard notation of N_{it} to stand for the areal density of charge centers per square centimeter (cm^{-2}), and Q_{it} for the areal density of charge in coulomb per square centimeter (C/cm^2).

Interfacial states are an unmitigated evil in semiconductor devices. Much effort, mostly empirical, has been invested to devise procedures that reduce the density of such states to an acceptably low level. In modern MOS technology, the density of these states is of the order of $10^9/cm^2$. To acquire a quantitative grasp of the significance of this surface-density value, we can point out that it approximates the density of electrons in the incipient inversion layer existing at the threshold condition, a very small areal density indeed. This statement is based upon the case arbitrarily taken in Sections 2-2 through 2-4, wherein net doping was $N_A - N_D \approx 1.6 \times 10^{15}/cm^3$.

The effect of states of the fast variety at the oxide–silicon interface in an MOS capacitor is to act much like an electrostatic shield, since electrons flowing from the interior of the crystal can readily gain access to the surface states. When positive bias, for example, is applied to the gate, such charge transport results in interface trapped charge. The trapped electrons terminate lines of force originating on positive charges at the field plate, and hence the lines do not penetrate into the silicon to form the depletion layer that precedes inversion-layer formation. This situation is illustrated qualitatively in Figure 2-15. Figure 2-15(a) shows the potential profile of an MOS capacitor that is ideal in the sense that no interfacial states are present. The squares in Figure 2-15(b) represent, this time, acceptorlike states that are negatively charged when occupied. Occupancy of those shown on the diagram was the result of the positive bias on the gate. Consequently, the capacitor with interfacial states shows much less band bending and depletion than the MOS capacitor that is free of such states. It was this kind of "shielding" behavior that John Bardeen described [35] to account for the fact that the many years of the field-effect experiments that led to the point-contact transistor had produced disappointing results. (See Section 1-1.) Thin semiconductor samples were employed in the field-effect experiments, but repeatedly the amount of depletion was far less than expected, and as a result the conductivity change in the thin layer was also much smaller than expected. The "surface states" he postulated to account for such observations we can legitimately label as interfacial states because of the inevitable presence in air of at least a spontaneous oxide layer on a semiconductor crystal. Elucidation of the interface-trapped-charge phenomenon and quantitative descriptor was achieved by extension of the capacitance-measurement techniques described in Sections 3-4 and 3-5, and also by surface-conductance measurements [36].

It has been found that an efficient method for reducing the density of interfacial states is low-temperature (~450°C) hydrogen annealing. So effective are the refined (but largely empirical) present-day procedures that the once vexing problem of interfacial states can be regarded as a thing of the past for terrestrial radiation-free applications and environments. Unfortunately, however, there are also applications where the problem returns full-blown [37]. For example, a communications satellite some-

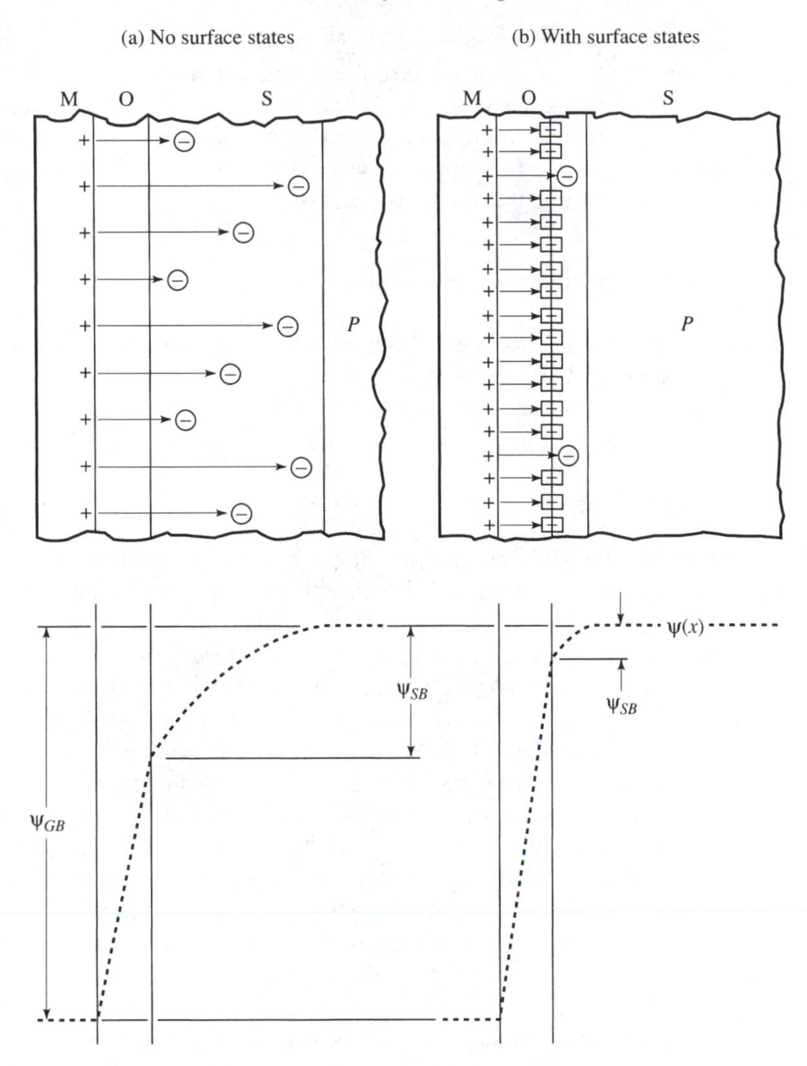

Figure 2–15 Tendency of fast surface states to shield the crystal interior from the effects of bias applied to the MOS capacitor. (a) Device with no surface states, showing termination of field lines by acceptor ions in the depletion layer. (b) Device with surface states that fill with electrons to terminate field lines, thus simulating an electrostatic shield that diminishes ψ_{SB}.

times finds itself in a sea of energetic particles and quanta, and such radiation creates numerous interfacial states and associated interface trapped charge. Ironically, the hydrogen that constitutes a "fix" for the problem of interfacial states under radiation-free conditions becomes itself a problem in a radiation environment, being responsible for yet another form of parasitic charge [38].

The specific nature and distribution in energy of the interfacial states (or traps) remain subjects of controversy. However, a model that is consistent with a substantial amount of experimental data is as follows [37]: When the Fermi level is at the middle of the gap (the threshold of weak inversion), the interfacial states overall tend to be neutral. This implies that all states in the upper half of the gap are acceptorlike, while all states in the lower half are donorlike. Hence, in an *N*-channel device at the threshold of strong inversion, the interface trapped charge is negative; in a *P*-channel device at threshold, on the other hand, the charge is positive.

EXERCISE 2–9 Explain the last observation.

■ The *N*-channel (*P*-substrate) device exhibits downward band bending at threshold, so that the Fermi level at the surface is above the middle of the gap. Thus, many of the acceptorlike states between the Fermi level and the middle of the gap are filled and negatively charged. Reverse conditions hold in the *P*-channel case.

The combination of high-lying acceptor states and low-lying donor states may seem odd, but is not without precedent. This concept, not examined heretofore, can be explained in a more familiar context by referring to Table 2, wherein it is pointed out that a gold atom introduced substitutionally in silicon introduces two states, only one of which is in evidence at a given time. When the Fermi level is below the gold donor state, located 0.35 eV above the valence-band edge, one sees the donorlike behavior of gold. But when the Fermi level is above its acceptor state, located 0.54 eV below the conduction-band edge (and hence very near the center of the gap), one sees acceptorlike behavior. In either case the gold atom is likely to be charged (ionized), but oppositely in the two cases. This *amphoteric* property is sometimes described by stating that the substitutional gold atom behaves like a *P*-type impurity in an *N*-type environment, and vice versa. Thus it has a tendency to raise resistivity in either case. The introduction of gold into silicon was used in past decades to reduce carrier lifetime—mainly by virtue of the acceptor state that lies so very near the center of the bandgap—thus creating "faster" devices.

Further comments on interfacial states and the effects of charge therein are found at the end of the next section. The balance of charges in interfacial states against other parasitic charges to be described there can also be affected by radiation.

2–6 Oxide Charge

When MOS technology had advanced far enough to eliminate, virtually, the states responsible for interface trapped charge, investigators then became aware of yet another nonideal property of the MOS capacitor. There appeared to be a rather constant and inescapable positive charge, and it, too, was in the neighborhood of the oxide–silicon interface. This charge, whose detailed identity and genesis has still to be determined in detail, has been named *fixed oxide charge* [34]. The accepted symbols for its areal charge density and areal charge-center density, respectively, are Q_f and N_f. (In much of the early literature the areal density of such a charge is identified as Q_{ss}, a symbol with the obvious shortcoming that it suggests "surface states" more readily than the fixed-charge concept.)

The properties of fixed oxide charge were described by an early team of investigators [39], who enumerated a number of its attributes: It is

- independent of surface potential ψ_S, at least over the central 0.7-V portion of the energy gap
- stable under moderate bias-temperature stress. In other words, it is a truly *fixed* charge in the sense that it is not *mobile* like the charge to be treated later in the present section.
- independent of oxide thickness for a given oxide-growth condition
- approximately independent of impurity type and density in the range from 10^{14} to 10^{17} cm^3
- located within, at most, 200 Å of the oxide–silicon interface
- reproducibly variable as a function of ambient gas (dry oxygen or water vapor) during oxidation
- reproducibly variable as a function of silicon temperature during oxidation
- capable of being increased by strong fields resulting from negative field-plate bias; in these cases its final value levels off at a value proportional to its initial value; the increase in N_f resulting from such treatment is accompanied by an apparent increase in "fast" surface states.
- dependent on crystal orientation at the silicon interface, following the same sequence as oxidation-rate constant and surface-atom density
- determined by the final step in the oxidation process.

In explanation of the last item it should be pointed out that during an oxidation process, SiO_2 is created at the oxide–silicon interface. That is, oxygen diffuses through the oxide that has already formed, with very little silicon diffusing to the exposed surface of the layer. Consequently, the SiO_2 most recently formed is deepest, or near the interface.

The picture that was offered in explanation for all these observations was that unreacted silicon is present near the interface and for a short distance into the oxide. That is, with a density that declines as one departs from the interface, there is excess (or unoxidized) silicon present. The problem has been studied more recently using a variety of techniques, including x-ray-photoelectron spectroscopy, Auger-electron spectroscopy, and electron microscopy [40–42]. These observations suggest the presence of a thin ($\sim$10 Å) layer of "nonstoichiometric" material near the oxide–silicon interface. Quantitative estimates of the excess (or deficiency) of a given component are not in agreement, partly because the measurements are difficult. Silicon–silicon bonding has also been suggested as an important feature of this layer, so the problem is more complicated than initially thought.

Although the precise nature of the oxide fixed charge has not yet been determined, there is wide agreement that this fixed charge resides in the oxide close to the interface. The term *fixed* conveys the double idea that first, unlike the interface-trapped charge, Q_f is independent of surface potential, and second, it is not mobile when the MOS capacitor is subjected to a bias at elevated temperature, as is the third kind of parasitic charge (which will be examined next, following a quantitative look at the effect of Q_f on capacitor properties). Also, the term *oxide* is included in the name of this charge component, in spite of its association with the interface, because the charge indeed resides in a region of finite thickness outside the pure-silicon portion of the MOS capacitor.

To examine the effect of Q_f on the MOS capacitor, let us initially assume $H_D = 0$. Then connect gate to bulk directly, so that $\phi_G = \phi_B$. With the sheet of positive charge existing at the interface, and with no barrier-height difference between gate and bulk, the potential profile will look like that in Figure 2-16(a). The silicon surface is depleted. In order to establish the flatband condition, one must bias the gate negatively, as shown in Figure 2-16(b). Let us compute V_{FB} for a typical value of N_f found on a (100) surface, namely, $10^{10}/\mathrm{cm}^2$. This crystal orientation is favored in MOS technology so that N_f (and statistical fluctuations in N_f) will be as small as possible. Assuming that Q_f is the only charge present, and using the same MOS-capacitor sample as in Sections 2-3 and 2-4, we have from Equation 2-6,

$$V_{FB} = \psi_{GB} = -\frac{Q_f}{C_{OX}} = -\frac{N_f q X_{OX}}{\epsilon_{OX}} = -\frac{(10^{10}/\mathrm{cm}^2)(1.6 \times 10^{-19}\ \mathrm{C})(10^{-5}\ \mathrm{cm})}{0.345 \times 10^{-12}\ \mathrm{F/cm}} = 0.046\ \mathrm{V}. \tag{2-40}$$

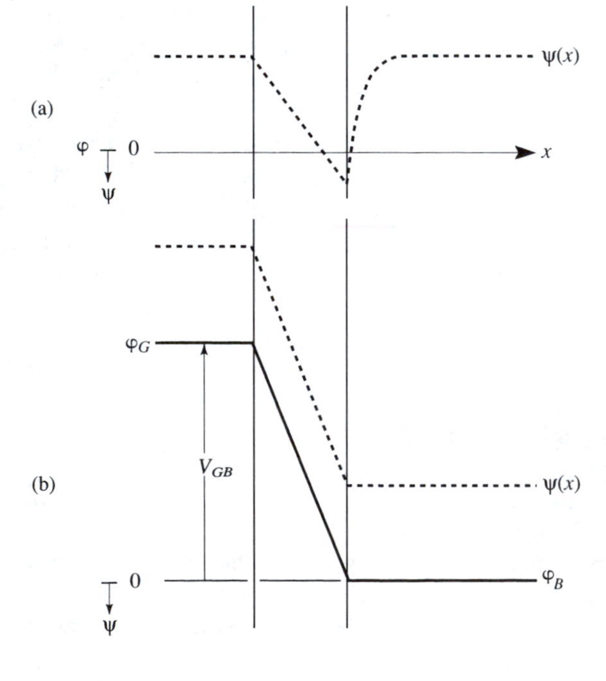

Figure 2-16 Qualitative potential profiles for MOS capacitor with thin layer of positive charge at the oxide–silicon interface, approximating the effect of Q_f. (a) Case of $\psi_{GB} = \varphi_{GB} = 0$. (b) Case of flatband condition, created by negative bias V_{GB}.

Hence the effect of Q_f on flatband voltage in this instance is less than 50 mV. For a (111) surface, by contrast, the contribution of Q_f to flatband voltage typically exceeds 100 mV.

The next kind of parasitic charge we shall consider is one that plagued early MOS technology. It proved to be the charge associated with positive ions, primarily alkali ions, present in the oxide. Identifying and correcting this problem was a crucial achievement on the way to practical MOS technology; it also constitutes a fascinating detective story.

In the early 1960s it was customary to deposit the aluminum field plate on MOS capacitors and MOSFETs by using a filament evaporator. In this apparatus, bits of aluminum wire that were placed on a tungsten filament melted when the filament was heated to incandescence. The molten aluminum adhered to the tungsten, or "wetted" it, and aluminum atoms that evaporated from the melt traveled in straight lines through the vacuum chamber to strike the silicon substrate, producing the desired layer of deposited aluminum. It was discovered in about 1963 that the tungsten filament was a significant source of sodium contamination. Sodium entered the molten aluminum, then the deposited layer, and finally the oxide layer upon which the aluminum was deposited. While general "cleaning up" of the MOS process from start to finish was necessary to eliminate the ubiquitous sodium contamination, especially that contributed by fingers, the elimination of filament evaporation was the greatest single advance in solving the sodium problem. The apparatus that was subsequently employed was the electron-beam evaporator, in which a focused electron beam strikes a pellet of aluminum, causing local melting and hence evaporation. The pellet itself is supported in a water-cooled cup that does not transmit contaminants to the aluminum charge. Furthermore, the electron beam is bent by a magnetic field so that the sample receiving the aluminum cannot "see" the electron source, and hence is shadowed from any contaminants that may originate there. For a significant year or two, this important technological change was held as a trade secret; but by the mid-1960s it had been transmitted by word of mouth throughout the intensely active and competitive MOS community.

It was essential to solve the sodium problem by eliminating sodium. That is, one cannot "live with" this parasitic charge in the MOS capacitor by compensating for it elsewhere, for at least two reasons: First, under conditions where such contamination is present, the amount of contamination is variable, and the electrical properties of the device vary as a result. Second, the sodium ions in the oxide are rather mobile; by applying a normal voltage to the MOS capacitor, one exerts an electrical force on these contaminating positive ions. With the temperature elevated to a few hundred degrees celsius, the ions can, with negative bias, be induced to move toward the field plate. Positive bias, in turn, drives them back toward the oxide–silicon interface. Such a positional change changes the flatband voltage V_{FB} and also the threshold voltage V_T. For these reasons, the presence of such contaminating alkali ions constituted a totally unacceptable condition.

To understand the effect of ionic contamination in the oxide, consider the situation shown in Figure 2-17. A sheet of charge having an areal density Q (in C/cm^2) is located at an arbitrary position within the oxide, as shown in the physical representation of Figure 2-17(a). Let the position be designated x, and let the spatial origin be

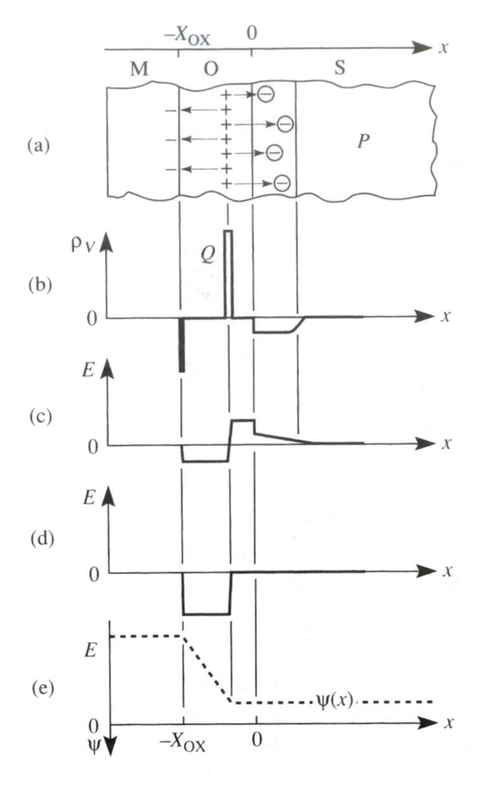

Figure 2-17 Hypothetical sheet of parasitic positive charge in oxide for analyzing problem of arbitrary one-dimensional ionic contamination of oxide. (a) Physical representation. (b) Charge profile with $H_D = \psi_{GB} = 0$. (c) Electric-field profile. (d) Electric-field profile after negative external bias is applied to create flatband condition. (e) Potential profile for flatband condition.

placed at the oxide–silicon interface. In this one-dimensional problem, lines of force emanating from the positive ionic charges must terminate on negative charges either in the gate electrode or in the silicon. The charge-density profile accompanying this situation, shown in Figure 2-17(b), assumes that $H_D = 0$ and that $\psi_{GB} = \varphi_{GB} = 0$ as well. The area of the delta function labeled Q must equal the sum of the areas of the two negative-charge profiles. The one on the left represents the sheet of charge in the field plate, and the near rectangle on the right, acceptor-ion charge in the depletion layer.

The field profile shown in Figure 2-17(c) corresponds to the charge profile of Figure 2-17(b). Let us now alter the situation by increasing the negative bias on the gate until just enough negative charge is delivered to the gate to cause all lines of force originating on contaminating ions to terminate on the gate electrode. Under these conditions, the silicon is free of charge and field, and has been brought to the flatband condition. The field and potential profiles accompanying this new condition are shown in Figure 2-17(d) and (e). This particular value of gate-to-bulk voltage is, once again, the *flatband voltage* for the device and structure assumed, and can be written

$$V_{FB} = -\frac{Q}{C(x)} = -\frac{(X_{OX} + x)Q}{\epsilon_{OX}}. \tag{2-41}$$

The capacitance per unit area $C(x)$ is a function of the arbitrarily assumed position for the ionic charge sheet, and $\epsilon_{OX} = 0.345 \times 10^{-12}$ F/cm is the absolute permittivity of the silicon oxide. Multiplying and dividing by the oxide thickness X_{OX} gives us

$$-\frac{X_{OX} + x}{X_{OX}} Q \frac{X_{OX}}{\epsilon_{OX}} = -\frac{X_{OX} + x}{X_{OX}} \frac{Q}{C_{OX}}, \tag{2-42}$$

where C_{OX} is the capacitance per unit area of the total-thickness MOS capacitor. Noting, then, that Q may be written $\rho_V(x)dx$, where ρ_V is volumetric charge density, it follows that the flatband voltage for an MOS capacitor having an arbitrary charge profile in the oxide may be written

$$V_{FB} = -\frac{1}{C_{OX}} \int_{-X_{OX}}^{0} \frac{X_{OX} + x}{X_{OX}} \rho_V(x)\, dx. \tag{2-43}$$

Figure 2-18 represents a charge distribution $\rho_V(x)$ extending through the oxide layer. When voltage is applied in accordance with Equation 2-43, the result is the charge-density profile shown there, wherein the positive and negative portions of the profile have equal areas.

EXERCISE 2–10 Justify the algebraic signs on the right-hand side of Equation 2-41.

■ The first minus sign is present because voltage is being applied to the left-hand plate of the capacitor in Figure 2-17(a), while the charge of interest Q resides on the right-hand "plate." Because position x is inherently negative for the origin assignment in Figure 2-17, and is smaller than X_{OX}, the quantity $X_{OX} + x$ is positive as desired.

The third and last major category of parasitic charge found in the oxide has been named *oxide trapped charge* [34], described by the symbols Q_{ot} for areal charge density and N_{ot} for areal charge-center density. It consists of carriers that have become "lodged" in the oxide.

Here it is worthwhile to combine two concepts presented earlier. Avalanche breakdown in a *PN* junction involves the presence of "hot" or energetic carriers in the space-

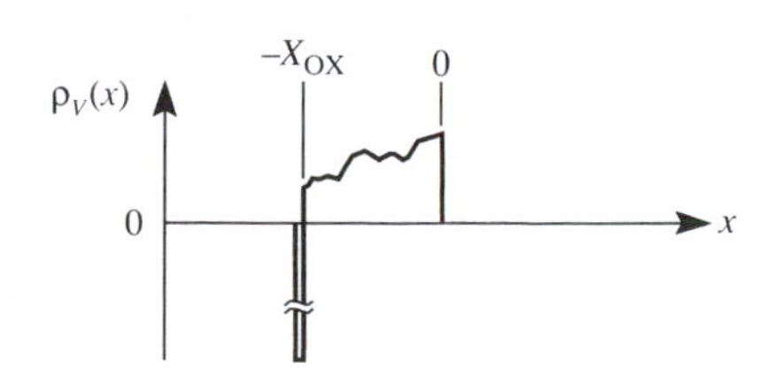

Figure 2–18 One-dimensional parasitic charge-density distribution in oxide, consisting of positive alkali ions. Negative external bias has equalized the negative and positive charge components, creating the flatband condition.

charge layer. Also, arbitrarily taking the case of electrons, by using the silicon conduction-band edge as energy reference, we find that the kinetic energy of the hot electron is equal to its total energy, which, in turn, is equal to its distance in energy above the band edge. Referring back to Figure 2-9(b), then, we can make the plausible observation that an electron with an energy somewhat over 3 eV (or 3 V, in the electrostatic-potential terms used there) will be capable of passing into the conduction band of the oxide layer. In an analogous manner, an energetic hole could enter the valence band of the oxide. Somewhat greater energy is required in the case of a hole, and also its mobility in the oxide is smaller than that of the electron, so let us continue to focus mainly on the electron case.

Once in the oxide conduction band, the electron can be caused to drift to a field plate that is "floating," or unconnected to any terminal. The necessary field in the oxide can be created in a MOSFET-like device by applying a source–drain bias that causes voltage division in the two oxide capacitances C_{SG} and C_{GD}, which are necessarily in series. The field, furthermore, in addition to causing electron drift in the oxide's conduction band, makes it possible for lower-energy electrons to enter the conduction band by means of tunneling.

EXERCISE 2–11 Explain the last statement.

■ Visualize the effect of a significant positive bias on the gate of the device in Figure 2-9. The conduction-band edge of the oxide will then drop steeply from right to left. This has the effect of thinning the potential barrier that keeps electrons in the silicon. With sufficient barrier thinning, electron tunneling becomes possible, so that an electron can enter the oxide with a fraction of the energy needed to surmount the barrier right at the interface.

Given this method for achieving electron penetration of the oxide layer, one can charge the field plate by delivering electrons to it. Furthermore, the necessary hot electrons can be supplied by causing avalanche breakdown in a junction that is near the interface, or that intersects the interface. This chain of phenomena, in fact, is the basis for an *erasable-programmable read-only memory* [43], or EPROM, to use its acronym. A particular MOSFET is either ON or OFF (representing either a 1 or a 0), depending on the state of charge of its floating field plate. Stored information can be erased by irradiating the device with ultraviolet light, thus giving energy to the stored electrons and permitting them to return to the silicon. This kind of EPROM is but one example of what is now a rather large group of nonvolatile memory options based upon MOS technology. The term *nonvolatile* in this context describes memories that retain their information (in this case, the charges on insulated field plates) when bias voltages are removed.

The reason for describing the EPROM here is to illustrate a case wherein carriers are deliberately introduced into the oxide. Some of the carriers so introduced become trapped in the oxide, and thus contribute to Q_{ot}. Especially with repeated write–erase cycles, the residual carrier charge increases.

But problems with such charge, unfortunately, are not confined just to cases wherein the carriers have been intentionally introduced into the oxide. With the continuous and rapid reduction of device dimensions or "feature sizes" and a slower reduction of operating voltages, the result has been growth in field values that are encountered in the normal operation of devices. As a result, hot carriers are increasingly present even when not intentionally created, and these can enter the oxide to become trapped there. The result is a shift in flatband voltage V_{FB}, and an equal shift in the very important threshold voltage V_T. Furthermore, the story does not end even here. Energetic radiation is once again a problem because it can create hole–electron pairs for which one or both members become trapped in the oxide.

EXERCISE 2–12 Explain the last statement.

■ Semiconductors and insulators differ only in degree, both having energy gaps. Hence, hole–electron pairs can be created in both, with greater energy being required in the insulator case. For example, the insulator diamond requires radiation energy above some 5.5 eV, while in the oxide case, higher energy is required, above approximately 8 eV.

Radiation-caused pair production in silicon oxide would seem to contribute self-neutralizing charges, but the electrons have a higher probability of escape from the oxide because of their higher mobilities [37]. As a result, the radiation-induced contribution to Q_{ot} tends strongly to be positive. There is evidence that under some fortuitous circumstances, this positive charge is partly canceled by the net negative interface trapped charge that the radiation may also cause. But even if this situation could be created consistently and reliably, it does not constitute a solution, because carrier mobility in the MOSFET channel declines concurrently.

EXERCISE 2–13 Why should mobility decline as Q_{it} increases?

■ Channel carriers are confined to a very thin region near the oxide–silicon interface. The presence of additional charge centers at the interface will result in more frequent scattering of carriers, by the same mechanism as in the case of scattering by impurity ions. The resulting shrinkage of the mean free path is responsible for the declining mobility.

EXERCISE 2–14 Why should one be concerned about declining mobility?

■ Equations 1-17 and 1-22 show that drain current I_D is directly proportional to electron mobility μ_n. As a result the important figure of merit, transconductance

g_m, is also proportional to μ_n, as Equation 1-24 shows. Hence the radiation leads to declining device performance.

Fluctuations in the population (of electrons) in the interfacial states causes fluctuations in channel-carrier scattering and, hence, in mobility. Such *mobility fluctuations* are currently under study as an additional source of $1/f$ or flicker noise in MOSFETs [44]. Beyond this, the transitions of electrons to and from the status of conducting carriers in the channel on the one hand, and trapped carriers in interfacial states on the other, give rise to *density fluctuations* that may also contribute to $1/f$ noise. These issues will probably be resolved in the near future [44].

2-7 Calculating Threshold Voltage

A good starting point for threshold-voltage determination is Equation 2-23, repeated here for convenience,

$$V_{GB} = H_D + \psi_{GS} + \psi_{SB}. \tag{2-44}$$

The term H_D involves only the barrier-height-difference effect. Assuming P-type silicon with a net doping of $10^{15}/\text{cm}^3$ and an aluminum field plate, we are informed by Figure 2-10 that $H_D = -0.9$ V. And the last term ψ_{SB}, the band-bending term, involves only a single phenomenon, treated accurately in Section 2-3. But the middle term ψ_{GS}, the potential drop through the oxide portion of the MOS capacitor, is fixed by the four kinds of charge outlined in Sections 2-5 and 2-6 and, in addition to these, by the charge in the silicon. All of these charge components are imposed "on" the fixed portion of the MOS capacitor, and therefore contribute to a potential drop through the oxide as a consequence of the basic law governing the parallel-plate capacitor. To start, let us assume that our technology and application will permit us to neglect the interface trapped charge altogether. In addition, let us assume that conditions that cause oxide trapped charge have been avoided. Then, combining Equations 2-44, 2-36, 2-40, and 2-43 yields this result:

$$V_{GB} = H_D - \left[\frac{1}{C_{OX}} \int_{-X_{OX}}^{0} \left(\frac{X_{OX} + x}{X_{OX}}\right) \rho_V(x)\, dx\right] - \frac{Q_f}{C_{OX}} - \frac{Q_S}{C_{OX}} + \psi_{SB}. \tag{2-45}$$

The first three terms on the right-hand side occurred previously in equations wherein flatband voltage V_{FB} was specified. This was true because each of these parasitic effects makes a fixed charge contribution to the MOS capacitor at a fixed spatial position (provided we rule out alkali-ion motion by maintaining room temperature). Hence each of these three terms makes a fixed contribution to flatband voltage V_{FB}. The last two terms of Equation 2-45 are by contrast variable and can change sign as well. But under flatband conditions, ψ_{SB} is by definition zero, which in turn rules out ionic and surplus-carrier charge in the semiconductor, so that Q_S also vanishes. Hence in the present case,

$$V_{FB} = H_D - \left[\frac{1}{C_{OX}}\int_{-X_{OX}}^{0}\left(\frac{X_{OX}+x}{X_{OX}}\right)\rho_V(x)\,dx\right] - \frac{Q_f}{C_{OX}}. \qquad (2\text{-}46)$$

Now let us make the realistic assumption that charge in the oxide has been eliminated, so that Equations 2-45 and 2-46 can each be simplified by dropping the integral term. Then, to develop an expression for the applied-voltage value that brings the MOS capacitor to the threshold of strong inversion V_T, note again that this is an applied voltage that causes the surface potential to be equal in magnitude and opposite in sign to the bulk potential. That is, in the present unnormalized notation, $\psi_S = -\psi_B$ (or in normalized notation, $U_S = -U_B$). By definition, then,

$$\psi_{SB} \equiv \psi_S - \psi_B = -2\psi_B, \qquad (2\text{-}47)$$

so that from Equation 2-44 and (modified) Equation 2-45,

$$V_T = H_D - \frac{Q_f}{C_{OX}} - \frac{Q_S}{C_{OX}} - 2\psi_B = V_{FB} - \frac{Q_S}{C_{OX}} - 2\psi_B. \qquad (2\text{-}48)$$

Let us choose $N_A = 10^{15}/\text{cm}^3$, for which $2\psi_B = -0.59$ V. The value $N_A = 1.6 \times 10^{15}/\text{cm}^3$ was employed in Sections 2-2 through 2-4 because it corresponds to the convenient value $U_B = 12.0$, but this small change in assumed doping has a negligible effect on H_D. Also, it makes no change at all in Q_f, which is doping independent. Hence the term $-Q_S/C_{OX}$ is the only one remaining to be evaluated. From Equations 2-29 and 2-30 we have

$$|Q_S| = \epsilon|E| - \left(\frac{kT}{q}qN\epsilon\right)^{1/2}\frac{dW}{d(x/L_D)}. \qquad (2\text{-}49)$$

But using Approximation C of Table 3 and noting that $|W_S| = q|\psi_{SB}|/kT$, gives us

$$\left.\frac{Q_S}{C_{OX}}\right|_{V=V_T} = -\frac{|Q_S|}{C_{OX}} = -\frac{|Q_S|X_{OX}}{\epsilon_{OX}} = -\frac{X_{OX}}{\epsilon_{OX}}\sqrt{2qN\epsilon|\psi_{SB}|} = -0.40\text{ V} \qquad (2\text{-}50)$$

for the present threshold condition, where the minus sign has been introduced because the space charge in the P-type silicon is negative. Hence from Equation 2-48 we find

$$\begin{aligned} V_T &= -0.9\text{ V} - 0.8\text{ V} - (-0.4\text{ V}) - (-0.6\text{ V}) \\ &= -1.7\text{ V} + 1.0\text{ V} = -0.7\text{ V}, \end{aligned} \qquad (2\text{-}51)$$

where $V_{FB} = -1.7$ V, the result of the two nonideal factors present.

It can be seen in Equation 2-51 that a corresponding ideal device would have $V_T = +1$ V. Hence the nonideal factors have caused a threshold-voltage shift of -1.7 V and a change of algebraic sign. That is, the capacitor will possess a beyond-threshold inversion layer when the external voltage is set at $V_{GB} = 0$. The *P*-type MOS capacitor possesses an inversion layer, or "channel," even with no voltage applied. As noted at the end of Section 2-4, the *N*-type MOS capacitor does not have a channel under similar conditions, and so the factors H_D and Q_f contributed to the early popularity of *N*-type starting material, and hence to *P*-channel devices, because it was desired that there be no inversion layer in the $V_{GB} = 0$ condition. Subsequent technological developments (particularly ion implantation) have made it possible to "tailor" doping and hence charge conditions near the surface, which has greatly increased the range of choice open to the MOS designer. Consequently, *N*-channel devices are now the dominant type, and *P*-channel devices are rarely encountered outside CMOS integrated circuits.

3 MOS-Capacitor Modeling

In the process of accurate MOS-capacitor modeling, Chapter 2 was able to make appreciable use of equations developed in Appendix F for the step junction. This was because the semiconductor portion of the former (which we shall assume to be uniformly doped) poses a problem identical to that posed by one side of the latter, with an important proviso. The MOS capacitor is virtually at equilibrium for any applied bias within a wide range of values (current through the insulator must be negligible), while the junction is at equilibrium for only one bias value, $V = 0$. Thus, for any arbitrarily chosen value of bias imposed on a particular MOS capacitor, one must specify a unique step-junction structure to have the kind of equivalence just noted.

From the practical point of view, an important difference between the MOS capacitor and the step-junction diode is that in the former a parallel-plate capacitor is inevitably and permanently in series with the semiconductor phenomena one wishes to study (often by capacitance measurements). As a result, comparisons of the two diodes pose subtle problems, with solutions that are sometimes not obvious. Such studies and comparisons, however, have contributed greatly to an understanding of the MOS system. Let us begin by extending the approximate-analytic modeling of Appendix F into the MOS context, where it has important features that are exact. Then device and equivalent-circuit comparisons will enable us to understand the bias-dependent capacitance descriptions that follow in this section.

3–1 Exact-Analytic Surface Modeling

Early treatments of the semiconductor-surface problem used a combination of analytic and numerical methods [45–49]. Later workers recognized the existence of asymptotic solutions of extremely simple form [50–52], and this had the effect of extending the scope of pure-analytic methods in treating such problems. Curiously, the latter group directed their attention toward the step-junction problem rather than the semiconductor-surface problem. The two problems are fully equivalent analytically, as is pointed out at the beginning of Appendix F, but this equivalence was not immediately recognized. The next expansion of analytic capabilities came with the realization that one of the asymptotic solutions defines an invariant spatial origin that permits the writing of explicit approximate expressions of simple form for both asymptotic and intermediate solutions [53]. This constitutes the depletion-approximation replacement (DAR) presented in Appendix F.

The Poisson–Boltzmann equation, Equation F11, can be integrated once analytically, yielding the electric-field expression given in differing forms in Equations F20 through F22 and F24. In each case, electric field is given as a function of electrostatic potential. Hence, given the value of a critical potential, such as that at the silicon surface (which is to say, at the oxide–silicon interface), we can calculate electric field in the silicon right at the surface directly and *exactly*. No approximations are involved. An application of Gauss's law then permits us to determine areal charge density in the silicon with equal accuracy. The charge so determined consists in the general case of ionic depletion-layer charge *and* of inversion-layer or channel charge for one polarity of applied bias, and of accumulation-layer charge for the other polarity. This straightforward procedure was applied early to the MOS-capacitor (and MOSFET) problem [54]. By combining this method with the DAR, however, one can reintroduce the spatial variable, thus generating analytic *profiles* of electric field and volumetric charge density [53, 55]. The necessary expressions will now be developed for subsequent use.

In Figures 2-4 through 2-9 we have oriented the MOS capacitor with the semiconductor portion at the right-hand side, and we shall continue to do so. Hence, to relate the MOS-capacitor problem with the junction problem defined in Figure F2, it is necessary to identify the present semiconductor (or, let us say, silicon) region with the right-hand region of Figure F2. Figure 3-1(a) carries over from Figure F2(a) the physical representation of a *PN* junction at equilibrium. The bulk potentials U_{01} and U_{02} are shown thereon, as is the junction position x_J, which enters into discussion in Appendix F. The corresponding "surface" problem is represented in Figure 3-1(b). The critical surface position is now designated x_S, and the equilibrium bulk potential of the silicon sample is U_B; the subscript B was introduced in Section 2-2 and is conveniently applied here to take advantage of the fact that there is now only one silicon region to deal with. Thus, letting $U_{02} \equiv U_B \equiv q\psi_B/kT$, and letting $W = U_B - U$, Equation F24 becomes

$$\frac{dW}{d(x/L_D)} = -\sqrt{2}\left[\frac{e^{U_B}(e^{-W} + W - 1) + e^{-U_B}(e^{W} - W - 1)}{e^{U_B} - e^{-U_B}}\right]^{1/2}, \tag{3-1}$$

and is consistent with Figure F2 in the following sense: There, U_B is positive (the right-hand side of the sample is N-type), W is positive-going, and electric field is negative; the

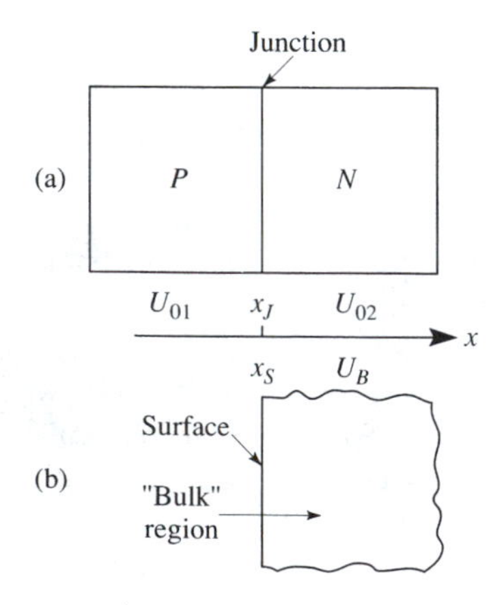

Figure 3–1 Relating the *PN*-junction and MOS-capacitor problems. (a) Physical representation of the general step-junction problem adopted for exact analysis. (b) Physical representation of the silicon portion of an MOS capacitor, or of the bulk sample for semiconductor-surface analysis. We have let $U_{02} \equiv U_B$.

lines of force are directed leftward from positive charge in the silicon to negative charge somewhere toward the left. We have here a depletion-inversion condition. It is best, however, to remove the sign ambiguity by a method similar to that used in Equation F27. Let U_B be replaced by $|U_B|$, agree to let W always be positive-going, and then for the depletion-inversion condition in *any* sample, N-type or P-type, we have

$$\left|\frac{dW}{d(x/L_D)}\right| = -\sqrt{2}\left[\frac{e^{|U_B|}(e^{-W} + W - 1) + e^{-|U_B|}(e^{W} - W - 1)}{e^{|U_B|} + e^{-|U_B|}}\right]^{1/2}, \qquad (3\text{-}2)$$

an expression that is valid without restriction. Absolute-value bars have been placed on normalized electric field as a reminder that its algebraic sign must be fixed by an appeal to physical reasoning.

For particular ranges of $|U_B|$ and W, Equation 3-2 can be replaced by certain of the approximate equations given in Table 3 [55, 56]. The range designations are overlapping so that one can select an approximation tailored to a particular problem. In the conditions of most interest, depletion and inversion coexist (Approximations A through C). But inversion is important only near and above threshold. Under conditions of extreme inversion, ionic charge can be neglected (Approximation D). It is interesting to note that the depletion–inversion expressions can be readily separated into terms that account for electric field arising from an inversion-layer charge and from a depletion-layer (ionic) charge. In the full and exact expression, Equation 3-1, the first major term in the numerator gives the potential dependence of ionic charge, and the second, of inversion-layer charge. The approximate expressions in Table 3 are labeled as to which charge they describe, and in a number of cases, inspection easily verifies their origin in the general expression. Ionic-charge expressions are given first (Approximations E and F). Subtraction of the ionic component from the total, of course, gives the inversion-layer component. Using Gauss's law, once again, either field component can be converted into a charge. For extremely small values of W, which is to say, for extremely small departures from neutrality, depletion and accumulation conditions yield the same electric field magnitude (Approximation G).

EXERCISE 3–1 In Table 3, how does Approximation A′ follow from A?

■ Substituting $W = 2|U_B|$ into Approximation A yields

$$\sqrt{(e^{-2|U_B|} + 2|U_B| - 1 + e^{-2|U_B|+2|U_B|} - 2|U_B|e^{-2|U_B|} - e^{-2U_B}}$$
$$= \sqrt{2(-2|U_B|e^{-2|U_B|} + 2|U_B|)} = 2\sqrt{|U_B|(1 - e^{-2|U_B|})}.$$

EXERCISE 3–2 In Table 3, how does Approximation B follow from A?

■ For the minimum value $|U_B| = 3$ indicated in B, larger than the minimum value $|U_B| = 2$ indicated in A, the subterms $(-W - 1)$ can be dropped from the second major term in A, resulting in an approximate doubling of the maximum error in the ranges specified when one goes from A to B.

EXERCISE 3–3 How is Approximation B′ obtained?

■ Dropping the exponential term in Approximation A′ causes only an approximate doubling of the maximum error in the specified ranges in going from A′ to B′.

The condition of *accumulation* is introduced in Appendix F. In this condition, one has negative space charge consisting of surplus electrons in an *N*-type region or positive space charge consisting of surplus holes in a *P*-type region. The expression for normalized electric field that is valid without restriction can be obtained from Equation 3-2 simply by replacing $|U_B|$ by $-|U_B|$:

$$\left|\frac{dW}{d(x/L_D)}\right| = \sqrt{2\left[\frac{e^{-|U_B|}(e^{-W} + W - 1) + e^{|U_B|}(e^{W} - W - 1)}{e^{-|U_B|} + e^{|U_B|}}\right]}. \tag{3-3}$$

The last two expressions in Table 3 deal with accumulation. The very mild restriction that $|U_B| > 2$ gives Approximation H. It is evident, then, that with increasing W, the result becomes the exponentially rising function independent of $|U_B|$ that is given as Approximation J. This explains why a single curve is plotted in Figure F3 to represent the accumulation condition.

The exact expressions, Equations 3-2 and 3-3, are summarized in Table 4 along with the simple result obtained for the MOS capacitor employing an intrinsic semiconductor. Finally, we should point out that one is usually interested in field in the semiconductor at the interface, and that the corresponding value of normalized potential in such a case is W_S. The subscript has been omitted in Tables 3 and 4, however, in the interest of generality.

EXERCISE 3–4 Point out the qualitative difference between Equations 3-2 and 3-3 that causes the depletion–inversion condition to involve two kinds of charge and a threshold phenomenon, while the accumulation condition does not.

■ In Equation 3-2 for the depletion–inversion case, the first major term in the numerator is associated with depletion-layer charge, while the second is associated with inversion-layer charge. The product $\exp(|U_B| - W)$ and the product $\exp(-|U_B| + W)$ cross over for $W = |U_B|$, the threshold of weak inversion. There the ionic and inversion charge densities at the plane of interest (usually the surface) are equal. In Equation 3-3 for the accumulation case, the product $\exp(|U_B| + W)$

exceeds the product $\exp(-|U_B| + W)$ for all allowable (positive) values of $|U_B|$ and W. Hence there is no threshold condition.

3–2 Comparing MOS and Junction Capacitances

Sorting out the coexisting phenomena in the *PN* junction was a major preoccupation of the 1950s, and even of the 1940s [57]. Small-signal junction capacitance was among these effects, and by the close of the 1950s, its voltage dependence was understood reasonably well; this is illustrated in Figure G2(b) and involves a crossover of diffusion and depletion-layer capacitance. The chief difficulty in the experimental observation of these capacitance functions is the massive conductance that accompanies them in the upper forward-bias regime. For good and practical reasons, understanding the nuances of current versus voltage received considerably more interest and attention than did junction capacitance. The *PN*-junction diode was an important device in its own right, as well as being the key constituent of devices such as the BJT.

Since the MOS capacitor normally exhibits negligible conductance at any bias, the capacitance–voltage observation was a vital method (and almost the only method) for experimental characterization of the device, a process completed about ten years later than for the junction. But the convenience of having conductance out of the picture was offset by other complications. First, as already noted, the parallel-plate oxide capacitor is permanently in series with the semiconductor phenomena in the bulk region that are of primary interest, so that one is required to "reach through" the oxide with any experimental observation. Second, the inversion layer has a negligible effect on dynamic *PN*-junction properties, but becomes a dominant factor in the case of the MOS capacitor. As a result, the physical arrangement for supplying (or not supplying) carriers to the inversion layer has an important bearing on dynamic behavior. An important option for such carrier delivery places a heavily doped region adjacent to the MOS sandwich. In other words, the MOS capacitor inevitably becomes a two-dimensional entity in one of its important embodiments. Third, as also noted, equilibrium analysis is appropriate for the bulk region of the MOS capacitor at any bias, while *only* at zero bias is equilibrium analysis valid for the *PN* junction. Hence to relate the two devices in the aspects wherein they are identical, one must "design" a different junction sample to correspond to every change of bias placed on the MOS capacitor. The immutable differences between the junction and MOS diodes obscured these identities, as did the fact that the time periods of the most intense study were about a decade apart. Not surprisingly, two sets of terminology evolved for certain identical critical conditions in the two devices. It is instructive to construct a series of examples that display these conditions.

Figure F2(b) presents a potential profile for an arbitrarily chosen asymmetric step junction at equilibrium. Analogous profiles are presented in Figure 3-2 for a set of other step junctions at equilibrium. The left-hand portion of each profile is dashed because our focus here is on the MOS capacitor (or, alternatively, on the "surface" prob-

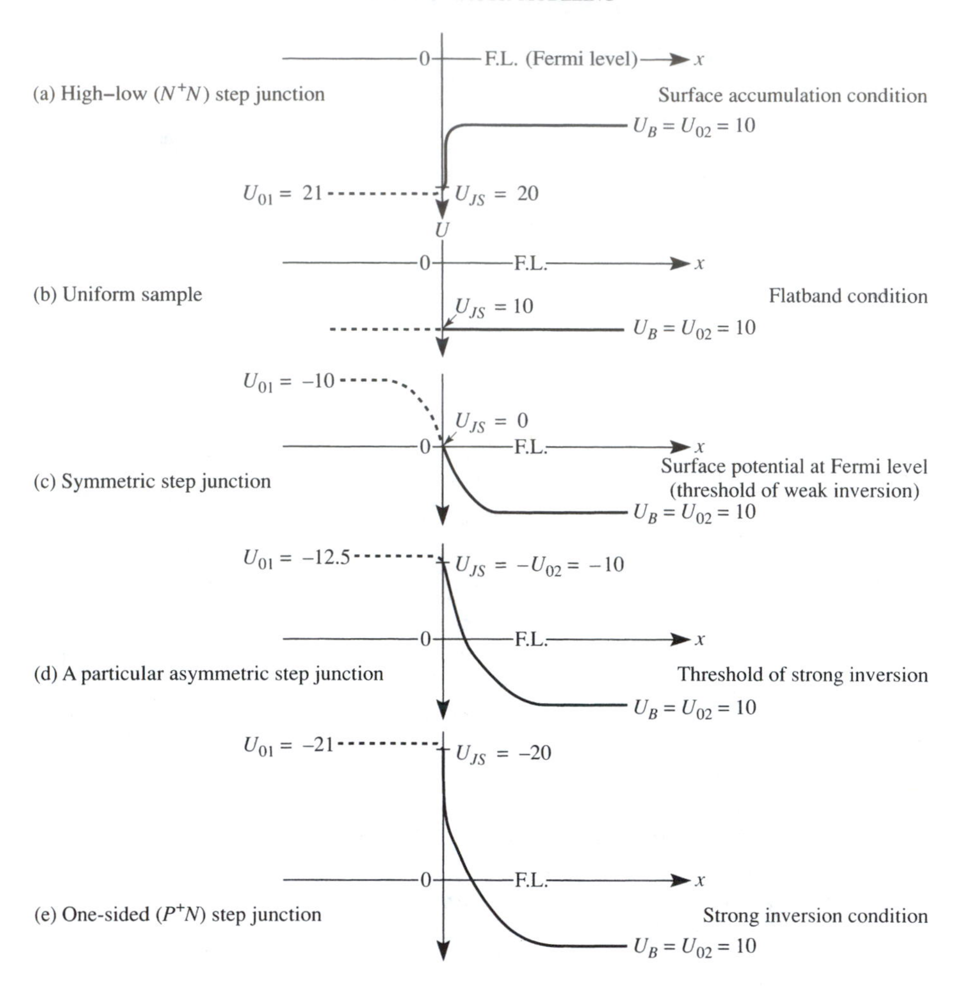

Figure 3–2 Relating the step-junction problem to the MOS (or surface) problem. Potential profiles are shown for a series of equilibrium step junctions that are identified at the left. These profiles are analogous to that of Figure F2(b). The right-hand side represents also the bulk region of an MOS capacitor at various bias conditions, with the condition named at the right. See text for discussion of the individual cases.

lem); thus the left-hand half of the junction problem is "virtual." We have designed these junction samples so that the conditions in their right-hand halves correspond identically to conditions in the bulk of a particular MOS capacitor at a particular bias. In the high–low or N^+N junction of Figure 3-2(a), there is a potential drop of one normalized unit on the high side. The trivial uniform-sample case in Figure 3-2(b) corresponds to the highly significant flatband condition in MOS terms. In Figure 3-2(c) we have the familiar symmetric step junction, corresponding to the threshold of weak inversion. The significance of this term is that at this point, the two carrier populations are balanced right at the surface. For further voltage increase, the incipient inversion layer is present, but with an areal density that contributes negligible conductance parallel to the surface.

The particular asymmetric junction depicted in Figure 3-2(d) is one in which the hole density at the junction equals the electron density far from the junction in the right-hand side. Further bias increase causes the high-side potential drop to fall, approaching one normalized unit, and causes a qualitative change in the potential profile. The curve proceeds upward more steeply than before, a consequence of the inversion layer right at the junction. Hence this is the important threshold condition. In Figure 3-2(e), then, the bias change has been carried to an extreme, yielding the one-sided or grossly asymmetric step junction, with a one-unit drop on its high side. In MOS terms, this is the condition of strong inversion in the MOSFET.

3–3 Small-Signal Equivalent Circuits

The *PN* step-junction sample shown in Figure 3-1(a) was the object of exact-analytic treatment, chosen for consistency with earlier literature. Also, the *N*-type surface sample in Figure 3-1(b) was chosen for consistency with that step-junction sample. Now we revert to the *P*-type substrate used throughout Chapters 1 and 2. (This is the most important and common case because of the mobility advantage of inversion-layer electrons.) Fortunately, the exact expressions written in the format of Equations 3-2 and 3-3 are valid for either case.

The MOS capacitor can be modeled usefully by means of small-signal equivalent circuits. The simplest case, of course, is the accumulation case. There the holes that have been attracted to the oxide–silicon interface form a thin, dense layer that closely resembles the sheet of charge on the positive plate of a conventional parallel-plate capacitor. Thus the equivalent circuit degenerates into a single capacitor, C_{OX}. (Recall that C_{OX} is stated by custom as a capacitance per unit area.) As a refinement, a small resistor can be placed in series with C_{OX}, but that is not necessary unless the silicon has unusually low net doping.

Only slightly more complex is the case depicted in Figure 2-5, wherein the MOS capacitor is given a bias of the depletion–inversion polarity, but well below threshold. Therefore the equivalent circuit consists of the capacitor C_{OX} in series with a depletion-layer capacitance that involves the same phenomenon as that of the *PN* junction. In MOS work, however, this feature is usually termed *bulk* capacitance. The charge involved can be described as ionic, but this term is unsuitable for subscript purposes because *inversion-layer* capacitance is to be introduced next. Hence the plain term "bulk" once again serves a useful role, giving us the symbol C_b for bulk capacitance. Note well that C_b is also measured in pF/cm^2 for consistency with the case of C_{OX}. This practice differs from that for the junction capacitances, wherein the depletion-layer capacitance C_t and the diffusion capacitance C_s (or C_u) are measured in pF. For this reason, the differing subscripts for the two depletion-layer capacitances (*b* and *t*) are justified, even though the same phenomenon is being modeled. Hence the appropriate equivalent circuit for this case can be drawn and labeled as in Figure 3-3.

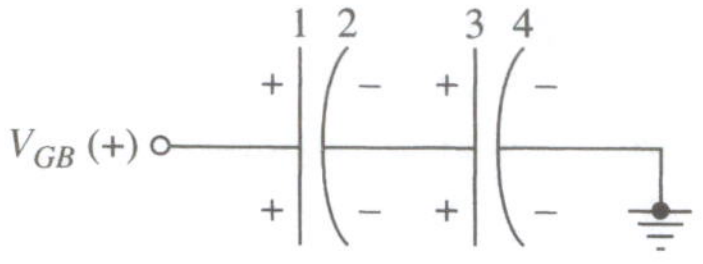

Figure 3-3 Equivalent circuit for the MOS-capacitor condition represented physically in Figure 2-5. The substrate is significantly depleted, but far below the threshold of strong inversion. (The numbered plates are discussed in Exercise 3-5.)

EXERCISE 3-5 How can one justify the presence of capacitor plates 2 and 3 in the equivalent circuit presented in Figure 3-3? The physical structure that is being modeled is shown in Figure 2-5, where there is nothing present having the physical properties of a pair of capacitor plates.

■ Plates 2 and 3 in Figure 3-3 are common, but mutually isolated from the balance of the circuit. On one side is the SiO_2 layer, an excellent insulator. On the other side is a silicon layer that is virtually devoid of carriers, the depletion layer. Hence the negative charge on plate 2 and the positive charge on plate 3 are the result of charge separation. The total charge on plates 2 and 3 remains zero for any value of V_{GB}. Since no net physical charge is required at the oxide–silicon interface, no physical capacitor plate or plates is needed in the physical structure to validate a model consisting of two capacitors in series.

Now let V_{GB} be increased more in the depletion–inversion (or positive) polarity, bringing the small-signal inversion-layer capacitance C_i into play. Like C_b, it involves a "virtual" plate at the oxide–silicon interface. The second plate of C_i is the inversion layer itself, tightly pressed against the oxide–silicon interface. But here a complication enters. In a one-dimensional MOS capacitor, the inversion-layer electrons have dropped into a potential well that in effect isolates them. As a result, C_i displays widely differing frequency dependences that are determined by the mode of communication of these electrons with the ohmic contact made to the substrate. In the simplest case, an N^+ region is placed adjacent to the field plate. This physical arrangement is shown in Figure 3-4(a). When a small signal is applied to a capacitor of this structure, electrons flow readily in and out of the inversion-layer well through the medium of the N^+ region, causing C_i to fit simply in parallel with C_d in the equivalent circuit. The N^+ region, in turn, is ohmically connected to the right-hand terminal. Observe that the configuration in Figure 3-4(a) is equivalent to the case of a MOSFET with the source and bulk terminals common. The velocity of charge transport in and out of the inversion layer approximates that in an ordinary conductor.

The other extreme situation exists when no special provision is made to deliver charge to and from the inversion layer in response to the applied signal. If the signal frequency is low enough, spontaneous carrier generation and recombination can yield a surplus carrier population that "follows" the signal. For typical lifetimes, however, such a frequency is very low indeed—of the order of 10 Hz. For this case too, C_i and C_b are in parallel, and mutually in series with C_{OX}. For a much higher frequency such as 100 kHz, however, C_i simply disappears from the equivalent circuit. As might be

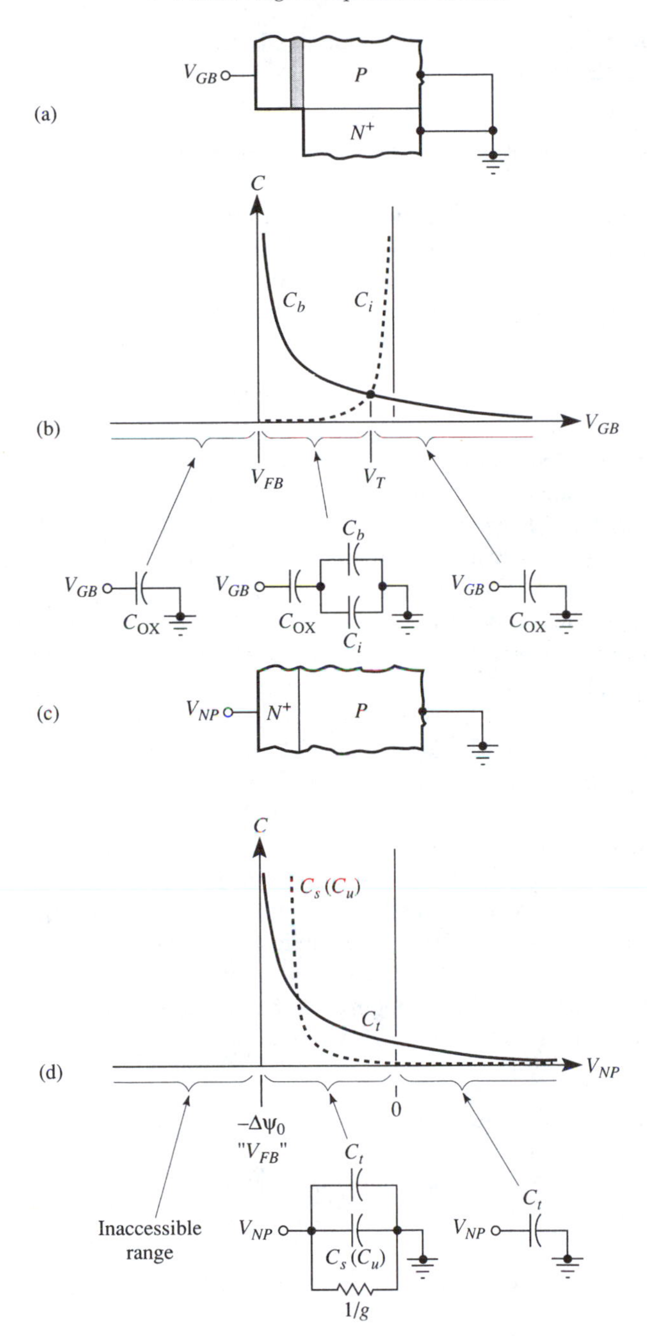

Figure 3–4 Comparing the MOS-capacitor and *PN*-junction diodes. (a) Physical structure of capacitor used for comparison. (b) Bulk capacitance C_b and inversion capacitance C_i versus applied voltage V_{GB}, showing appropriate equivalent circuit by operating regime. (c) Physical structure of junction diode used for comparison. (d) Depletion-layer capacitance C_t and diffusion capacitance $C_s(C_u)$ versus applied voltage V_{NP}, showing appropriate equivalent circuits by operating regime.

expected, there exists a wide frequency range wherein C_i is frequency-dependent, an unsatisfactory state of affairs.

EXERCISE 3–6 What is unsatisfactory about a frequency-dependent C_i?

■ The point of the kind of equivalent circuit being discussed here is to model the device using passive components that are *fixed* in value, and hence frequency-independent. One expects complications at frequency extremes, but when the complications exist at intermediate frequencies, the value of the equivalent circuit is diminished.

The gross dependence of C_b and C_i on bias, V_{GB}, is summarized in Figure 3-4(b), along with an indication of which equivalent circuit is appropriate for each regime of operation. Note that the voltage axis does not have an origin, because its location depends upon the parasitic effects treated in Chapter 2. As a result, the capacitance profiles are valid for a real-life MOS capacitor. The depletion-layer capacitance diverges at the flatband voltage V_{FB}. On the other hand, the C_b and C_i curves intersect precisely at the threshold of strong inversion V_T. The physical basis of this fact has an explanation that is straightforward, but not obvious. It is the subject of Section 3-6, where the phenomenon is termed MOS-capacitance crossover. (Section 3-7 treats the same topic analytically.) In principle, the inversion-layer population and C_i both increase without limit for $V_{GB} > V_T$, so that C_i shunts out C_b, and the equivalent circuit reverts simply to C_{OX}.

For clarifying contrast, we take the familiar N^+P diode shown in Figure 3-4(c) and plot in Figure 3-4(d) a set of capacitance profiles analogous to those in Figure 3-4(b) for the MOS capacitor. The differences are intriguing and informative. This time there *is* an origin, the only bias condition for which the device is at equilibrium. When the imposed junction voltage (which for simplicity we are taking to be equal to V_{NP}) is given the inaccessible value $-\Delta\psi_0$, we have a condition analogous to the flatband voltage V_{FB}. For equivalent P-type dopings in the junction diode and MOS capacitor, the C_t curve is nearly the same as the C_b curve.

EXERCISE 3–7 Why are the C_b and C_t curves not identical?

■ The bias voltage on the junction is measured from the N^+ region to the P region, and thus includes the small voltage drop on the N^+ side of the junction, which amounts to 26 mV and is nearly bias-independent [58]. In the MOS capacitor, on the other hand, the voltage drop relevant to C_b is that from the oxide–silicon interface to the ohmic contact on the P-type region. Because the N^+-side drop in the junction case is both small and nearly constant, the two capacitance profiles are similar. The element C_b is often compared to C_t for this reason. In spite of the small difference in the two capacitances, it is important to realize that the silicon solutions, $\psi(x)$ and $E(x)$, are *identical* when $\psi(x_J) = \psi(x_S)$.

EXERCISE 3–8 Why doesn't C_s, diffusion capacitance, appear in Figure 3-4(b)?

■ Diffusion capacitance is absent from the MOS device for the trivial reason that the SiO_2 layer prevents carrier injection into the silicon.

The fact that C_i does not appear in Figure 3-4(d) is nontrivial, however. After all, the N^+P junction possesses an inversion layer at equilibrium. But the behaviors under bias of MOS-capacitor and *PN*-junction inversion layers are totally different. With a growing positive bias (V_{GB}) on the *P*-substrate MOS capacitor, surface potential ψ_S follows closely behind gate potential ψ_G, and electron density at the interface, $n(x_S)$, increases *exponentially* with ψ_S.

In the *PN*-junction case, junction potential ψ_J is very insensitive to bias [59]. For the sake of a specific example, let us take the sample shown physically in Figure 3-5(a), having the equilibrium electron-density profile shown in Figure 3-5(b). This sample is chosen to have respective net-doping values of $N_D^+ = 10^{20}/\text{cm}^3$ and $N_A = 10^{16}/\text{cm}^3$. This makes it parallel to the high–low junction of Figure F9. In either sample, potential drop on the high side is very nearly one normalized unit. Hence the junction potential is

$$U_J = |U_{01}| - 1 = \ln (N_D^+/n_i) - 1 = 22.0. \qquad (3\text{-}4)$$

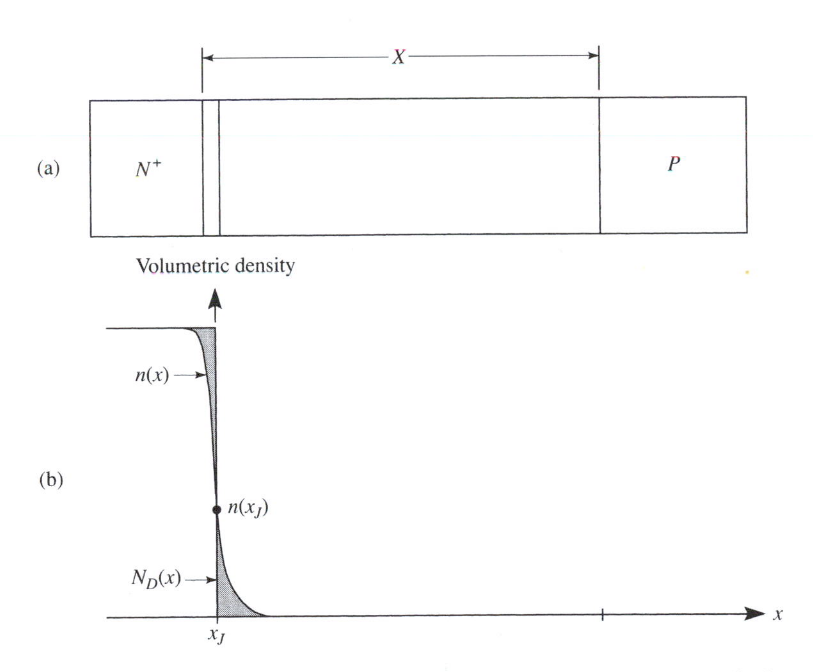

Figure 3–5 An equilibrium one-sided junction with the same respective net-doping magnitudes as the high–low junction of Figure F9. (a) Physical representation, showing approximate depletion-layer boundaries. (b) Net-doping profile $N_D(x)$ and electron profile $n(x)$.

In a grossly asymmetric junction of either type, the junction potential is uniquely related to (peak) electric field E_J by the expression [58]

$$E_J = (8.914\text{ V/cm})e^{U_J/2} = 119\text{ kV/cm}. \tag{3-5}$$

The numerical result is for the present example. (The quantity E_J can also be obtained from Approximation B in Table 3.) For a first-order estimate of the effect of a small positive bias increment ΔV_{NP} applied to this one-sided junction, let us assume that the profile shown in Figure 3-5(b) simply translates leftward. The amount of this leftward translation would be 1 Å for every micrometer of rightward translation for the right-hand depletion-layer boundary! This is because the doping asymmetry is 10^4, and charge balance on the two sides must be preserved. In the process, the population of the inversion layer is diminished. In summary we may observe that not only are inversion-layer responses to bias on the MOS capacitor and the *PN* junction grossly different in magnitude, but also they are different in *algebraic sign*! The miniscule inversion-layer capacitance of the one-sided junction is a negative capacitance.

EXERCISE 3–9 Continue to assume a simple leftward translation of the profile in Figure 3-5(b) in response to an increment in reverse bias. Doesn't the presence of the inversion layer upset the approximate calculation just given?

■ No. A more detailed description of the process is this: When the profile shifts leftward from its equilibrium position by a small amount Δx, the change in areal density of donor ions on the N^+ side is

$$+\,N_D^+\left(1 - \frac{1}{e}\right)\Delta x.$$

The change in the areal density of inversion-layer electrons is

$$-\,N_D^+\left(\frac{1}{e}\right)\Delta x.$$

This change eliminates the need for an equal number of donor ions, which were "anchoring" these electrons with their lines of force. Thus, subtracting this number from the positive increment in donor–ion density yields

$$N_D^+\left(1 - \frac{1}{e}\right)\Delta x + N_D\left(\frac{1}{e}\right)\Delta x = N_D^+\Delta x.$$

This areal-density change precisely balances that on the right,

$$N_A(10^4)\Delta x.$$

EXERCISE 3–10 Estimate the percentage change in $n(x_J)$ when the profile in Figure 3-5(b) translates leftward 1 Å.

■ The reverse current in a reverse-biased junction is so near zero that the balance of drift and diffusion currents in the transition region of the junction must be preserved. Taking specifically the position x_J and equating the current components there yields

$$\left.\frac{dn}{dx}\right|_{x=x_J} = n(x_J)\,\frac{E_J}{kT/q}.$$

Here

$$n(x_J) = \frac{N_D^+}{e} = \frac{10^{20}/\mathrm{cm}^3}{e} = 3.68 \times 10^{19}/\mathrm{cm}^3.$$

Using the approximation

$$\left.\frac{dn}{dx}\right|_{x=x_J} = \frac{\Delta n}{\Delta x}$$

yields

$$\Delta n = n(x_J)\,\frac{E_J}{kT/q}\,\Delta x,$$

or

$$\Delta n = (3.68 \times 10^{19}/\mathrm{cm}^3)\left(\frac{119 \times 10^3\ \mathrm{V/cm}}{0.02566\ \mathrm{V}}\right)(-10^{-8}\ \mathrm{cm}).$$

Thus

$$\Delta n = -1.71 \times 10^{18}/\mathrm{cm}^3,$$

and the percentage change in $n(x_J)$ is

$$\left(\frac{-1.71 \times 10^{18}/\mathrm{cm}^3}{3.68 \times 10^{19}/\mathrm{cm}^3}\right)(100) = -4.6\%.$$

In fact, the profile $n(x)$ is not quite preserved under reverse bias. A more detailed examination of this matter is found in Problems A12 and A13.

3-4 Ideal Voltage-Dependent Capacitance

We will illustrate the combination of depletion and inversion effects in an MOS capacitor with varying bias by starting with a simple case. Assume the device to be ideal in the sense of Sections 2-1 through 2-3. In addition to setting aside parasitic effects temporarily, we shall adopt a specific structure. Take the oxide thickness to be $X_{OX} = 0.1\ \mu m$, and the net doping of the silicon to be $N_A = 10^{15}/cm^3$. Because of ideal properties in the device assumed, the flatband condition will exist at $V_{GB} = 0$ V.

As the bias V_{GB} is quasistatically varied through wide ranges on either side of zero, let us observe the differential capacitance $C = dQ/dV_{GB}$, where Q is the net static charge on the field plate. The differential-capacitance meter delivers a small-signal voltage, assumed to have a frequency of 100 kHz, to the MOS capacitor and observes the resulting current. Bias voltage is delivered simultaneously through a high-impedance device (in principle, simply an inductor), which renders the bias circuit open to the ac signal. However, improved and refined methods for observing differential capacitance have evolved, as noted later, because of the importance of such characterization.

We will proceed through a sequence of bias conditions parallel to those examined in Section 3-3. With V_{GB} strongly negative, the silicon surface is heavily accumulated, so C_{OX} is observed. In Figure 3-6 a normalized ordinate is employed, so that C/C_{OX} has the value of unity at its left-hand branch. Of particular importance are the amount and distribution of incremental charge stored and recovered in the silicon in response to the ac signal applied, a matter treated in Figure 3-7, which shows the location and shape of the silicon "plate" of the MOS capacitor. The accumulation case just cited

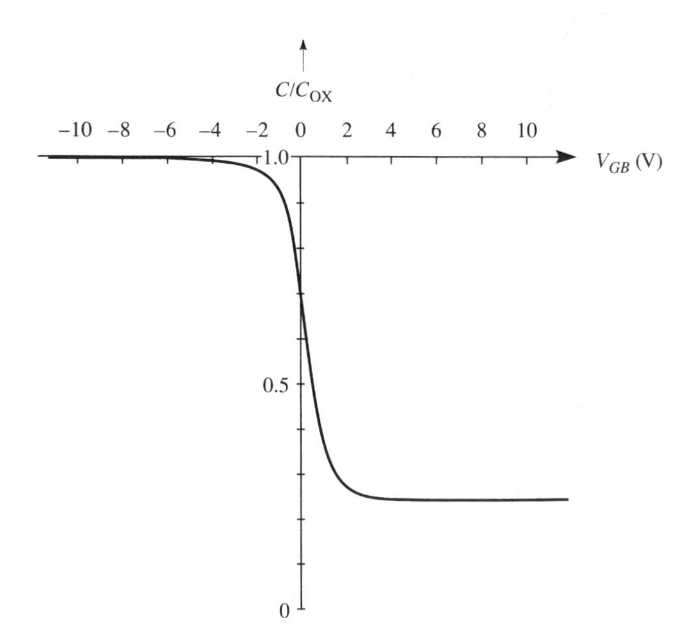

Figure 3-6 Normalized capacitance versus voltage for a device with $X_{OX} = 0.1\ \mu m$ and $N_A = 10^{15}/cm^3$, measured at 100 kHz.

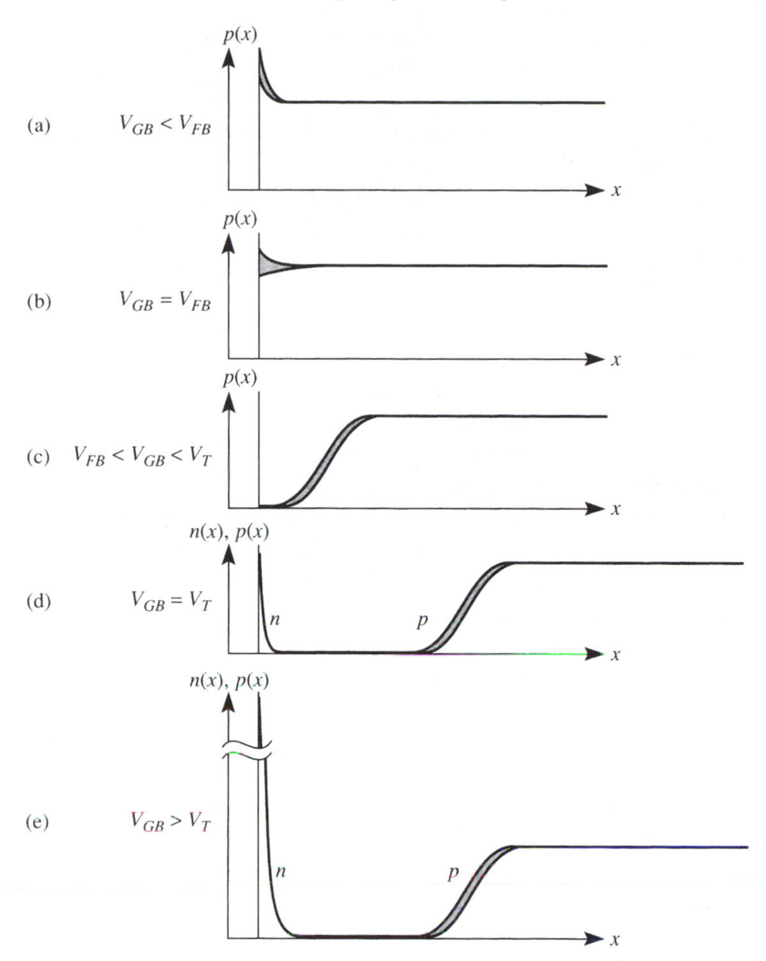

Figure 3–7 Carrier profiles in P-type substrate of MOS capacitor, showing incremental charge-storage locations in presence of a 100-kHz small signal under certain bias conditions. (a) Accumulation. (b) Flatband. (c) Depletion. (d) Depletion and threshold of strong inversion. (e) Depletion and strong inversion.

is shown in Figure 3-7(a), and the nearness of the incremental charge to the oxide–silicon interface is evident.

Now move to the flatband case, for which the analogous charge distribution is shown in Figure 3-7(b). The "centroid" of the incremental charge is now farther removed from the interface. There is a nonnegligible voltage drop in the silicon, and overall capacitance has declined somewhat through the resulting series-capacitor effect. Let us be quantitative, assuming an applied small-signal voltage that causes a surface-voltage amplitude of 50 mV, or about two normalized units. (For a fixed applied-signal amplitude, the excursions of surface potential will vary from one bias regime to another, but that fact does not invalidate the quantitative illustration presented here.)

Figure F4 shows that for $W = W_S = 2$ on the accumulation curve (the curve farthest right in the set), the distance to the point where $W(x/L_D)$ drops to zero is about 1 Debye length. This can be taken as a measure of line-of-force penetration into the silicon; where $W = 0$, neutrality still exists. This distance is about 0.13 μm, appreciably greater than $X_{OX} = 0.1$ μm. On the opposite voltage excursion, the common $W(x/L_D)$ curve is relevant, and for $W_S = 2$ this time, the line-of-force penetration is about 1.5 Debye lengths. Increasing signal amplitude leads to increasing asymmetry. Because the time-dependent lines of force penetrate so deeply at the flatband-bias condition, C/C_{OX} exhibits about a 30% drop from its former value in the device arbitrarily selected. This can be seen in Figure 3-6.

As bias change is continued in a positive direction, a depletion layer forms, as illustrated in Figure 3-7(c). The representation here is fully equivalent to that in Figure G3(b) for a step junction. With further positive change in bias, the depletion-layer boundary retreats farther from the interface, the depletion capacitor becomes "thicker," or smaller in value, and so the series combination of C_{OX} and C_b drops in value. This process saturates as the threshold voltage is approached. The saturation is evident in Figure F4. Visualize a curve for $U_{02} = U_B = 11.5$, the case for $N_A = 10^{15}/\text{cm}^3$. At the place where the near-vertical curve diverges from the common curve, one has the threshold condition. After that, line-of-force penetration (which is to say, depletion-layer thickness) stabilizes, growing only about one additional Debye length. Recall that the spatial zero in Figure F4 corresponds to the point on the $p(x)$ curve where $p(x) \approx 0.55$ p_0, placing that spatial position right at the depletion-layer boundary. The reason for saturation, of course, is that lines of force originating on the field plate can now terminate on the inversion-layer charge right at the interface. Figure 3-7(d) portrays conditions right at threshold, $V_{GB} = V_T$. No incremental charge is represented in the inversion layer because a relatively high frequency (100 kHz) was assumed, and no "fast" source of inversion-layer electrons was assumed. The recombination-generation process is much too slow to follow the ac signal. Figure 3-7(e), then, shows the depletion layer at its maximum thickness because the surface is now heavily inverted. At this point, Figure 3-6 shows that C/C_{OX} has fallen to its minimum value of about 0.27.

The C–V curve in Figure 3-6 is taken from an extensive set of such curves calculated numerically by Goetzberger [59]. The relevant subset of his curves is given in Figure 3-8. The dashed portions of these curves correspond to the assumptions of the foregoing discussion, with a high-frequency ac signal and no electron-supplying mechanism. The solid continuing curves rising from the dashed curves show the effect of inversion-layer capacitance C_i when it has been included in the circuit. In other words, the capacitive function represented by each solid curve amounts to that resulting from putting C_{OX} in series with the sum of a pair of functions like those plotted in Figure 3-4(b). The points identified by lowercase letters correspond to the five parts of Figure 3-7. At and beyond threshold there are two relevant points for each bias, with the lower points literally corresponding to Figure 3-7 and the upper points including the C_i presence. At threshold, points d, capacitances C_i and C_b make equal contributions to net capacitance, although this is not obvious by inspection; a simple calculation is required to demonstrate it.

The theoretical extremes that are represented by the dashed and solid curves at the right in Figure 3-8 can be approached practically in several ways. One method cited

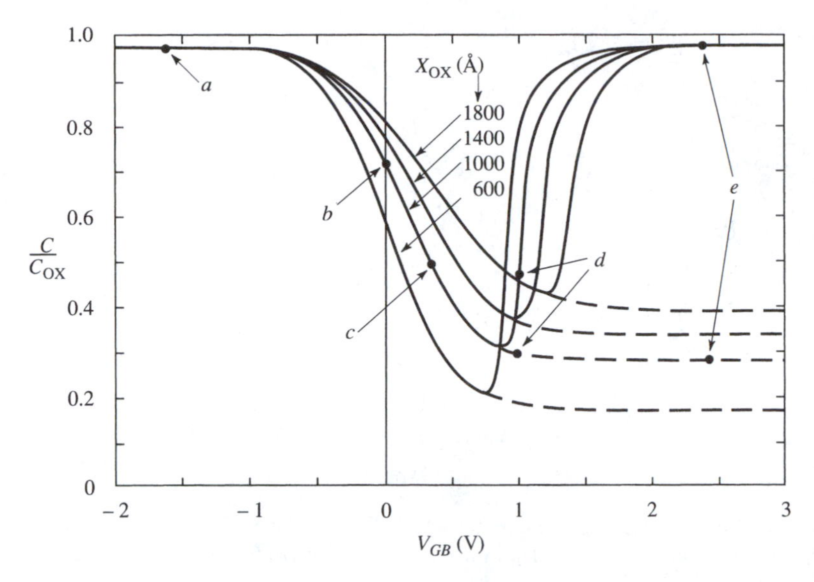

Figure 3–8 Ideal MOS curves of small-signal capacitance versus voltage for $N_A = 10^{15}/\text{cm}^3$ and several values of oxide thickness X_{OX}. The points indicated are related to the corresponding parts of Figure 3-7. (After Goetzberger [59], with permission from *Bell System Technical Journal*, copyright 1966, AT&T.)

earlier for activating C_i is a provision for supplying inversion-layer electrons. The structure of Figure 3-4(a) does this efficiently, because the signal voltage creates instantaneously a potential well at the surface. Into the well, electrons from the N^+ region are driven by an electric field. The mechanism here is identical to that of the charge-coupled device, or CCD [60]. The inversion-layer electrons are every bit as responsive as holes driven into and out of the depletion-layer boundary region by an electric field.

Another controlling variable cited earlier is signal frequency. In illuminating experiments carried out by Grove et al. [61], it was shown that the frequency dependence of *C–V* behavior leads to a continuum of curves. Several of these are shown in Figure 3-9. But to come even closer to the desired quasistatic conditions would require a frequency lower than 10 Hz, and differential-capacitance measurements there become extremely difficult. For this reason, quasistatic methods have been developed. A very slowly rising voltage ramp is applied to the MOS capacitor, and displacement current is observed. This method was first described by Berglund [62], and was then independently developed by Castagné [63], Kerr [64], and Kuhn [65]. Goetzberger et al. [66] give a good review of these and other *C–V* methods.

Grove and his colleagues experimentally illustrated other ways to manipulate the *C–V* curves between their theoretical extremes [67]. One method subjected the MOS capacitor to varying levels of illumination, thus producing excess minority carriers in the silicon that were available for transport in and out of the time-dependent potential well at the interface. The result is shown in Figure 3-10. Qualitatively very similar is the temperature dependence of the curves displayed in Figure 3-11. Because n_i and

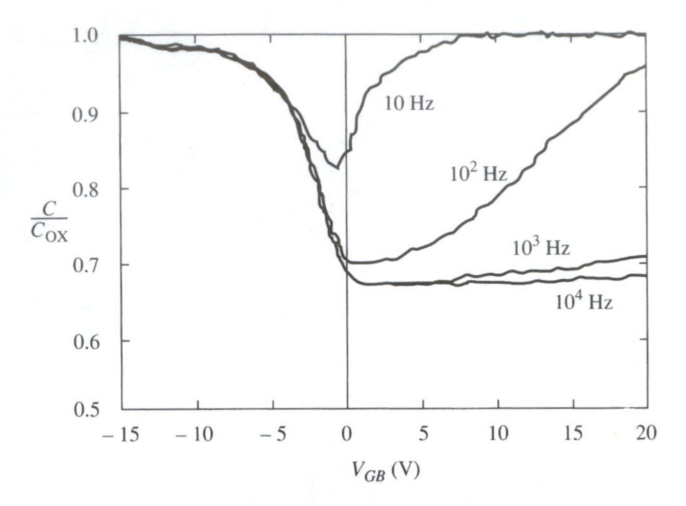

Figure 3–9 Capacitance versus voltage for an MOS capacitor with small-signal frequency as a parameter. (After Grove et al. [61], with permission, copyright 1964, American Institute of Physics.)

hence minority-carrier density are exponentially dependent on temperature, the supply of these carriers, essential in the C_i mechanism, is sensitive to temperature.

The quasistatic extreme achieved with a very low-slope ramp voltage has an opposite extreme, which is the application of a voltage pulse with a rise time far smaller than carrier lifetime. In a device with no electron-supplying provisions, the depletion layer can grow much thicker than the inversion-determined limit. This is termed a condition of *deep*

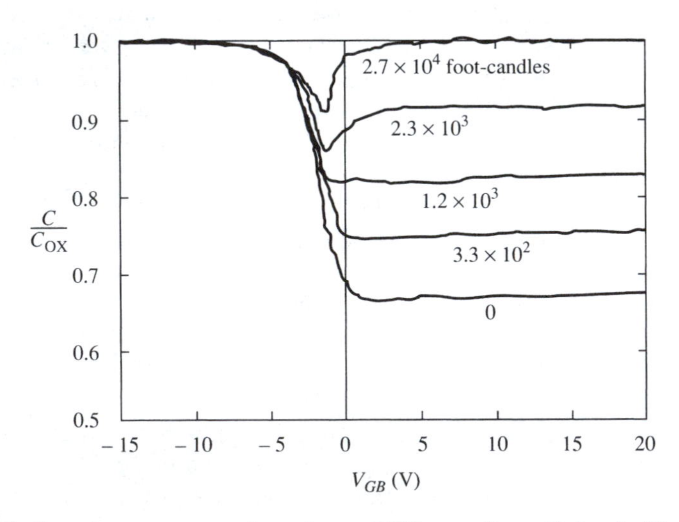

Figure 3–10 Capacitance versus voltage for an MOS capacitor with level of incident illumination as a parameter. (After Grove et al. [67], with permission, copyright 1965, Pergamon Press Ltd.)

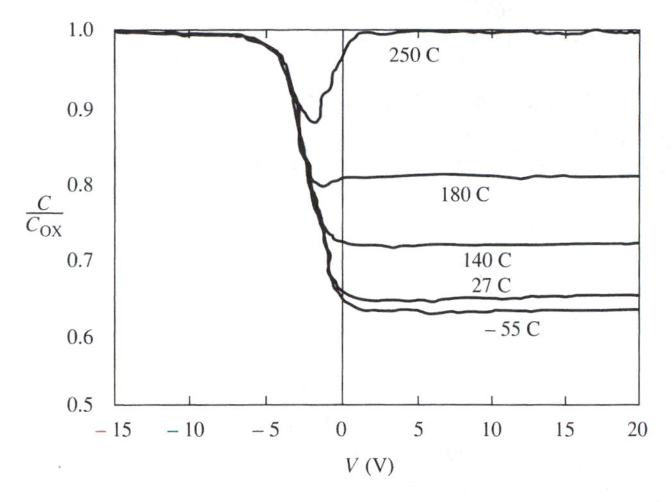

Figure 3–11 Capacitance versus voltage for an MOS capacitor with temperature as a parameter. (After Grove et al. [67], with permission, copyright 1965, Pergamon Press Ltd.)

depletion. Figure 3-12 shows the resulting characteristic, a continuation of the low-voltage depletion capacitance curve. For comparison, the inversion capacitance and the inversion–limited depletion-capacitance curves are also shown. Avalanche breakdown, or impact ionization, by carriers in the high-field region near the interface sets a limit to the depth of depletion. This is also displayed in Figure 3-12 and is marked by a singularity in the slope of the characteristic. The avalanche-produced electrons drop quickly into the inversion-layer potential well, and no further drop in capacitance is possible.

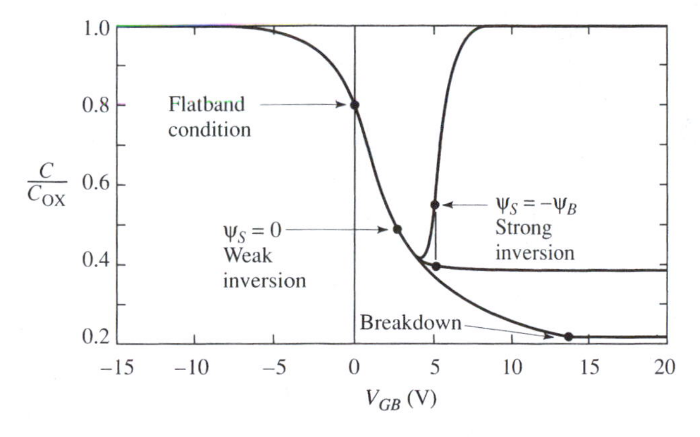

Figure 3–12 Capacitance versus voltage for an ideal MOS capacitor. The three curves at the right-hand side are, from top to bottom, that obtained with the inversion layer fully active, the inversion–limited depletion-capacitance characteristic, and the deep-depletion characteristic, with the last having an extreme-voltage limit set by avalanche breakdown. (After Grove et al. [67], with permission, copyright 1965, Pergamon Press Ltd.)

3-5 Real Voltage-Dependent Capacitance

The nonideal features of the MOS capacitor treated in Sections 2-4 through 2-7 affects its C–V properties markedly. Let us make an assumption that is plausible for present-day conditions, where both mobile ionic charge and interface trapped charge are negligible. With this assumption, we can write from Equation 2-46,

$$V_{FB} = H_D - \frac{Q_f}{C_{OX}} = -1.7\ \text{V}, \tag{3-6}$$

where the first two terms of Equation 2-51 were used to obtain the numerical value of V_{FB}. Here we have again taken the case of a P-type silicon substrate of $N_A = 10^{15}/\text{cm}^3$, an oxide thickness of $X_{OX} = 0.1\ \mu\text{m}$, and an aluminum field plate, all as in Section 2-4. The flatband point for an ideal MOS capacitor falls at $V_{GB} = 0$, as was noted in Section 3-4 and is illustrated in Figure 3-13(a). Also shown there is the C–V characteristic for the real device of $V_{FB} = -1.7$ V that was just determined. The flatband point must fall at the same value of C/C_{OX} on the C–V curve for the real device as on that for the ideal device, because the two capacitors are assumed to differ only in flatband voltage. The leftward shift of the real curve with respect to the ideal curve is displayed in Figure 3-13(a); it is the only difference in the two curves.

Converting from the P-type case of Figure 3-13(a) to the N-type case of Figure 3-13(b) produces a mirror reversal of the ideal characteristic. This fact is a consequence of the symmetry in the analytic solution that was stressed in Appendix F and Section

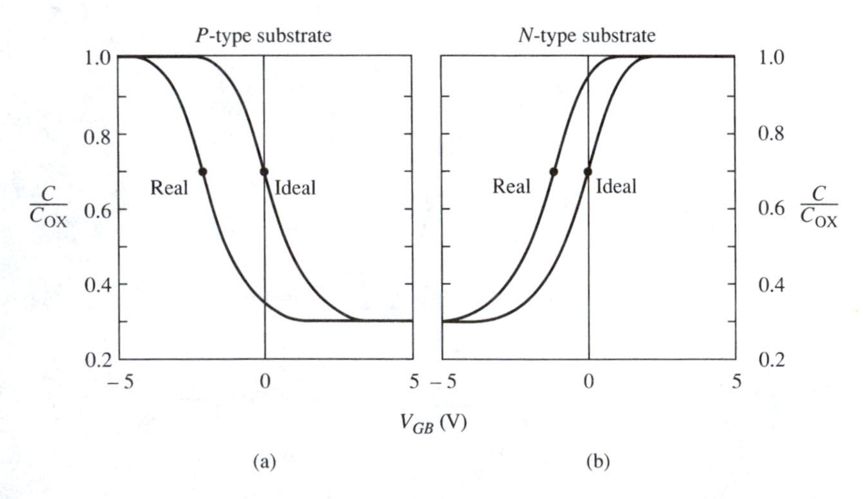

Figure 3–13 Comparing the C–V characteristics of an ideal MOS capacitor and a capacitor that is real but free of variable bias-dependent phenomena. (a) For the P-type substrate, Q_f and H_D cause a leftward shift of the real curve. (b) For the N-type substrate, Q_f causes the same leftward shift as before, while H_D contributes a smaller but still negative shift. (After Grove et al. [67] with permission, copyright 1965, Pergamon Press Ltd.)

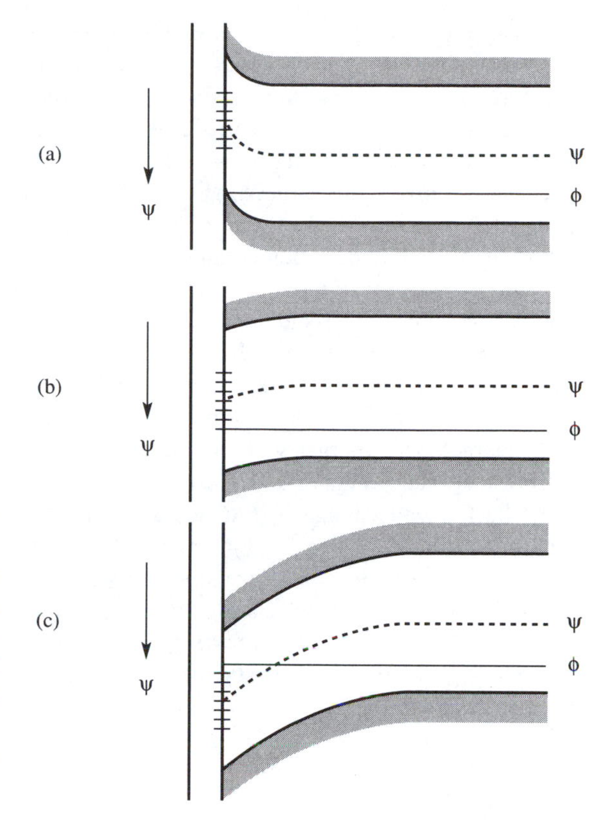

Figure 3–14 Band diagrams for various bias conditions on a *P*-substrate MOS capacitor with donorlike "fast surface states." (a) Negative bias, accumulation, and empty (positively charged) states. (b) Small positive bias and depletion, and states still charged. (c) Greater positive bias and depletion, with states now empty and hence neutral.

3-1. This time Figure 2-10 informs us that for an *N*-type substrate doped to $10^{15}/\text{cm}^3$, we have $H_D = -0.32$ V. Assuming Q_f to be the same as before, then

$$V_{FB} = H_D - \frac{Q_f}{C_{OX}} = -1.1 \text{ V}. \tag{3-7}$$

The translation of the real curve is again negative, but by a smaller amount than before.

The parallel shift of the *C–V* characteristic illustrated in Figure 3-13 occurs only when the parasitic effects present are bias-independent. As Section 2-5 and Figure 2-14 explain, the phenomenon responsible for interface trapped charge is bias-dependent or, more precisely, surface-potential-dependent. As a result, it causes shape distortion in the *C–V* characteristic. As surface-potential change and band bending occur, the responsible states are pushed above or below the Fermi level (depending upon the sense of the change) and give up or take on electrons, the essence of Bardeen's classic model of surface states [35].

One particular case of such states is illustrated in Figure 3-14. Donorlike states are assumed to exist uniformly through a range of potentials (or energies) located near the middle of the gap at the *P*-type silicon surface. For negative bias (let us assume that $V_{FB} = 0$), the accumulation direction, these states are well above the Fermi level, as can be seen in Figure 3-14(a), and hence are empty of electrons. The assumption that these states are donorlike means that their charge is positive, and hence that a negative shift of the *C–V* characteristic results. (Note the negative signs in front of all terms of Equation 2-45 that contain the factor $1/C_{OX}$. The negative signs are present because negative voltage on the gate is needed to respond to positive charge in other parts of the capacitor.) The state of constant charge is preserved even as band bending in the depletion direction commences, a condition illustrated in Figure 3-14(b). As a result, the portion of the *C–V* characteristic thus explored remains common in shape with the ideal characteristic, a fact confirmed in the leftmost curve of Figure 3-15, an experimental result from 1967 [39]. Finally, sufficient positive bias pulls the surface states well below the Fermi level, as shown in Figure 3-14(c), where they fill and become neutral, or "invisible," by virtue of their donorlike character. At this point the *C–V* characteristic merges with that for a device that is identical, but permanently free of interface trapped charge.

The flatband voltage V_{FB}, or the amount of parallel shift separating the real and ideal *C–V* characteristics for a given MOS capacitor, is also very useful in MOSFET analysis. Hence let us emphasize a relationship that may not have been evident heretofore. The key equation for relating voltage difference and potential difference (not the same!) with only the nonideal barrier-height effect present was Equation 2-21, $V_{GB} = H_D + \psi_{GB}$. With the generalization that allows other parasitic effects to be present also, the key equation becomes

$$V_{GB} = V_{FB} + \psi_{GB}. \tag{3-8}$$

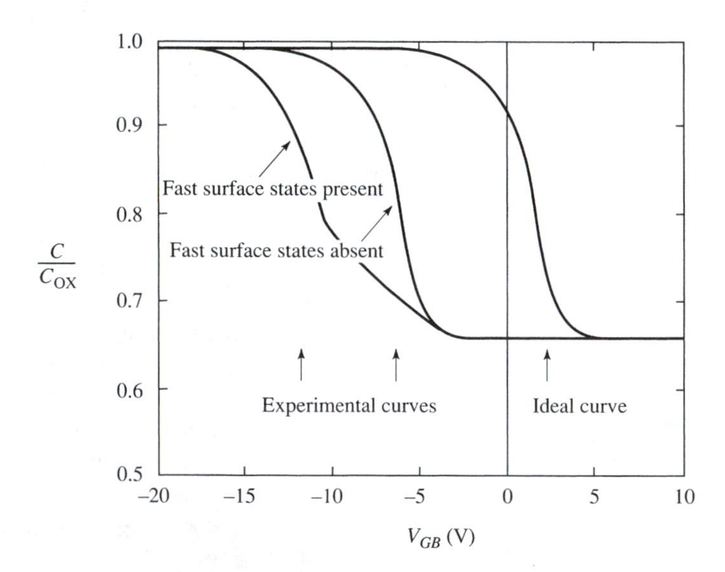

Figure 3–15 The effect of bias-dependent interface trapped charge residing in "fast surface states" at the oxide–silicon interface. (After Deal et al. [39] with permission, copyright 1967, The Electrochemical Society, Inc.)

Hence in the real device, ψ_{GB} can be written

$$\psi_{GB} = V_{GB} - V_{FB}. \tag{3-9}$$

To convert any of the capacitor equations of Chapter 2 into terms of V_{FB}, it is simply necessary to substitute $(V_{GB} - V_{FB})$ for ψ_{GB}.

3–6 Physics of MOS-Capacitance Crossover

Figure 3-4(b) noted the fact that bulk capacitance C_b and inversion capacitance C_i are equal at the threshold voltage. That is, the two C–V functions intersect at a point we have named the crossover point, for which $V_{GB} = V_T$. A clear but nonobvious physical explanation has been given for the crossover phenomenon [68], an explanation that we now outline.

In what follows, we employ the customary definitions for the relevant differential capacitances:

$$C_b \equiv \frac{dQ_b}{d\psi_S}, \tag{3-10}$$

where Q_b is bulk charge per unit area and ψ_S is surface potential, and

$$C_i \equiv \frac{dQ_n}{d\psi_S}, \tag{3-11}$$

where Q_n is inversion-layer charge per unit area, with the subscript carried over from Section 1-3. Also, the threshold condition is defined as that for which surface potential is equal in magnitude and opposite in sign to the bulk potential, with the Fermi level taken as potential reference.

Let us consider such an MOS capacitor with increasing gate voltage of the depletion–inversion polarity. The decline in bulk capacitance that results is governed by a power law, while the growth in inversion-layer capacitance is governed by a stronger and more complicated function, as we have seen. Curiously, these two differing functions appear to intersect at the threshold of strong inversion. This intersection is more striking when one considers that the charge storage in the inversion layer occurs very close to the surface (interface) in question; by contrast, the incremental charge storage involved in bulk capacitance (1) has a position that is remote from the surface, (2) has a position that is a function of surface potential, and (3) has a spatial profile that is "fuzzy" in the sense that the incremental charge is stored in a region of appreciable thickness, as illustrated in Figure 3-7. All three issues are relevant because potential calculations are very sensitive to charge position. We confine ourselves once again to a one-dimensional geometry, since the MOS capacitor poses a problem that can be made one-dimensional to any arbitrary degree of accuracy by increasing its area. Also, we consider only the case of a uniformly doped substrate. The question to be answered is this: Why do two mechanisms with such differing characters yield identical capacitance values at the threshold condition?

In addition to the considerations just outlined, there are alternative physical descriptions of the charge entering into bulk capacitance. To be specific, let us continue with the case of a *P*-type silicon bulk material. The increment of bulk charge that accompanies a positive increment of surface potential can be regarded (a) as the charge of the acceptor ions that are "uncovered," or (b) as the local decrease in hole population near the depletion-layer boundary, with the negative sign involved in "decrease" providing the necessarily negative sign for the charge increment. Note that the inversion-layer charge description is, by contrast, singular.

The descriptions just offered are consistent with the practice presented in Figure 3-16(a). There we affix to the x-axis a spatial origin located at the silicon surface, as is usually done. Then we translate to the right a charge profile by an amount depend-

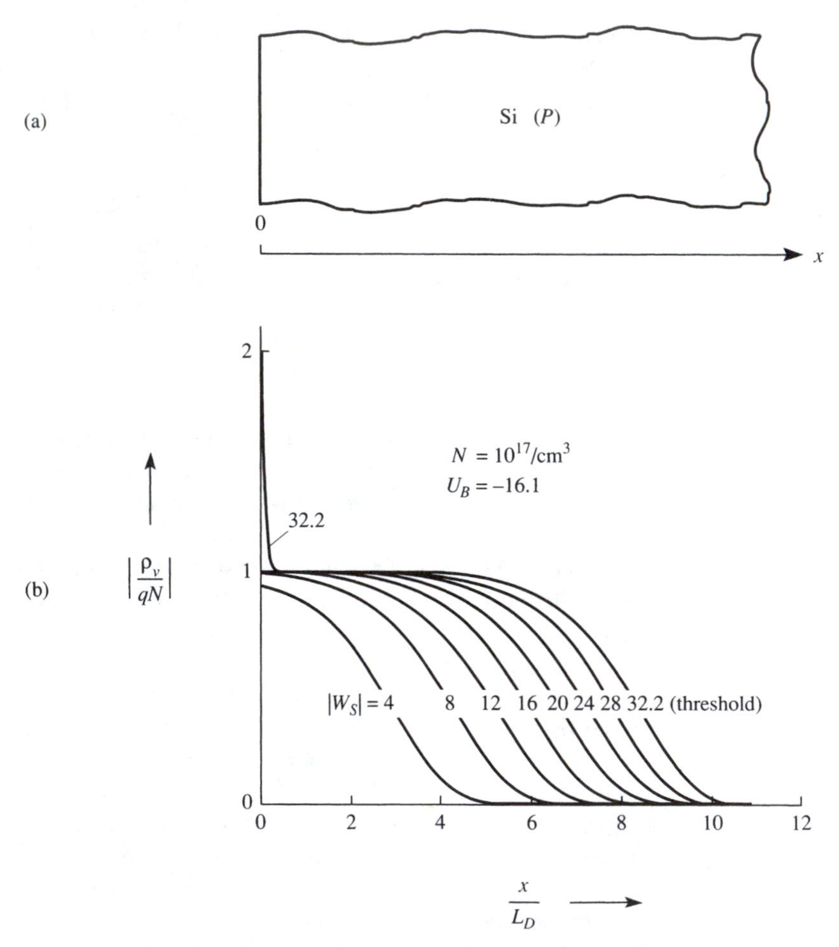

Figure 3-16 Traditional view of the silicon-surface or MOS-capacitor problem. (a) A *P*-type silicon sample with the spatial origin placed at the surface. (b) A series of numerically determined volumetric charge-density profiles for progressively increasing normalized surface potential W_S.

ing upon surface potential. Figure 3-16(b) shows a series of charge-profile positions for equal increments of surface potential, expressed in normalized terms, with the bulk potential (rather than the Fermi level) taken as reference. Definitions here are consistent with those given in Appendix F. Once bulk doping is specified, the charge profile, which comprehends both bulk charge and inversion-layer charge, is unique and well known.

By shifting conceptual gears, however, we find that the physical basis for the equality of bulk and inversion-layer capacitance at equilibrium becomes evident. Let us first deal with the issue of "fuzziness." In a one-dimensional problem, to reiterate, electric field is unlike electrostatic potential in that it is insensitive to charge position. Only the *total* charge per unit area beyond a certain position matters. The constancy of charge density through a range of distances that is evident in Figure 3-16(b) means that electric field will be linear through the same range, as shown in Figure 3-17(a) for a specific though arbitrary case. If the linear field profile is extrapolated to the x axis, it defines a position whose significance is explained in Appendix F. It is the position at which the abrupt space-charge boundary will occur when the depletion approximation is applied to the same MOS-capacitor problem, and when electric field at the silicon surface attributable to ionized impurities is the *same* for the depletion approximation as for the actual case. The shaded area in Figure 3-17(a) amounts to one normalized potential unit, kT/q, and accounts for the one-unit difference between the depletion-approximation potential-profile parabola and the asymptotic parabola of the exact numerical solution.

Now to be more specific, choose any position x in Figure 3-17(b) in the positional range where space-charge density is constant. Then the total charge per unit area to the right of that position (toward the bulk) will be the same for the depletion-approximation profile and for the actual profile. As a result, we may point out that the two crosshatched areas in Figure 3-17(b) are equal. This position (defined by extrapolating the linear portion of the field profile) is, of course, the origin of the depletion-approximation replacement (DAR) discussed in Appendix F.

With respect to this origin, let us define the surface position x_S. Furthermore, and this is where the conceptual shift enters, let us consider the surface potential ψ_S to be an independent variable. Then, in a thought experiment, the surface position becomes the dependent variable, $x_S = f(\psi_S)$. The silicon crystal exhibits positive or negative "growth" at its surface to accommodate any arbitrary value of ψ_S. This approach takes advantage of the invariance that exists at the depletion-layer boundary: The origin bears a fixed positional relationship to the invariant charge, field, and potential profiles that exist there.

Consider a further illustration of the convenience in this approach. Choose any position for x_S that is left of the position in Figure 3-17(a) where the actual and depletion-approximation-field profiles have merged. Then the ionic charge per unit area in the silicon can be written *accurately* as:

$$Q_B = qNx_S, \tag{3-12}$$

where N is net impurity density in the uniformly doped substrate, whether one employs the actual *or* the depletion-approximation formulation.

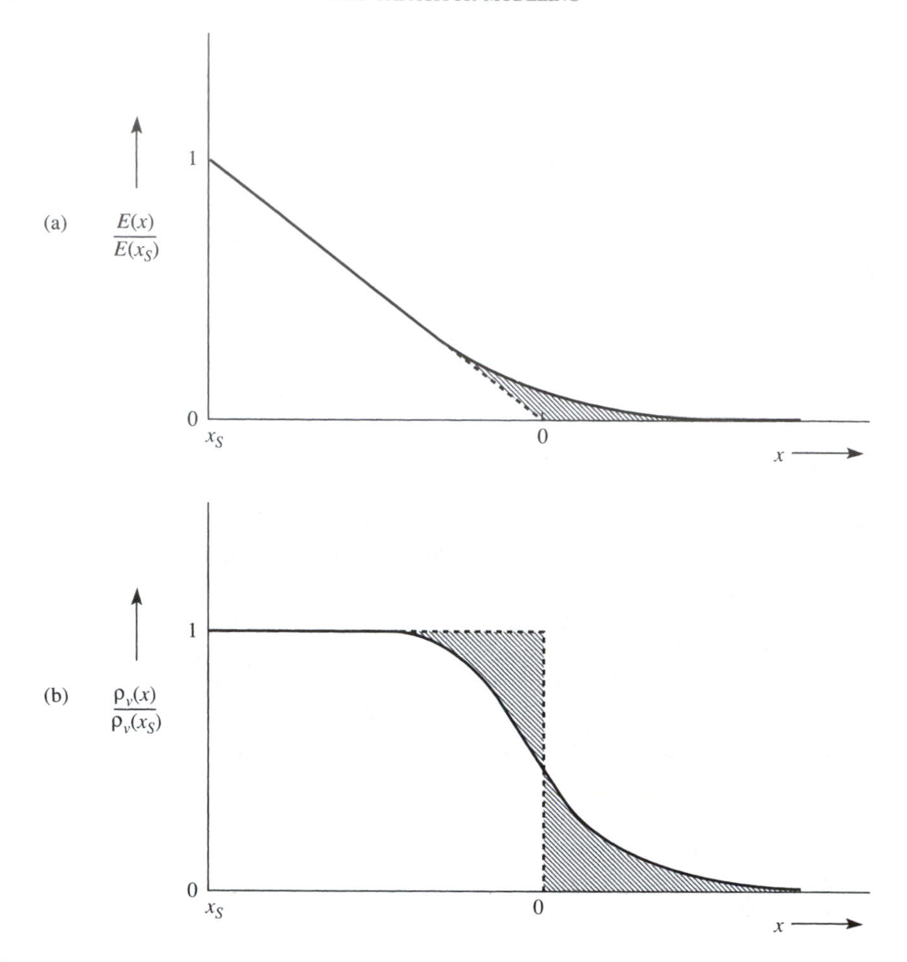

Figure 3–17 Physical meaning of the invariant spatial origin employed in the depletion-approximation replacement. (a) Electric-field profile in the neighborhood of the depletion-layer boundary, showing the actual profile (solid line) and the depletion-approximation profile (dashed line) that defines the new spatial origin. (b) The new spatial origin in relation to the volumetric charge-density profile. The two crosshatched regions have equal area in view of part (a) above and Gauss's law.

The depletion approximation was invoked previously because it aids the physical interpretation of the DAR origin located in the vicinity of the depletion-layer boundary. But now let us set the depletion approximation aside and think in terms of the actual field, charge-density, and potential profiles in a real MOS capacitor, with the first two shown in Figure 3-17(a) and (b), respectively. Then focus on the origin plane in Figure 3-17 and "freeze" its position in the silicon crystal, anchoring it to a particular atomic plane. The next step is to apply the thought experiment just described. Suppose ψ_S has been chosen at a value that corresponds to a moderate amount of inversion. Then apply a small positive increment to ψ_S. The result is the addition of a few monolayers of silicon at the surface, incorporating the right number of acceptor ions and the right number of inversion-layer electrons. Now the "equality" matter at *threshold* has a self-evident explanation. By definition, at threshold the volumetric electron density at the surface equals the volumetric ionic density there. Hence a tiny surface-potential increment at threshold involves equal amounts of inversion-layer charge and bulk charge in the thought-experiment layer of infinitesimal thickness; the silicon monolayers just "grown" are located where the two density functions intersect. This precise equality holds within the limits of the usual assumptions and approximations—those involving Boltzmann statistics, band symmetry, equivalent densities of states, and complete ionization.

Note that we are not offering the thought experiment as a description of the storage process. Rather, it is being offered as a description of before-and-after conditions (before and after the application of a surface-potential increment) that are *identical* to those in the real MOS capacitor. Consequently, the change in bulk charge as a result of the applied potential increment was the same in the thought experiment and the actual device. Finally, then, the bulk-charge and inversion-layer-charge increments are equal in the actual MOS capacitor, and equal charge increments for a given potential increment will lead by definition to equal inversion-layer and bulk capacitances at threshold. In this way the thought experiment provides the necessary bridge to physical plausibility for an initially obscure equality between two very different functions in the MOS-capacitor problem, and reemphasizes the critical character of the threshold condition.

3–7 Analysis of MOS-Capacitance Crossover

An early analytic demonstration of C_b and C_i equality at threshold was given by Tobey and Gordon [69]. They used two significant approximations to do so, however. One was the depletion approximation, and the other was the assumption that the inversion-layer charge is negligible at threshold in comparison to the bulk charge. In fact, the former typically amounts to several percent of the latter in terms of areal density. A more accurate approximate treatment was given subsequently by Tsividis [70], verifying the earlier conclusions. Later a still more accurate assessment was given [71], using the methods that are the subjects of Appendix F and Section 3-1. This analysis, instructive in several ways, proceeds as follows.

The exact expression for normalized electric field is given as the first equation in Table 4. For the $|U_B|$ range of interest to us, it readily simplifies to Approximation A in Table 3, which in turn simplifies to Approximation C.

EXERCISE 3–11 Explain these simplifications.

■ The ratio of the two terms in the denominator of the first expression, or $(\exp|U_B|)$: $[\exp(-|U_B|]$, amounts approximately to 10^{10} for $|U_B| = 11.5$, corresponding to a net doping of $10^{15}/\text{cm}^3$. Even for $|U_B|$ as low as 2.0, the ratio is about 55. Hence, dropping the second term in the denominator and performing the indicated division yields Approximation A. Then, for $W \geq 3$, the exponential term in the first parentheses can be dropped, and the terms $-W$ and -1 can be dropped in comparison with e^W in the second parentheses, yielding Approximation C.

Because we are specifically interested in electric field in the silicon at the interface, let us write Approximation C in terms of W_S:

$$\left| \frac{dW}{d(x/L_D)} \right|_{W=W_S} = \sqrt{2(W_S - 1 + e^{W_S - 2|U_B|})}. \tag{3-13}$$

The first two terms in parentheses are associated with the depletion-layer or the bulk charge, while the last (exponential) term is associated with the inversion-layer charge. The fact that the two charge components can be so resolved was noted at the end of Section 3-1 for the general (exact) electric-field expression that was the starting point for the present section. Then, following through the chain of approximations outlined in Exercise 3-11 verifies the assertion just made about Approximation C.

The small-signal capacitances C_b and C_i are in parallel, and therefore are additive. Let us use the conventional symbol Q_s to represent the total charge per unit area stored in the silicon. (The present MOS context and lowercase subscript should prevent confusion of this symbol with the similar symbol used in Appendix G for charge stored in the diffusion capacitance of the junction diode.) Accepting these symbols gives us the definition

$$C_b + C_i \equiv \frac{dQ_s}{d\psi_S}. \tag{3-14}$$

(We follow convention here by omitting the negative sign that a strict construction would require in this and the following expressions.) The value of Q_s as a function of ψ_S (or W_S) can be obtained via Gauss's law directly from the expression for electric field in the silicon at the oxide–silicon interface, Equation 3-13.

Letting Q_b stand for bulk (ionic) charge stored in the depletion-layer or bulk capacitance gives us in a parallel way,

$$C_b \equiv \frac{dQ_b}{d\psi_S}. \tag{3-15}$$

Once again, the value of Q_b as a function of ψ_S (or W_S) can be obtained via Gauss's law for conditions sufficiently below threshold, directly from the bulk-charge field component of Equation 3-13. Finally, we are able to calculate the most important and interesting capacitance C_i as follows:

$$C_i = \frac{dQ_s}{d\psi_S} - \frac{dQ_b}{d\psi_S}. \tag{3-16}$$

Note carefully that one may *not* compute C_i for any range of operation directly from the inversion-layer-charge component of electric field in Equation 3-13 in a manner parallel to that used to determine C_b in Equation 3-15.

EXERCISE 3–12 Why does this departure from parallelism exist? Why cannot one use the last term of Equation 3-13 and Gauss's law to calculate C_i in at least a limited range?

■ The important difference between C_b and C_i is that C_b can exist in the absence of C_i, and C_i can exist *only* in the presence of C_b. In a one-dimensional problem electric field is independent of the charge position, but the electrostatic potential is highly sensitive to the charge position. Hence, even though a one-to-one relationship exists between the components of Equation 3-13 and the components of stored charge (via Gauss's law), if one attempts to calculate C_i from the prescription $\Delta Q_n/\Delta W_S$ (where Q_n is the areal density of the inversion-layer charge), one obtains the wrong answer because ΔW_S is incorrect.

Normalized electric-field magnitude $|dW/d(x/L_D)|$ can be converted to unnormalized form by writing

$$|E| = \frac{1}{L_D}\frac{kT}{q}\left|\frac{dW}{d(x/L_D)}\right|. \tag{3-17}$$

Thus Equation 3-13 in partly unnormalized terms becomes

$$|E| = \frac{1}{L_D}\frac{kT}{q}\sqrt{2(W_S - 1 + e^{W_S - 2|U_B|})}. \tag{3-18}$$

An expression for Q_s, the total areal charge density in the silicon, can hence be found from Equation 3-18 by applying Gauss's law (or by noting that electric displacement possesses the desired dimensions):

$$Q_s = \frac{\epsilon\sqrt{2}}{L_D}\frac{kT}{q}\sqrt{W_S - 1 + e^{W_S - 2|U_B|}}. \tag{3-19}$$

Total silicon-based capacitance thus becomes

$$C_b + C_i = \frac{dQ_s}{d\psi_S} = \frac{q}{kT}\frac{dQ_s}{dW_S}$$

$$= \frac{\epsilon\sqrt{2}}{L_D}\frac{d}{dW_S}\sqrt{W_S - 1 + e^{W_S - 2|U_B|}}. \tag{3-20}$$

Or,

$$C_b + C_i = \frac{\epsilon\sqrt{2}}{L_D} \frac{1 + e^{W_S - 2|U_B|}}{\sqrt{W_s - 1 + e^{W_S - 2|U_B|}}}, \tag{3-21}$$

where the unity term in the numerator of the last factor arose from the bulk-charge phenomenon, and the exponential term from the inversion-layer phenomenon. Thus it is now evident that C_b and C_i are equal at threshold.

EXERCISE 3–13 Explain the last statement.

■ At threshold, $W_S = 2|U_B|$. Hence the exponential term in the numerator is unity at threshold and the two phenomena make equal contributions.

To summarize, the capacitance expressions having the wide-ranging validity of Approximation C in Table 3 are

$$C_b = \frac{\epsilon\sqrt{2}}{L_D} \frac{1}{\sqrt{W_s - 1 + e^{W_S - 2|U_B|}}}, \tag{3-22}$$

and

$$C_i = \frac{\epsilon\sqrt{2}}{L_D} \frac{e^{W_S - 2|U_B|}}{\sqrt{W_s - 1 + e^{W_S - 2|U_B|}}}. \tag{3-23}$$

Improved MOSFET Theory

A real MOS capacitor can be brought arbitrarily close to ideal one-dimensional conditions simply by increasing its gate area. Although we have in some cases introduced a junction near the gate edge to supply carriers to the inversion layer, that feature can equally well be omitted. The MOSFET, by contrast, possesses inevitable complications that remove it a long way from ideal properties. Its MOS-capacitor portion must interact with junctions in order for the device to work. Its drain–source bias results in an *IR* drop in the channel, which causes surface potential to vary from drain to source. Because of this variation the MOSFET will pose, at best, a two-dimensional problem. Further, the presence of its drain–source current, or channel current, means that nonequilibrium conditions exist there. Quasi Fermi levels are useful for treating such conditions. Finally, technological evolution has brought a steady shrinkage of MOSFET feature sizes. Most present-day MOSFETs have submicrometer critical dimensions. These tiny dimensions cause some aspects of MOSFET analysis to constitute a three-dimensional problem.

Immediately below we wish to look at the surface inversion layer in the presence of an orthogonal junction. Then we factor into MOSFET theory the depletion-layer and bulk-charge behavior that received emphasis in Chapters 2 and 3. (Recall that in Section 1-3 we treated inversion-layer charge as the only component of charge in the silicon.) The effect of reverse bias from source to substrate, known as *body effect*, will concern us then, leading up to numerical treatments of the MOSFET.

4–1 Channel-Junction Interactions

Admitting the inevitability of a two-dimensional problem when junctions have been introduced, we shall proceed to idealize the geometry as much as remains possible. As the first idealization, let us assume that the source and drain junctions (both taken to be N^+P junctions) are plane and parallel to one another, while normal to the oxide–silicon interface. This relationship is displayed in Figure 4-1(a), a cross-sectional representation. For the second simplification, we have chosen the source–drain spacing L to be large, 5 μm, as also shown in Figure 4-1(a). In the resulting *long-channel* MOSFET, the dimension L considerably exceeds the thickness of all depletion layers, those of source and drain junctions and the surface (or oxide–silicon interface). The extreme position of the boundary for the last depletion layer, approximately that at threshold, is shown in Figure 4-1(a) by a dashed line, assuming that source and bulk are electrically common and assuming the absence of drain–source bias. As a final sim-

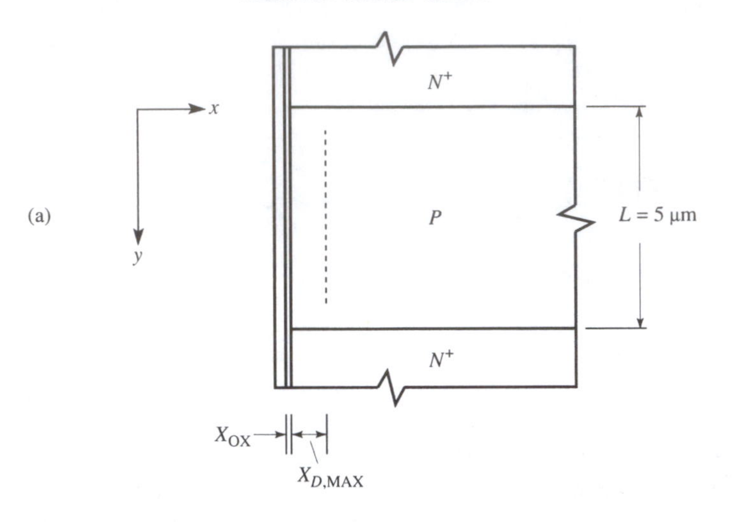

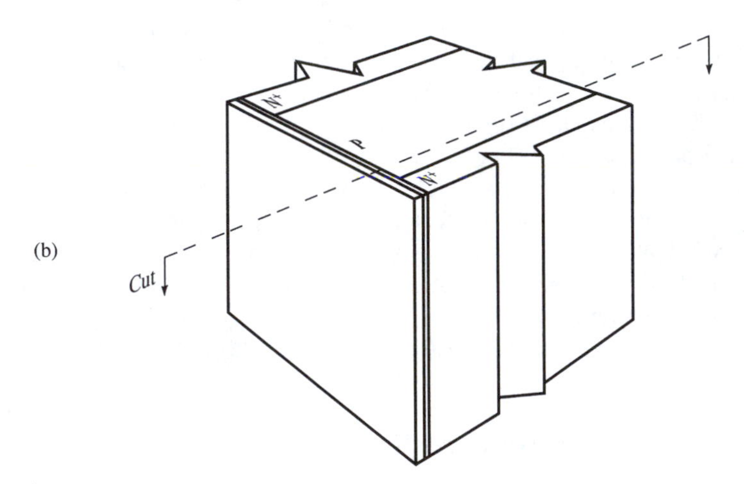

Figure 4–1 Idealized MOSFET having plane source and drain junctions normal to the oxide–silicon interface and having gate overlap of the source and drain regions. (a) Cross-sectional representation, showing dimensions and the extreme position of the depletion-layer boundary (dashed line). (b) Perspective representation, showing cut that will permit display of band diagrams in orthogonal directions. (After Warner and Grung [72].)

plification, we have permitted the oxide and field plate (gate electrode) to overlap the source and drain regions appreciably. This is manifestly a departure from practice that has been standard for many years. One seeks to terminate the gate structure precisely at the metallurgical junction, thus minimizing troublesome parasitic capacitances. But with overlap, some features of the explanation that follows become clearer. In spite of this and the other departures from present-day realism, the distorted structure that results is capable of functioning as an actual MOSFET and will facilitate some instructive descriptions.

In Figure 4-1(b) the same structure has been rendered in perspective. Our aim is to approximate the two-dimensional problem we face by using two orthogonal one-dimensional solutions. As the next step toward that goal, let us cut the sample in the manner indicated in Figure 4-1(b) and discard the right-hand portion. With this accomplished, and with the further step of rendering the oxide and field plate "invisible," we arrive at the uncluttered sample depicted in Figure 4-2(a). (Although we have chosen not to represent the field plate and oxide here, we assume that they are capable of normal function.)

Let us set $V_{GB} = V_{FB}$, so that the band diagram from surface to bulk is as shown in Figure 4-2(b). This is the same x-axis profile as the one used throughout Chapters 2 and 3. The orthogonal one-dimensional band diagram in the y or source-to-drain, direction is that of a one-sided step junction. A significant proviso is needed, however.

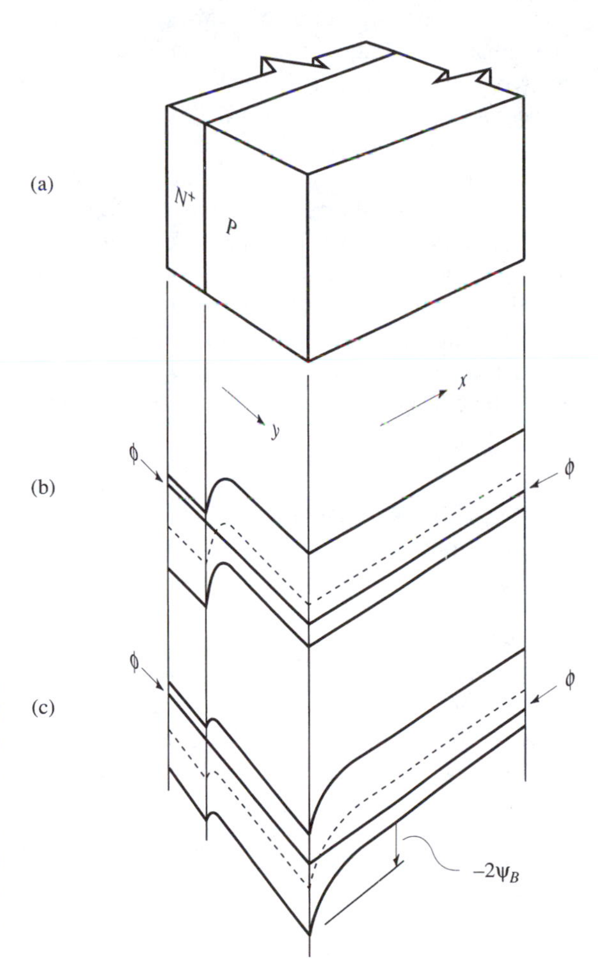

Figure 4–2 Effect of gate bias on idealized MOSFET. (a) Device of Figure 4-1(b) with field plate and oxide invisible and with drain region discarded. (b) Orthogonally intersecting one-dimensional band diagrams adjusted to the threshold condition. (c) Two-dimensional band picture for the threshold condition. (After Warner and Grung [72].)

The bias V_{GB} that creates flatband conditions in the P-type bulk region will *not* do so in the N^+ region. But this complication is one that we can set aside for now. We can show that the surface potential of a heavily doped region is quite unresponsive to change induced through an MOS capacitor, so we shall neglect any change there resulting from the application of V_{FB} from gate to bulk.

To see the reason, refer back to Figure 2-6(d)—a weak-depletion band diagram based upon the depletion approximation. For the conditions arbitrarily assumed, ψ_{GS} and ψ_{SB} are roughly equal, meaning that the rectangular and triangular electric-field profiles in Figure 2-6(b) have comparable areas. Now, in a thought experiment, hold E_S constant while increasing substrate doping. As a result, the triangular area shrinks monotonically while the rectangular area remains constant. Since the triangular area is identically ψ_{SB}, this means the surface potential progressively approaches the bulk potential. (Note that to make this qualitative point, an explanation using the depletion approximation is fully adequate.) For the accumulation condition, line-of-force penetration is inherently small, so the property of surface-potential insensitivity holds for either polarity of applied bias on a heavily doped substrate.

As the next step, let us bias the device at the threshold voltage. The effect on the orthogonal band diagrams is shown in Figure 4-2(c). Look first at the band diagram in the x or depth direction. It shows the familiar threshold-condition band bending of $-2\psi_B$, or about $+0.59$ V for $N_A = 10^{15}/\text{cm}^3$. Now examine the band diagram on the "front" of the silicon sample, involving the N^+P source junction. Once again surface potential on the N^+ region is little affected by the bias change. But the overall band diagram now resembles that of an N^+N (high–low) junction. The presence of the incipient inversion layer makes the surface region of the substrate appear N-type, but not as N-type as the N^+ region. Because the silicon remains at equilibrium, comparing the situation to that in an N^+N junction at equilibrium is appropriate.

The band diagram on the front of the block superficially resembles that for an N^+P junction under forward bias. This comparison is much less apt, however, because that would be patently a nonequilibrium situation, and bias has not yet been applied to the junction.

EXERCISE 4–1 Assuming threshold conditions, expand upon the comparison of the "front-surface" band diagram to an N^+N junction and to a forward-biased N^+P junction, commenting especially on hole populations near the surface.

■ The inversion-layer electrons are *surplus* carriers, and their enhanced density is accompanied by depressed hole density, with the product being $pn = n_i^2$. The same would hold in the N^+N junction at equilibrium. On the other hand, if the electron elevation in the P-type region were the result of forward biasing an N^+P junction, those *excess* electrons would be accompanied by an equal density of excess holes.

For the next examination of the idealized two-dimensional MOSFET, let us return to the structure of Figure 4-1. Looking directly at the "front" of the substrate and once again letting field plate and oxide be invisible yields the picture in Figure 4-3(a). An added feature now is an indication of equilibrium depletion-layer boundaries for the source and drain junctions. The most meaningful positions for the channel ends are at those boundaries, as also indicated by 0 and L on the y axis. Once again we choose the flatband condition of Figure 4-2(b), with the result displayed in Figure 4-3(b). The surface potential $\psi_S(y)$ (the mid-gap line) is evidently aligned with the bulk potential ψ_B from $y = 0$ to $y = L$, confirming flatband conditions. Next let us apply a gate–source bias V_{GS} (equal to V_{GB}, since we have short-circuited the source junction). Quite arbitrarily, we have brought the upper band edge for the P-type substrate into alignment with ψ_B, as can be seen in Figure 4-3(c). To avoid cluttering the diagram, we have discarded the balance of the upper band edge and all of the lower band edge.

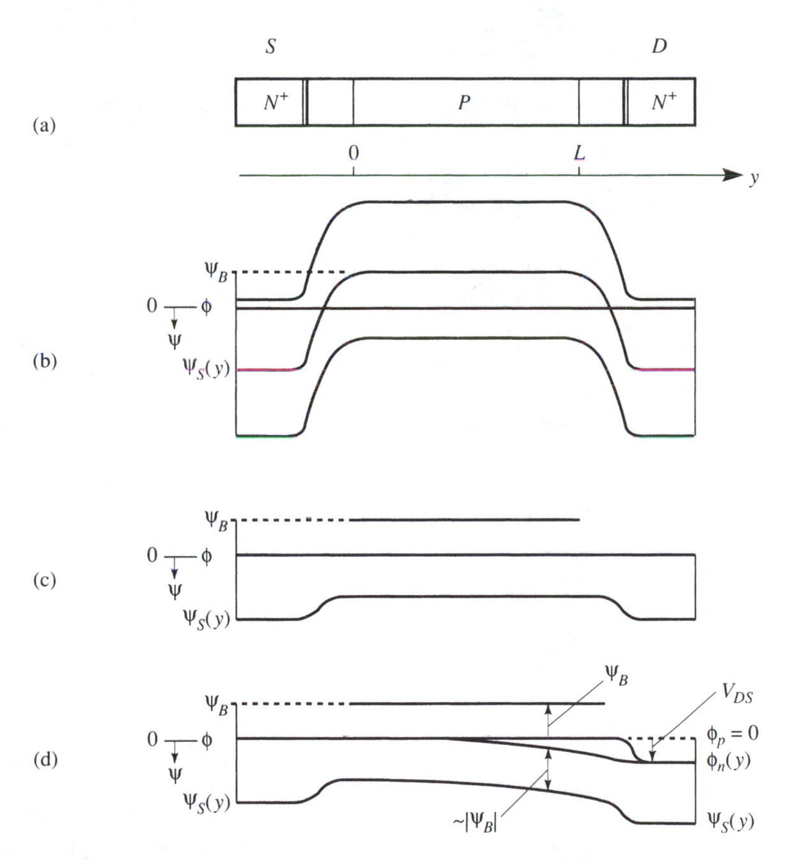

Figure 4–3 Idealized MOSFET of Figure 4-1(c) with biases applied. (a) Physical representation of structure, looking normal to substrate surface. (b) Band diagram for flatband condition in bulk region. (c) Potential profile $\psi_S(y)$ for V_{GS} having arbitrary positive value and $V_{DS} = 0$. (d) Potential profiles for small V_{DS} and for the same value of V_{GS} as in part (c). (After Warner and Grung [72].)

EXERCISE 4–2 Does the condition represented in Figure 4-3(c) constitute the threshold condition? If not, how far does it depart and in which direction?

■ We saw that in Figure 4-2(c), at threshold, we had band bending of $-2\psi_B = +0.59$ V. Since half the gap is 0.56 V, it follows that the band bending at threshold exceeds half the gap by 0.03 V, bringing the upper band edge that far below ψ_B in Figure 4-2(c). In Figure 4-3(c), then, the band bending is below threshold, with surface potential being more positive than the bulk potential by +0.56 V rather than by +0.59 V.

The next step is to apply a small drain–source bias V_{DS} and then examine its effect on the band diagram at the sample surface. This constitutes a nonequilibrium condition, as noted before, so quasi Fermi levels must enter the discussion. The effect of V_{DS} is to forward bias the source junction and reverse bias the drain junction.

EXERCISE 4–3 Does V_{DS} divide evenly between the two junctions? If not, how does it divide?

■ The asymmetry of forward and reverse properties of the junctions, assumed to be identical junctions, ensures that reverse bias on the drain junction exceeds forward bias on the source junction. Since the two junctions carry the same current, neglecting leakage components, one could consult an accurate diagram of forward and reverse current versus voltage for the silicon junction of interest to assess approximately the voltage disparity at a particular current.

The further effect of V_{DS} is to cause ψ_S to vary from source to drain, as shown by the lowest curve in Figure 4-3(d). The quasi Fermi level for electrons $\phi_n(y)$ shows a similar curvature. Through the properties of the electrochemical potential, or quasi Fermi level, the electron current in the channel (and, as we shall see, the total channel current) has a value of $d\phi_n/dy$ times $n(0, y)$. The symbol $n(0, y)$ is convenient for designating volumetric electron density at the interface in the present two-dimensional problem. By contrast to the electron case, we find that $d\phi_p/dy \approx 0$, so we conclude that holes contribute negligibly to conduction along the surface.

EXERCISE 4–4 The slope of $\phi_n(y)$ in Figure 4-3(d) increases toward the drain, implying a surface-electron density that declines toward the drain, since channel current is y-independent. But it is also evident that $\psi_S - \psi_B = \psi_{SB}$, the band bending, increases toward the drain. Doesn't greater band bending imply higher electron density?

■ In the *equilibrium* problem posed by the MOS capacitor, greater band bending indeed implies higher inversion-layer density. But in the *nonequilibrium* problem posed by the MOSFET, this is not so. The important spacing here is $\psi_S(y) - \phi_n(y)$, by definition of the quasi Fermi level. For the bias condition shown in Figure 4-3(c), this spacing is roughly constant because it depends logarithmically upon $n(0, y)$, which declines slowly from source to drain.

EXERCISE 4–5 How do we know that $n(0, y)$ declines from source to drain?

■ We know that $n(0, y)$ vanishes at $x = L$, the left-hand boundary of the drain junction. (See Problems A10 and A18 for more detail.)

It is appropriate to point out here that the quantity $V(y)$ described as "voltmeter voltage in the channel" in Section 1-3 is actually $\phi_n(y)$, because a voltmeter measures Fermi-level difference. This thought must not be carried beyond the conceptual level, however; a voltmeter probe must make an ohmic contact to a region where Fermi level is to be sensed, and an ohmic contact to a region exists only if the quasi Fermi levels have merged at the point of contact. While this is clearly not the case in the channel region as shown in Figure 4-3(d), it does hold in the N^+ source and drain regions. Thus one can use a voltmeter to measure $V_{DS} = (\phi_n)_{DS} = (\phi_p)_{DS}$.

4–2 Ionic-Charge Model

The elementary MOSFET analysis of Chapter 1 employs a series of simplifying assumptions. The most extreme of these is the assumption that the areal density of inversion-layer charge at position y is proportional to $\psi_{GS}(y)$, or to $E_{OX}(y)$, to use a symbol of self-evident meaning. (Recall that the subscript S on the symbol ψ refers to *surface*.) The dominance of ionic or bulk charge over inversion-layer charge in wide operating ranges causes the rudimentary analysis to be substantially in error. What is surprising and nonobvious, however, is that the simply derived current equations can be rendered extremely useful and quite accurate by applying empirical factors to them [72].

An early treatment of the effect of bulk-charge presence on MOSFET characteristics was carried out by Ihantola and Moll [73]. In presenting this analysis, however, we shall preserve several important simplifying assumptions, worth reviewing at this point. The first is the gradual-channel or long-channel assumption, which validates the use of one-dimensional equations for x-direction dependences. This is simply another way of describing the superposition of two one-dimensional solutions, put into effect in Figure 4-2. One way of specifying gradual-channel conditions is to set a low upper limit on the ratio of $|E_S(y)|$ to $|E_S(x)|$. For any given bias condition, one can reduce the ratio at a particular position by increasing channel length L, thus validating the equivalence of the two terms (gradual-channel and long-channel) that denote the basic assumption we shall designate (1). Further assumptions are (2) uniform bulk doping, (3) transport of electrons in the channel by drift only, (4) a constant mobility μ_n governing that transport throughout the channel, (5) negligible junction leakage current, and (6) electrically common source and bulk regions.

Figure 4-3(d) will be a helpful guide in the following discussion. Let us assume, though, that instead of having V_{GS} slightly below threshold as in that figure (see Exercise 4-2), we have a value somewhat above threshold. The profile $\psi_S(y)$ is quite insensitive to increases in V_{GS} beyond threshold because inversion-layer charge is so close to the silicon surface, a fact that will be exploited in the present analysis. Accepting the assumptions just given, we can take Equation 3-9 as a starting point. Because Fermi level has a constant value throughout the gate region and a different constant value throughout the bulk region, we may write

$$V_{GS} - V_{FB} = \psi_{GB} = \psi_{GS}(y) + \psi_{SB}(y) = \text{constant}. \tag{4-1}$$

From the capacitor law,

$$\psi_{GS}(y) = -\frac{Q_s(y)}{C_{OX}}. \tag{4-2}$$

Here, as in the rudimentary analysis, a minus sign enters the capacitor law because voltage is applied to the field plate, while the charge of interest is on the opposite "plate" of C_{OX}. Thus we have the first term of Equation 4-1. An approximate second term can be written by consulting Figure 4-3(d):

$$\psi_{SB}(y) \approx \phi_n(y) - 2\psi_B. \tag{4-3}$$

This expression can be explained as follows: All potentials are referred to $\phi_p = 0$. Toward the drain end of the channel, $\phi_n(y)$ is somewhat positive. Choosing an arbitrary point in the channel and requiring it to be right at the threshold condition requires the further positive band bending of $|2\psi_B|$. But we can generalize and write the band bending as $-2\psi_B$. For the present P-type substrate, ψ_B is negative, so that all terms of Equation 4-3 are thus rendered positive. For an N-type substrate, on the other hand, all terms would be negative, so that Equation 4-3 is valid in either case. The approximation that Equation 4-3 involves is neglect of the small amount of band bending connected with variations in channel strength above threshold.

This brings us to

$$Q_s(y) = -C_{OX}[\psi_{GB} + 2\psi_B - \phi_n(y)] = Q_n(y) + Q_b(y). \tag{4-4}$$

The first of these two equations results from substituting Equations 4-2 and 4-3 into Equation 4-1. The second is from the definition of Q_s. Solving for the charge component of primary interest, Q_n, yields

$$Q_n(y) = -C_{OX}[\psi_{GB} + 2\psi_B - \phi_n(y)] - Q_b(y). \tag{4-5}$$

Applying the depletion approximation (Table 1) yields an estimate of the other charge component:

$$Q_b(y) = -qN_AX_{D,\text{MAX}} = -\sqrt{2\epsilon qN_A[\phi_n(y) - 2\psi_B]}. \tag{4-6}$$

Substituting Equation 4-6 into Equation 4-5 gives

$$Q_n(y) = -C_{OX}[\psi_{GB} + 2\psi_B - \phi_n(y)] + \sqrt{2\epsilon q N_A[\phi_n(y) - 2\psi_B]}. \tag{4-7}$$

The analogous expression from Section 1-3 is

$$Q_n(y) = -C_{OX}[V_{GS} - V_T - V(y)], \tag{4-8}$$

and comparing the two reveals the difference in the two analyses. The electron Fermi level ϕ_n now replaces "channel voltage" V. Subsequent logic in the present analysis is the same as before, however, so we write

$$I_{CH}\int_0^L dy = Z\mu_n \int_0^{V_{DS}} Q_n(\phi_n)\, d\phi_n, \tag{4-9}$$

from which

$$I_{CH} = -\mu_n C_{OX}\frac{Z}{L}\int_0^{V_{DS}}\left[(\psi_{GB} + 2\psi_B - \phi_n) - \frac{1}{C_{OX}} 2\epsilon q N_A(\phi_n - 2\psi_B)^{1/2}\right]d\phi_n. \tag{4-10}$$

Because $I_{CH} = -I_D$, as noted before, we have after integration

$$I_D = \mu_n C_{OX}\frac{Z}{L}\left[(\psi_{GB} + 2\psi_B)\phi_n - \frac{\psi_n^2}{2} - \frac{2}{3}\frac{\sqrt{2\epsilon q N_A}}{C_{OX}}(\phi_n - 2\psi_B)^{3/2}\right]_0^{V_{DS}}. \tag{4-11}$$

Letting

$$\gamma \equiv \frac{\sqrt{2\epsilon q N_A}}{C_{OX}}, \tag{4-12}$$

and using Equation 4-1 to eliminate ψ_{GB} from Equation 4-11 yields

$$I_D = \mu_n C_{OX}\frac{Z}{L}\left[\left(V_{GS} - V_{FB} + 2\psi_B - \frac{V_{DS}}{2}\right)V_{DS} - \frac{2}{3}\gamma\left((V_{DS} - 2\psi_B)^{3/2} - (-2\psi_B)^{3/2}\right)\right]. \tag{4-13}$$

Later we will show that

$$-V_{FB} + 2\psi_B = -V_T + \gamma\sqrt{-2\psi_B}, \tag{4-14}$$

where V_T is the threshold voltage. With this substitution, Equation 4-13 becomes

$$I_D = \mu_n C_{OX}\frac{Z}{L}\left[\left(V_{GS} - V_T + \gamma\sqrt{-2\psi_B} - \frac{V_{DS}}{2}\right)V_{DS} - \frac{2}{3}\gamma\left((V_{DS} - 2\psi_B)^{3/2} - (-2\psi_B)^{3/2}\right)\right]. \tag{4-15}$$

In the SPICE program for modeling the MOSFET [74], the rudimentary analysis of Section 1-3 is known as *Level-1* characterization. The SPICE *Level-2* characterization, on the other hand, considers ionic charge. Symbols for basic and physical quantities in the Level-2 model are listed and explained in Tables 5 and 6. This SPICE model employs a somewhat more general version of Equation 4-15 for drain current, given in Table 6. In particular, the Level-2 equation permits a finite bulk–source bias V_{BS}, which we have previously assumed to be zero.

It is evident that with increasing V_{DS}, the term $-V_{DS}^2$ causes I_D to reach a maximum value, or to saturate, in a manner qualitatively similar to that displayed in Figure 1-5(a). These saturating curves are plotted as solid lines in Figure 4-4, and certain findings from the rudimentary theory are shown with dashed lines. The values $N_A =$

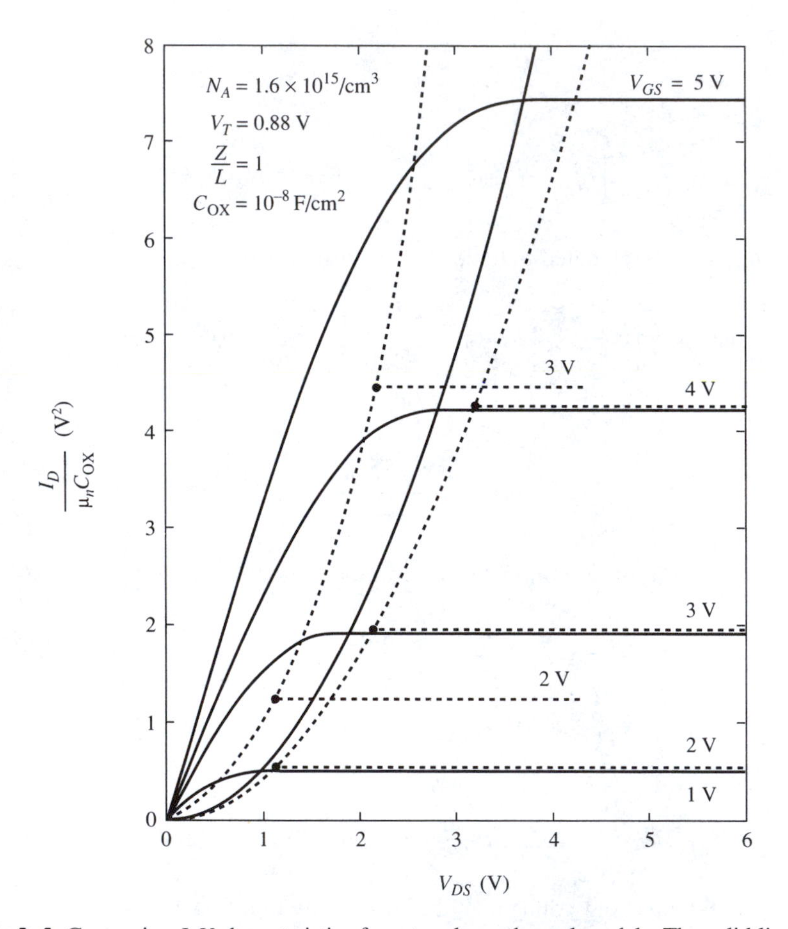

Figure 4–4 Comparing *I–V* characteristics from two long-channel models. The solid lines come from the Ihantola–Moll model, which includes bulk-charge effects and becomes the Level-2 SPICE model. The dashed lines are from the rudimentary model of Chapter 1, the Level-1 model. The left-hand dashed locus is for the same device as the solid lines, showing a current prediction about twice as great. The right-hand locus is obtained by empirically adjusting the Level-1 model to yield the same saturation-regime current at $V_{GS} = 5$ V. (After Warner and Grung [72].)

$1.6 \times 10^{15}/\text{cm}^2$, $V_T = 0.88$ V, $Z/L = 1$, and $C_{OX} = 10^{-8}$ F/cm^2 were used for calculating these curves, as is indicated in Figure 4-4. Notice that in terms of saturation-regime current, the two models differ by about a factor of two.

EXERCISE 4–6 Account qualitatively for the differing current predictions of the two models.

■ In the rudimentary model, areal charge in the inversion layer is assumed to match that in the gate. In the more accurate model, appreciable charge is "wasted" in that it belongs to a depletion layer, which means that there are fewer electrons in the channel available for conduction.

The question of what V_{DS} value yields current saturation at a given V_{GS} can also be addressed analytically. Rewriting Equation 4-7 with the aid of Equation 4-1 yields

$$Q_n(y) = -C_{OX}[V_{GS} - V_{FB} + 2\psi_B - \phi_n(y)] + \sqrt{2\epsilon q N_A[\phi_n(y) - 2\psi_B]}. \quad (4\text{-}16)$$

At the drain end of the channel, $\phi_n(L) = V_{DS}$. With the onset of saturation, the channel "pinches off" at the drain end, and hence $Q_n(L) = 0$. With these substitutions, then, we have

$$0 = -C_{OX}[V_{GS} - V_{FB} + 2\psi_B - V_{DS}] + \sqrt{2\epsilon q N_A(V_{DS} - 2\psi_B)}. \quad (4\text{-}17)$$

Solving this expression for V_{DS} yields

$$V_{DS} = V_{GS} - V_{FB} + 2\psi_B + \frac{\epsilon q N_A}{C_{OX}^2}\left[1 - \sqrt{1 + \frac{2(V_{GS} - V_{FB})C_{OX}^2}{\epsilon q N_A}}\right], \quad (4\text{-}18)$$

which was used to plot the solid locus of saturation points in Figure 4-4. (This equation is the subject of Problem A25.) The dashed locus at the left in Figure 4-4 was plotted using the analogous simple-theory expression, $V_{DS} = (V_{GS} - V_T)$.

Next, an empirical adjustment factor was applied to the simple-theory current to achieve a match of saturation currents for $V_{GS} = 5$ V. Note that the adjusted currents for lower V_{GS} are very close to those from the more complex analysis! The dashed locus at the right corresponds to these adjusted curves. (Curved-regime characteristics have been omitted for both the adjusted and unadjusted simple-theory curves to avoid cluttering the diagram.) MOSFET operation in saturation is by far the most important mode. The fact that rudimentary theory and empirical adjustment yield such a useful fit to more exact results means that the computer design of MOS integrated circuits began early, when only simple theory was available, in the early to mid-1960s. Computer-aided design became an important factor in the rapid advance of MOS technology, a point made in Section 1-6. To repeat, the reason for the supreme importance of circuit optimization by computer in the MOS case is that the MOSFET is relatively deficient in "drive" capability, or transconductance.

A useful expression for characteristic slope, or channel conductance, in the curved regime can be written by differentiating Equation 4-13:

$$g = \frac{dI_D}{dV_{DS}}$$
$$= \mu_n C_{OX} \frac{Z}{L}\left[(V_{GS} - V_{FB} + 2\psi_B - V_{DS}) - \frac{\sqrt{\epsilon q N_A}}{C_{OX}}(V_{DS} - 2\psi_B)^{1/2}\right]. \qquad (4\text{-}19)$$

Substituting Equation 4-18 into this expression of course causes g to vanish. But in the linear regime, near the origin, V_{DS} may be neglected, yielding

$$g(0) = \mu_n C_{OX} \frac{Z}{L}\left[V_{GS} - \left(V_{FB} - 2\psi_B + \frac{\sqrt{2\epsilon q N_A(-2\psi_B)}}{}\right)\right]. \qquad (4\text{-}20)$$

Now refer back to Equation 4-6, which gives bulk charge as a function of y. Near the source, the quasi Fermi levels merge, causing $\phi_n(0) = 0$. Thus

$$Q_b(0) = -\sqrt{2\epsilon q N_A(-2\psi)}. \qquad (4\text{-}21)$$

Dividing this expression by C_{OX} yields voltage drop caused by bulk charge at the source end of the channel, a term in the threshold voltage that was neglected in simple theory, and the correct interpretation of the last term in Equation 4-20. In fact, the three terms in parentheses in Equation 4-20 constitute the threshold voltage V_T in the present theory:

$$V_T = V_{FB} - 2\psi_B + \frac{\sqrt{2\epsilon q N_A(-2\psi_B)}}{C_{OX}}. \qquad (4\text{-}22)$$

Recall from Problem A26(b) that the simple-theory analog of Equation 4-20 is

$$g(0) = \mu_n C_{OX}\left(\frac{Z}{L}\right)(V_{GS} - V_T), \qquad (4\text{-}23)$$

and then note the direct correspondence between V_T and Equation 4-22. This is the desired verification of Equation 4-14.

Drain current in saturation is a matter that can be approached as before. Substituting Equation 4-18 into Equation 4-15 yields a complicated but important result. A shorthand version has been given by Brews [75] as

$$I_D = m\mu_n C_{OX}\left(\frac{Z}{L}\right)(V_{GS} - V_T)^2. \qquad (4\text{-}24)$$

The coefficient m is doping-dependent. For a lightly doped substrate, m approaches 1/2, its value in the rudimentary theory, Equation 1-17.

4–3 Body Effect

A large fraction of the devices in an MOS integrated circuit have source regions that are not electrically common with the bulk, or substrate, terminal. In normal operation, such source regions may "move toward V_{DD}" in voltage. For N-channel devices, this means the source becomes more positive than the P-type bulk, a case of reverse bias on the source–bulk junction. As a result, the MOSFET in question exhibits an increased threshold voltage. Because MOSFET voltages are conventionally stated with the source as reference, the relevant voltage is V_{BS}, now negative.

To see one way that a negative V_{BS} can arise, refer to Figure 4-5(a). The two right-hand devices, M_{12} and M_{22}, have a common and large value of V_{GG}, causing both of them to be ON, or conducting. The lower device, M_{22}, will exhibit a finite ON voltage, $V_{DS(ON)}$. To visualize this, refer back to Figure 1-15(b). Because M_{12} has a high and fixed gate voltage, its I–V characteristic will to first order resemble the load characteristic presented there. Let us assume that $V_{DS(ON)} = 0.5$ V, as indicated in Figure 4-5(a). This then constitutes the reverse voltage V_{BS} on the source junction of the upper device on the left-hand side, M_{11}. Figure 4-5(b) presents the two left-hand devices

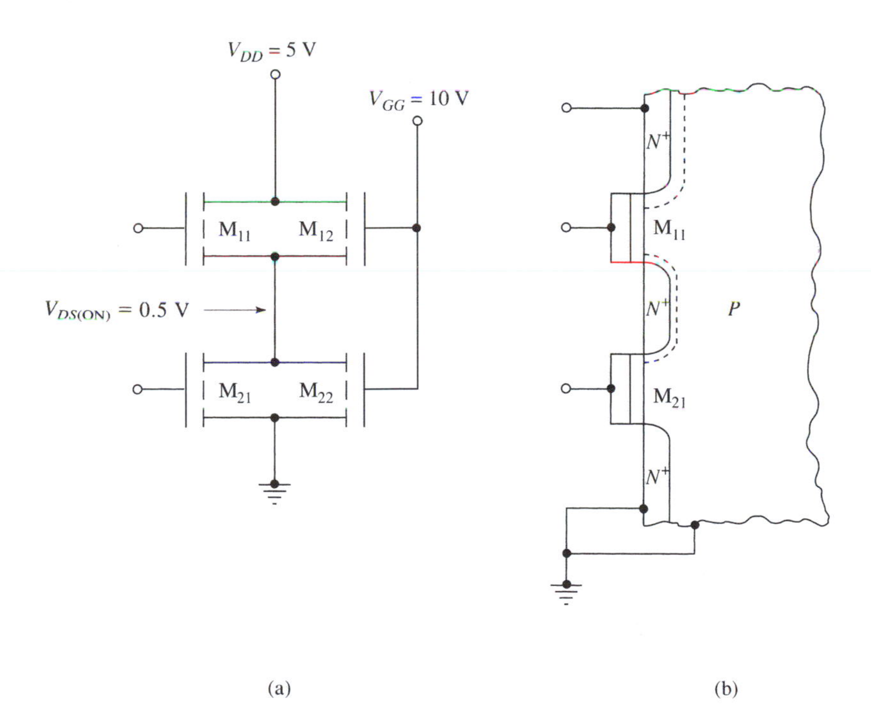

Figure 4–5 Sample situation leading to body effect. (a) The two right-hand devices are biased ON and conduct. Voltage drop $V_{DS(ON)}$ in M_{22} causes the source junction of M_{11} to be reverse-biased. (b) Cross section through left-hand device pair, showing by depletion-layer indication the reverse bias on their common region, the source of M_{11}, with respect to the substrate.

in cross section. The qualitative indications of depletion-layer thickness are intended to convey that the lowest junction shown has zero bias, the highest junction has a large reverse bias, 10 V, and the middle junction, which is common to M_{11} and M_{21}, has a small reverse bias, 0.5 V.

To examine how a negative V_{BS} on M_{11} causes an increase in its threshold voltage, we can profitably return to the kind of perspective representation used in Figure 4-2. Once again let us start with a flatband condition in the x direction, as shown in Figure 4-6(a). The application of reverse bias to the source junction, represented on the "front" of the block, can be seen toward the left. In reverse bias, splitting of the quasi Fermi levels does indeed occur, with ϕ_n dropping below ϕ_p, but they rejoin just a short distance from the junction on either side [76]. This occurs, basically, because the absolute departure from equilibrium minority-carrier density is highly constrained in reverse bias (as compared to forward bias). This kind of splitting over a short distance is represented in perspective in Figure 4-6(a).

Recall, now, that the channel is strongest, and the surface potential ψ_S is farthest below the relevant Fermi level ϕ_n, at the source end. Indeed, for an ON device, the in-

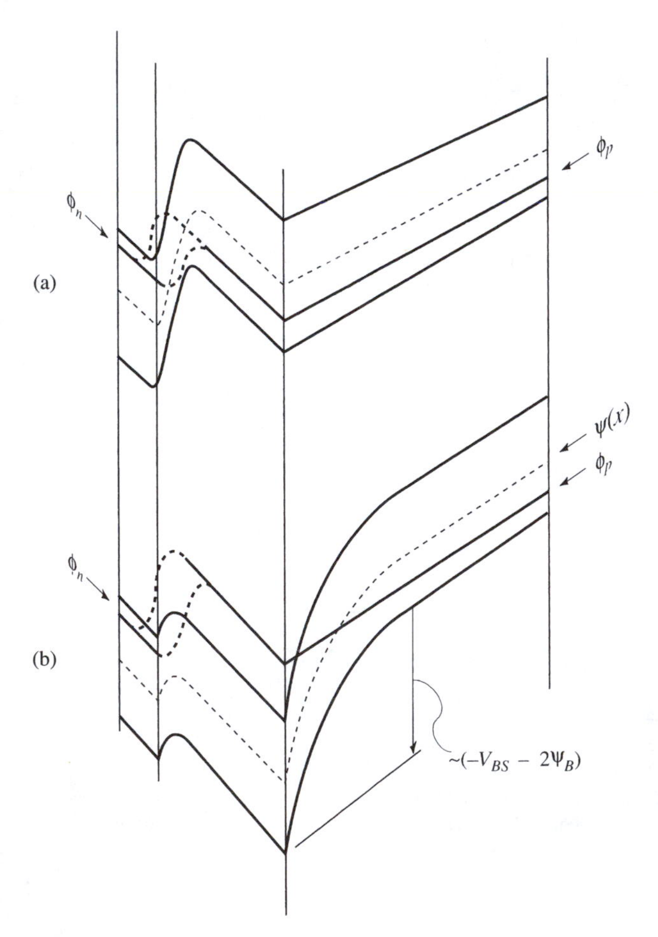

Figure 4–6 Body effect with intersecting one-dimensional solutions in the manner of Figure 4-2. (a) Flatband condition in bulk region and reverse bias V_{SB} on source–bulk junction. Fermi levels ϕ_p and ϕ_n split but merge again near junction. (b) Gate voltage produces sufficient band bending to bring $\psi_S(0)$ below ϕ_n by $|\psi_B|$, the threshold condition at the source end of the channel. (After Warner and Grung [72].)

terval ψ_S–ϕ_n must exceed $|\psi_B|$ there. It is evident in Figure 4-6(a) that the splitting of ϕ_p and ϕ_n is such that substantially more band bending will be necessary than was the case with $V_{BS} = 0$. Figure 4-6(b) shows the situation after gate bias V_{GS} has been increased enough to provide the requisite band bending. The character of the divided Fermi levels, with both their points of merger close to the junction, suggests that the incremental band bending needed will be somewhat less than V_{BS}. Hence as a high-side estimate we can write

$$\text{source-end band bending at threshold} \approx -V_{BS} - 2\psi_B, \tag{4-25}$$

recalling that both ψ_B and V_{BS} are negative at present. This amount of band bending is indicated and labeled in Figure 4-6(b). The need for this incremental measure of band bending in order to establish threshold is the reason for the *body effect*, or threshold increase.

The increment ΔV_{GS} needed to reach threshold is accompanied by an increment in bulk charge ΔQ_b. The threshold-voltage increment, therefore, amounts to

$$\Delta V_T = \frac{\Delta Q_b}{C_{OX}}. \tag{4-26}$$

It is important to realize that V_{BS} itself does not directly enter ΔV_T, but rather, it is the added bulk charge ΔQ_b imposed upon the oxide capacitor that is responsible.

EXERCISE 4–7 Why doesn't V_{BS} enter ΔV_T directly?

■ Our convention refers all voltages to the source of a MOSFET. With $V_{BS} < 0$, V_{GS} is *still* referred to the source terminal and cannot directly include V_{BS}. (If all voltages were referred to the bulk-terminal voltage, then V_{BS} would enter ΔV_T.)

It is evident that the new expression for threshold voltage in the presence of finite V_{BS} can be written by adding a term to Equation 2-48:

$$V_T = H_D - 2\psi_B - \frac{Q_f}{C_{OX}} - \frac{Q_S}{C_{OX}} - \frac{\Delta Q_b}{C_{OX}}. \tag{4-27}$$

In order to carry out an order-of-magnitude determination of ΔV_T, let us assume that Fermi-level splitting at the source end of the channel equals source–bulk bias, or

$$\phi_n - \phi_p = -V_{BS}. \tag{4-28}$$

Then,

$$\Delta V_T = -\frac{\Delta Q_b}{C_{OX}} = -\frac{(-q)N_A \Delta X_D}{C_{OX}}. \tag{4-29}$$

In functional notation,

$$\Delta V_T = \left(\frac{qN_A}{C_{OX}}\right)[X_D(-V_{BS} - 2\psi_B) - X_D(-2\psi_B)]. \tag{4-30}$$

Taking the depletion-approximation expression for the one-sided junction from Table 1 gives us

$$\Delta V_T = \left(\frac{\sqrt{2\epsilon q N_A}}{C_{OX}}\right)(\sqrt{-V_{BS} - 2\psi_B} - \sqrt{-2\psi_B}). \tag{4-31}$$

Empirical values of ΔV_T for a variety of MOSFET geometries were observed by Penney and Lau [77], with the resulting median curves plotted as solid lines in Figure 4-7. These curves are for $N_A = 7 \times 10^{15}/\text{cm}^3$ and for two basic aspect-ratio cases. Load devices qualify as long-channel MOSFETs, and inverter (or driver) devices as short-channel MOSFETs. The latter are less susceptible to body effect. It is evident in Figure 4-6(b) that the surface depletion layer will interact in complex three-dimensional fashion with those of the source and drain junctions as channel length is reduced. While the resulting details are not evident, one can claim in a vague way that the substantial band bending associated with a considerably reverse-biased drain junction will "contribute" some of the necessary band bending at the surface. The same idea is sometimes expressed in terms of bulk-charge "sharing" by the drain-junction and surface depletion layers.

For comparison with the solid curves, the dashed curve was obtained by plotting Equation 4-31. It falls between the experimental curves and qualitatively resembles the

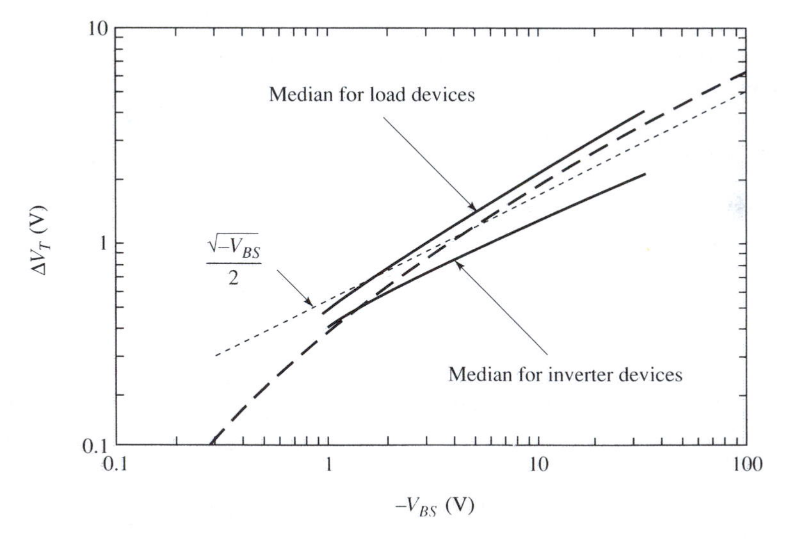

Figure 4-7 Solid lines show experimental threshold-voltage shift for long-channel (load) MOSFETs and short-channel (inverter or driver) MOSFETs with $N_A = 7 \times 10^{15}/\text{cm}^3$. (After Penny and Lau [77] with permission, copyright 1972, Van Nostrand Reinhold.) Dashed curve is plot of Equation 4-31 for the same doping. Dotted curve is plot of empirical Equation 4-32.

long-channel case, assumed for its development. The dotted curve, then, is a plot of the empirical expression

$$\Delta V_T = \tfrac{1}{2}\sqrt{-V_{BS}}, \tag{4-32}$$

sometimes used for modeling the body effect. It is a power-law expression, and hence is linear in log–log presentation. For more accurate modeling, one can deal with bulk charge in the manner employed in Section 4-2 to develop the Level-2 current expression. The resulting expressions are complex, which accounts for the popularity of Equation 4-32.

4–4 Advanced Long-Channel Models

An early and accurate long-channel model was contributed by Pao and Sah [54]. While it used two superposed one-dimensional solutions, thus making it by definition a long-channel model, it eliminated the depletion approximation and employed numerical integrations in the x and y directions. Further, in the x (depth) direction, it made use of certain expressions equivalent to those developed in Appendix F and Section 3-1. The Pao–Sah analysis was subsequently recast with our notation and with the extension of the analysis from equilibrium to nonequilibrium conditions [72]. Significantly, the Pao–Sah model permits carrier transport in the channel by drift *and* by diffusion, rather than by drift alone, as was the case in Section 4-2 and Chapter 1. A nonobvious consequence of this generalization is that the model is able to predict saturation-regime characteristics, with the valid result that true saturation (or constancy) of the drain current is not observed. Recall that previous models predicted constant current in saturation by appeal to a physical argument (such as that in Problem A10).

The Pao–Sah model is extremely accurate, but is of limited practical utility in its original form because of the kind of numerical procedures it requires. This motivated a search for an approximate analytic model of greater accuracy than the earlier ones. Subsequently, a *charge-sheet* model was developed by Brews [78] and by Baccarani et al. [79]. In a simplification over the Pao–Sah model, it reverts to the depletion approximation. In a second simplification, accounting for the model's name, it assumes a channel of zero thickness. However, like the Pao–Sah model, it permits both drift and diffusion of channel carriers, and consequently predicts saturation-regime characteristics. The approximations combined to yield the Brews model contribute partly compensating errors that account for its surprising fidelity, a matter that is examined elsewhere in some of its aspects [80].

Another model was offered by Dang [81], a bit before the Brews and Baccarani et al. models. It is intermediate in approximation between the Brews and Baccarani et al. models on the one hand, and the Ihantola–Moll model, on the other hand. These four models have been examined in some detail in relation to one another [72].

SPICE Models

Originally there were three SPICE MOSFET models, identified as Levels 1, 2, and 3 [32, 82–84]. Level 1 employs the rudimentary analysis of Chapter 1, with empirical adjustment. The most important correction is a coefficient applied to the current expressions to bring saturation-regime current into congruence with experiment and more accurate analyses. The result of such adjustment is shown in Figure 4-4, where the rudimentary model is brought into good agreement with the ionic-charge model.

The Ihantola–Moll, or ionic-charge, analysis is taken as the Level-2 SPICE model. It is used in a form more general than that developed in Section 4-2, permitting nonzero values of bulk–source voltage. A notable departure of the Level-2 from the Level-1 analysis is a three-halves-power dependence of current on voltage in the second major term of Equation 4-13, rather than the square-law dependence in the Level-1 case, Equation 1-14. Because the numerical handling of a fractional power is more time-consuming than that of an integral power, circuit designers were motivated to seek empirical modifications of the Level-2 model. In the resulting SPICE Level-3 model, they not only removed this shortcoming but were able to deal empirically with many other effects as well, such as short-channel and channel-shortening effects. The latter, addressed in Problem A10g, is attributable to the variations of drain–junction depletion-layer thickness, and is analogous to the Early effect in the BJT. The former implies accounting for some three-dimensional complications, alluded to in Section 4-1. These inevitably arise when channel length approaches depletion-layer dimensions.

5-1 Level-2 Parameters

The semiempirical Level-1 and Level-3 models are especially advantageous for circuit analysis. But because our present interest is equally on device physics and circuit analysis, we will focus primarily on the Level-2 Ihantola–Moll model [73]. The starting point is a generalized form of Equation 4-13 that takes into account the effect of finite bulk–source voltage,

$$I_D(V_{GS}, V_{BS}, V_{DS}) = 2K\left\{\left[V_{GS} - V_{FB}^* + 2\psi_B - \left(\frac{V_{DS}}{2}\right)\right]V_{DS} + \left(\frac{2\gamma}{3}\right)[(V_{DS} - 2\psi_B - V_{BS})^{3/2} - (-2\psi_B - V_{BS})^{3/2}]\right\}, \tag{5-1}$$

where

$$K \equiv \left(\frac{1}{2}\right)\frac{\mu_n C_{OX} Z}{L^*}, \tag{5-2}$$

and L^* is an elaborated version of channel length L that takes into account a technological channel-shortening effect. Impurities implanted to form the source and drain regions diffuse laterally during subsequent heat treatments by the amount L_D^*, and as a result, the channel is shortened by the amount $2L_D^*$. Hence,

$$L^* = L - 2L_D^*. \tag{5-3}$$

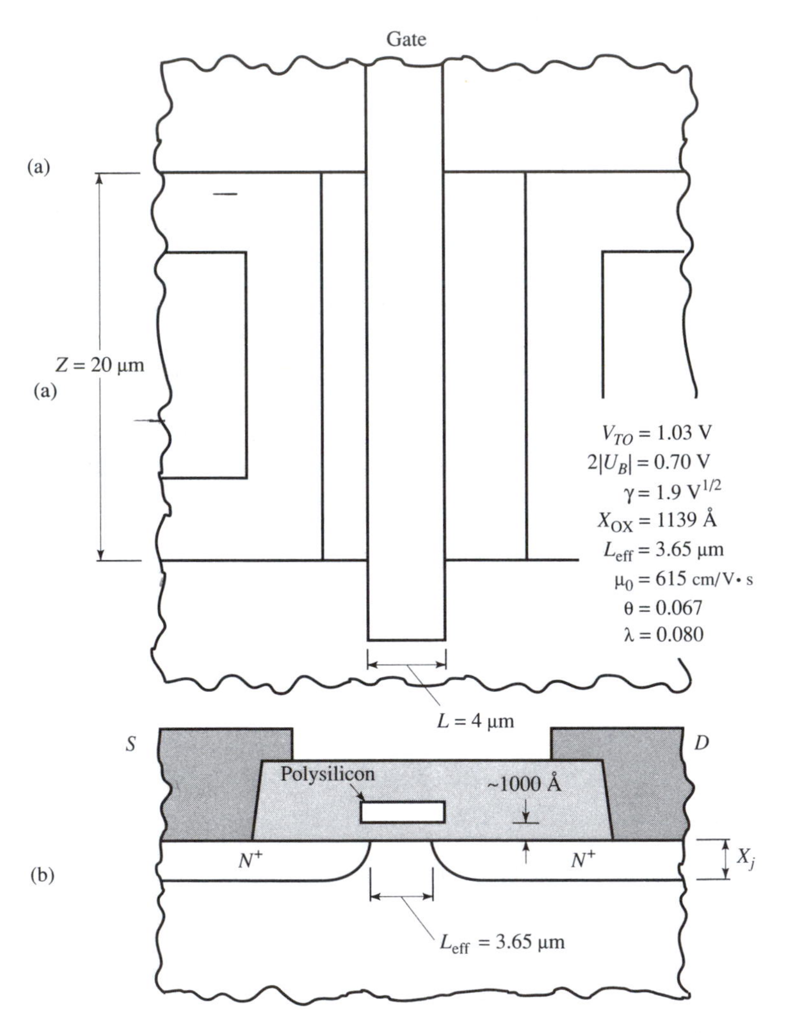

Figure 5–1 MOSFET structure chosen for study. (a) Parameter values and plan-view dimensions, emphasizing gate length of 4 μm. (b) Cross-sectional structure, emphasizing effective gate length of 3.65 μm. (After Warner and Grung [72].)

Our purpose here is to introduce the large number of possible input parameters for the Level-2 SPICE model. To compare the model with experiment, we will employ the device geometry shown in Figure 5-1 and the current–voltage characteristics shown in Figure 5-2. Also given in the latter figure are the characteristics calculated using the Ihantola–Moll model described in Section 4-2. Figure 5-3(a) presents a static equivalent-circuit model for the ideal MOSFET described by Equation 5-1. The current generator I_D is a function of the three variables V_{GS}, V_{BS}, and V_{DS}. Figure 5-3(b) adds to the equivalent circuit the parasitic elements present in a real MOSFET, needed for satisfactory dynamic analysis of the device.

Tables 5 and 6 give one possible set of SPICE parameters that are most directly related to the structure of the MOSFET. The oxide thickness X_{OX} determines the capacitance per unit area C_{OX}, and along with the surface mobility μ_n, enters into the current coefficient K. The substrate doping N_A and the specific capacitance C_{OX} enter the body-effect parameter that we repeat here:

$$\gamma = \frac{\sqrt{2\epsilon q N_A}}{C_{OX}}. \tag{5-4}$$

The substrate doping N_A also determines the bulk potential ψ_B since

$$\psi_B = \left(\frac{kT}{q}\right) \ln \left(\frac{N_A}{n_i}\right). \tag{5-5}$$

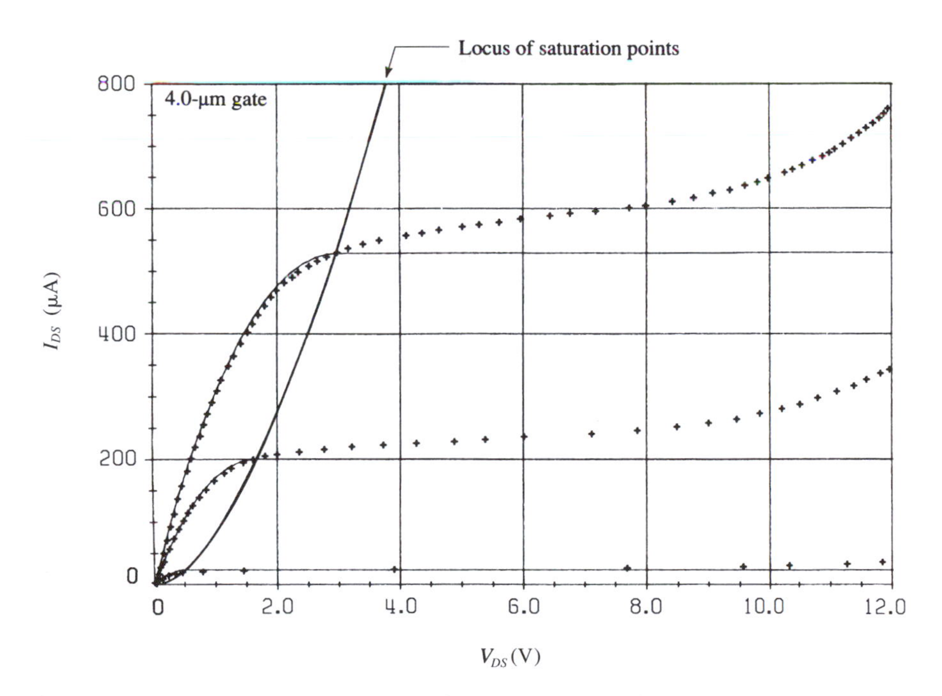

Figure 5–2 Three output characteristics for the MOSFET of Figure 5-1. The plotted points were obtained experimentally, and the solid curves were calculated from the Ihantola–Moll model. (After Warner and Grung [72].)

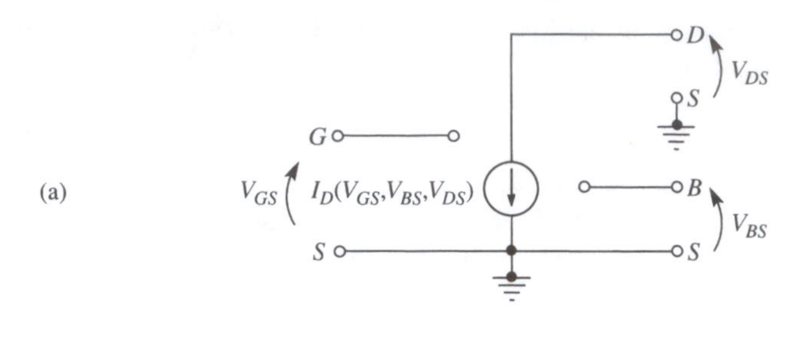

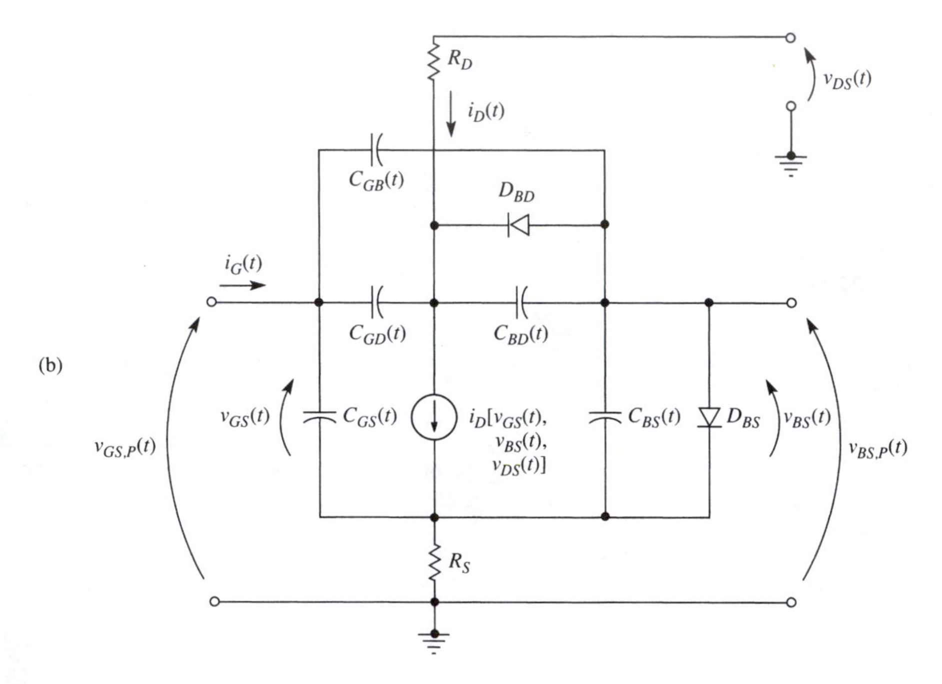

Figure 5–3 Level-2 modeling of the MOSFET. (a) Equivalent-circuit model based upon the Ihantola–Moll static model [73]. (b) Equivalent-circuit model that includes parasitic elements needed for accurate Level-2 static and dynamic modeling.

The mobility parameters μ_n, U_{EXP}, and E_{CRIT} are used to define an effective mobility as will be described later. In addition, the channel-length modulation factor λ is similar to the base-thickness modulation parameter in the BJT model. The effective oxide fixed charge density N_f empirically adjusts the zero-order threshold voltage determined by H_D, the barrier-height difference. For metal field plates, the SPICE values agree exactly with the experimental values given by Deal and Snow [26], so that

$$H_D = \pm\psi_B + 0.61 \text{ V}. \tag{5-6}$$

But for polysil field plates, we will use "effective" values N_f^* and H_D^* because the SPICE values differ slightly from the experimental results of Werner [27]. Figure 2-10(b) shows that the approximate SPICE values were chosen for convenient symmetry, with the barrier-height difference stated as

$$H_D^* = \pm[\psi_B - (0.5575\ \text{V})T_G]. \tag{5-7}$$

The positive sign represents a P-type substrate, and the negative sign an N-type substrate. The type-of-gate parameter T_G is positive for a polysil gate when the gate-doping type is opposite to that of the substrate, and negative when the two types are the same. The effective flatband voltage V_{FB}^* is given by

$$V_{FB}^* = H_D^* - \frac{qN_f^*}{C_{\text{OX}}} \tag{5-8}$$

and finally the threshold voltage is given by

$$V_T = V_{FB}^* - 2\psi_B + \gamma\sqrt{-2\psi_B}. \tag{5-9}$$

To compensate for the departure of the SPICE Equation 5-7 from the more exact Equation 2-20, we add a constant value of $3.3 \times 10^{10}/\text{cm}^2$ to the actual value of N_f. For example, if the actual fixed-oxide charge N_f equals $3.2 \times 10^{10}/\text{cm}^2$, then N_f^* equals $6.5 \times 10^{10}/\text{cm}^2$. The internal SPICE program also employs an ancillary parameter that it terms the effective barrier height, defined as

$$V_{BI}^* = V_{FB}^* - 2\psi_B = V_T - \gamma\sqrt{-2\psi_B}. \tag{5-10}$$

This parameter does simplify the form of some equations, but we will not use it.

Now let us return to Figure 5-3(b) to describe more fully the various elements it introduces. The source and drain implants have a sheet resistance R_{SH} that causes parasitic series source resistance and drain resistance, given by

$$R_S = N_{RS}R_{\text{SH}}, \tag{5-11}$$

and

$$R_D = N_{RD}R_{\text{SH}}, \tag{5-12}$$

where N_{RS} and N_{RD} are the numbers of resistive squares associated with each implanted region. At this point we conform to SPICE convention by letting R_S stand for parasitic series source resistance, although it previously stood for sheet resistance. Similarly, R_D stands for parasitic series drain resistance. To accommodate the change, we now let R_{SH} stand for sheet resistance.

The source and drain implants also produce PN junctions with a saturation-current density (extrapolated to zero bias) of J_0. The saturation currents for the corresponding junctions are given by

$$I_{BS} = A_S J_0, \tag{5-13}$$

and

$$I_{BD} = A_D J_0, \tag{5-14}$$

where A_S and A_D are areas of the source and drain implants. In normal operation, the source and drain junctions are reverse-biased so that these diodes contribute only relatively unimportant leakage currents. However, the associated depletion-layer capacitances are important, and are divided into sidewall portions and bottom portions. The zero-bias values of the bottom parts are given by

$$C_{0BS.B} = A_S C_{0.B}, \tag{5-15}$$

and

$$C_{0BD.B} = A_D C_{0.B}, \tag{5-16}$$

where $C_{0.B}$ is the junction capacitance per unit area. The sidewall portions of the drain and source junctions contribute capacitances given by

$$C_{0BS.\mathrm{SW}} = P_S C_{0.\mathrm{SW}}, \tag{5-17}$$

and

$$C_{0BD.\mathrm{SW}} = P_D C_{0.\mathrm{SW}}, \tag{5-18}$$

where $C_{0.\mathrm{SW}}$ is the sidewall capacitance per unit length, P_S is the length of the sidewall for the source region, and P_D is the corresponding length for the drain region.

5-2 Level-2 Model

The MOSFET model to be presented here contains seventeen equations in twenty variables. To define the boundary between the curved and saturation regimes, we employ the drain–source saturation voltage, a time-dependent form of Equation 4-18:

$$\begin{aligned} v_{DS.\mathrm{SAT}}(t) = {} & v_{GS}(t) - V_{FB} + 2\psi_B \\ & + \frac{\gamma^2}{2}\left[1 - \sqrt{1 + 4\,\frac{v_{GS}(t) - V_{FB} - v_{BS}(t)}{\gamma^2}}\right]. \end{aligned} \tag{5-19}$$

Also, to define the boundary between subthreshold and normal operation, we will use the ON voltage defined by

$$v_{\mathrm{ON}}(t) = V_{FB} - 2\psi_B + \gamma\sqrt{-2\psi_B - v_{BS}(t)} + \left(\frac{kT}{q}\right)N(t), \tag{5-20}$$

where

$$N(t) = 1 + \frac{C_n}{C_{OX}} + \frac{\gamma}{2\sqrt{(-2\psi_B) - v_{BS}(t)}}. \tag{5-21}$$

Notice that the time dependence of the ON voltage results from the time-dependent bulk-source voltage $v_{BS}(t)$. For $N(t) = 1$, the ON voltage equals the normal threshold voltage V_T. For other values of $N(t)$, there is a slightly more positive value of effective threshold voltage—the ON voltage. The adjustable parameter C_n is used in modeling the subthreshold regime of operation.

We define four regimes of operation. A good way to visualize these regimes is to consider a MOSFET inverter with the bulk-source junction short-circuited and with a resistive load. Then increase the gate-source voltage from 0 to a high value, moving upward along the loadline. Initially the silicon surface is in the accumulation condition (because of parasitic features of the MOS capacitor and "threshold-tailoring" steps). Next we encounter depletion, and the term "subthreshold" is used to refer to both the accumulation and depletion regimes. As the device turns on, it is in the saturation regime, and then finally the curved regime.

The accumulation regime is defined by $v_{GS}(t) < [v_{ON}(t) + 2\psi_B]$; the depletion regime, by $[v_{ON}(t) + 2\psi_B] < v_{GS}(t) < v_{ON}(t)$; the saturation regime, by $v_{ON}(t) < v_{GS}(t) < [v_{ON}(t) + v_{DS.SAT}(t)]$; and finally the curved regime, by $[v_{ON}(t) + v_{DS.SAT}(t)] < v_{GS}(t)$. To identify an equation valid in a particular regime, we will append one of four letters—a, d, s, or c—to the equation number. These symbols stand, of course, for accumulation, depletion, saturation, and curved, respectively.

The effective K is given by

$$K_{EFF}(t) = K\left[\frac{1}{1 - \lambda v_{DS}(t)}\right]\left[\frac{E_{CRIT}\epsilon/C_{OX}}{v_{GS}(t) - v_{ON}(t)}\right]^{U_{EXP}}, \tag{5-22}$$

where E_{CRIT} and U_{EXP} are empirical parameters and λ is the channel-length-modulation factor. There exist many empirical models for effective mobility and, in turn, for effective K. Here we selected only one of the possible Level-2 models and will regard E_{CRIT} and U_{EXP} as curve-fitting parameters that provide better agreement between theory and experiment.

The terminal voltages that comprehend parasitic effects are denoted by the subscript P. They are defined in terms of the intrinsic, or "internal," voltages used exclusively heretofore:

$$v_{GS.P}(t) = v_{GS}(t) + R_S[i_G(t) + i_B(t) + i_D(t)], \tag{5-23}$$

$$v_{GD.P}(t) = v_{GD}(t) + R_D[i_D(t)], \tag{5-24}$$

$$v_{BS.P}(t) = v_{BS}(t) + R_S[i_G(t) + i_B(t) + i_D(t)], \tag{5-25}$$

$$v_{BD.P}(t) = v_{BD}(t) + R_D[i_D(t)]. \tag{5-26}$$

The terminal currents are defined by

$$i_G(t) = C_{GS}(t)\,\frac{\partial v_{GS}(t)}{\partial t} + C_{GD}(t)\left[\frac{\partial v_{GS}(t)}{\partial t} - \frac{\partial v_{DS}(t)}{\partial t}\right] + C_{GB}(t)\,\frac{\partial v_{GB}(t)}{\partial t}, \tag{5-27}$$

$$\begin{aligned} i_B(t) = &+ I_{BS}\left[\exp\left(\frac{q v_{BS}(t)}{kT}\right) - 1\right] + I_{BD}\left[\exp\left(\frac{q[v_{BS}(t) - v_{DS}(t)]}{kT}\right) - 1\right] \\ &+ C_{BS}(t)\,\frac{\partial v_{BS}(t)}{\partial t} + C_{BD}(t)\left[\frac{\partial v_{BS}(t)}{\partial t} - \frac{\partial v_{DS}(t)}{\partial t}\right] - C_{GB}(t)\,\frac{\partial v_{GB}(t)}{\partial t}, \end{aligned} \tag{5-28}$$

and

$$\begin{aligned} i_D(t) = &\, i_D^R(t) - I_{BD}\left[\exp\left(\frac{q[v_{BS}(t) - v_{DS}(t)]}{kT}\right) - 1\right] \\ &- C_{GD}(t)\left[\frac{\partial v_{GS}(t)}{\partial t} - \frac{\partial v_{DS}(t)}{\partial t}\right] - C_{BD}(t)\left[\frac{\partial v_{BS}(t)}{\partial t} - \frac{\partial v_{DS}(t)}{\partial t}\right]. \end{aligned} \tag{5-29}$$

In the last equation, the resistive component of drain current $i_D^R(t)$ is defined in the four regimes of operation as follows:

$$i_D^R(t) = i_D^R(v_{ON}(t), v_{BS}(t), v_{DS}(t))\exp\left[\frac{v_{GS}(t) - v_{ON}(t)}{N(t)kT/q}\right], \tag{5-30a}$$

$$i_D^R(t) = \text{same as above}, \tag{5-30d}$$

$$i_D^R(t) = i_D^R(v_{GS}(t), v_{BS}(t), v_{DS.\mathrm{SAT}}(t)), \tag{5-30s}$$

$$i_D^R(t) = i_D^R(v_{GS}(t), v_{BS}(t), v_{DS}(t)), \tag{5-30c}$$

where

$$\begin{aligned} i_D^R(v_{GS}(t), v_{BS}(t), v_{DS}(t)) = 2K_{\mathrm{EFF}}(t)\Bigg\{&\left[v_{GS}(t) - V_{FB}^* + 2\psi_B - \frac{v_{DS}(t)}{2}\right]v_{DS}(t) \\ &+ \left(\frac{2\gamma}{3}\right)[(v_{DS}(t) - 2\psi_B - v_{BS}(t))^{3/2} \\ &- (-2\psi_B - v_{BS}(t))^{3/2}]\Bigg\}. \end{aligned} \tag{5-31}$$

The three gate capacitances, which will be described in Section 5-4, are defined in the four regimes of operation as:

$$C_{GS}(t) = C_{0GS}W, \tag{5-32a}$$

$$C_{GS}(t) = C_{0GS}W + \frac{2}{3}\,C_{OX}L^*W\left[\frac{v_{ON}(t) - v_{GS}(t)}{2\psi_B} + 1\right], \tag{5-32d}$$

$$C_{GS}(t) = C_{0GS}W + \frac{2}{3}\,C_{OX}L^*W, \tag{5-32s}$$

$$C_{GS}(t) = C_{0GS}W + C_{OX}L^*W\left[1 - \frac{v_{GS}(t) - v_{ON}(t) - v_{DS}(t)}{2[v_{GS}(t) - v_{ON}(t)] - v_{DS}(t)}\right]; \tag{5-32c}$$

$$C_{GD}(t) = C_{0GD}W, \tag{5-33a}$$

$$C_{GD}(t) = C_{0GD}W, \tag{5-33d}$$

$$C_{GD}(t) = C_{0GD}W, \tag{5-33s}$$

$$C_{GD}(t) = C_{0GD}W + C_{OX}L^*W\left[1 - \frac{v_{GS}(t) - v_{ON}(t)}{2[v_{GS}(t) - v_{ON}(t)] - v_{DS}(t)}\right]; \tag{5-33c}$$

and

$$C_{GB}(t) = C_{0GB}L^* + C_{OX}L^*W, \tag{5-34a}$$

$$C_{GB}(t) = C_{0GB}L^* + C_{OX}L^*W\left[\frac{v_{ON}(t) - v_{GS}(t)}{2\psi_B}\right], \tag{5-34d}$$

$$C_{GB}(t) = C_{0GB}L^*, \tag{5-34s}$$

$$C_{GB}(t) = C_{0GB}L^*. \tag{5-34c}$$

Finally, the source and drain junction capacitances are given by

$$C_{BS}(t) = \frac{C_{0BS.B}}{\sqrt[M_B]{1 - \dfrac{v_{BS}(t)}{\Delta\psi_0}}} + \frac{C_{0BS.SW}}{\sqrt[M_{SW}]{1 - \dfrac{v_{BS}(t)}{\Delta\psi_0}}}, \tag{5-35}$$

and

$$C_{BD}(t) = \frac{C_{0BD.B}}{\sqrt[M_B]{1 - \dfrac{v_{BS}(t) - v_{DS}(t)}{\Delta\psi_0}}} + \frac{C_{0BD.SW}}{\sqrt[M_{SW}]{1 - \dfrac{v_{BS}(t) - v_{DS}(t)}{\Delta\psi_0}}} \tag{5-36}$$

This completes the seventeen equations of the SPICE model. These equations will be referred to here as the Level-2 SPICE model for the MOSFET or, briefly, as the SPICE model.

Using the SPICE model and the parameter values listed in Table 7, we can generate the *I–V* characteristics shown in Figure 5-4, which are consistent with the experimental data given in Figure 5-2. Exactly the same results can be obtained by using the five electrical parameters noted in Table 8 in place of the physical parameters N_A, T_G, N_f^*, R_{SH}, and J_0. Table 9 lists the equivalent sets of parameter values.

For a drain–source voltage of 2.47 V, the square root of the drain current is plotted in Figure 5-5(a) as a function of V_{GS}. In Figure 5-5(b) the drain–source current is plotted using a logarithmic current scale. If the MOSFET were exactly a square-law device, the curve in Figure 5-5(a) would be a straight line and would intersect the x axis at the threshold voltage, $V_T = 1.03$ V. Figure 5-5(b) shows the subthreshold behavior of the drain current. The slope of the curve in the subthreshold range from 10^{-11} to 10^{-7} A is controlled by the variable N, which in turn is stated in Equation 5-21.

The transconductance $g_{m.P}$ of the MOSFET is plotted in Figure 5-6(a), along with the output conductance $g_{o.P}$. These are defined , respectively, as $\partial I_D/\partial V_{GS.P}$ and $\partial I_D/\partial V_{DS.P}$. The voltage gain $A_{V.P}$, Figure 5-6(b), is defined as $g_{m.P}/g_{o.P}$, which is the maximum gain that can be achieved in most circuits using the MOSFET. The voltage gain has a maximum value near the threshold voltage and decreases substantially as $V_{GS.P}$ in-

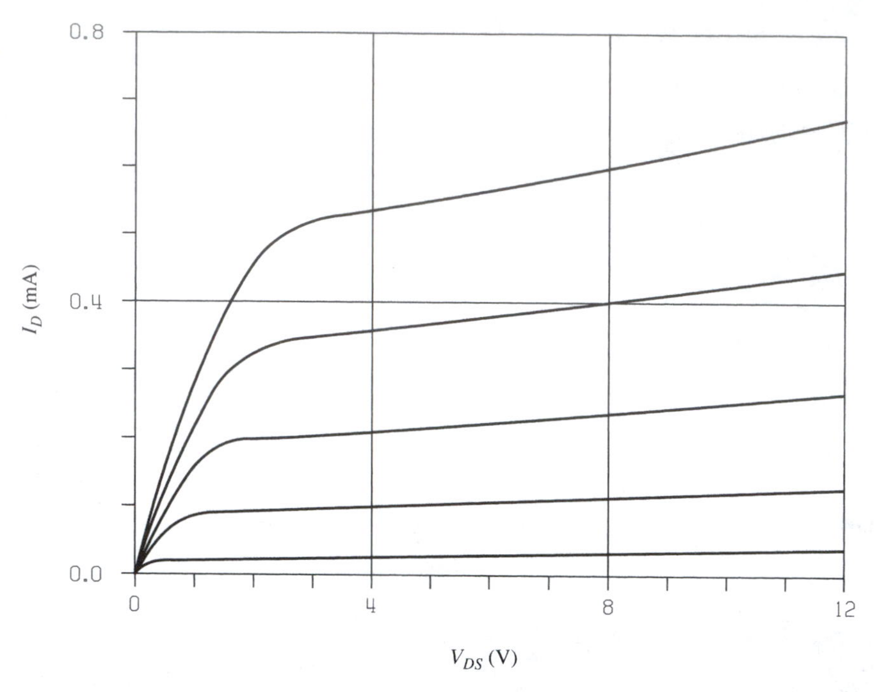

Figure 5-4 Current–voltage characteristics for the MOSFET of Figure 5-1, calculated using the Level-2 model. Comparison with the experimental results in Figure 5-2 shows good agreement.

creases. Thus for high gain, the MOSFET must be biased at low current levels. Figure 5-7 plots the same curves as Figure 5-6 using a logarithmic current scale. Figure 5-7(a) clearly shows that the transconductance of the MOSFET is about qI_D/kT at low current levels, a fact exploited in low-power (especially CMOS) circuitry. A constant voltage-gain magnitude of more than 600 is achieved over a considerable range in the subthreshold regime.

5-3 Small-Signal Applications of Model

One of the simplest small-signal voltage amplifiers is the common-source circuit shown in Figure 5-8. The load resistor and dc supply voltages were selected to be compatible with the depletion-load circuit to be discussed in the next section. Using the SPICE-parameter values given in Table 7, we can calculate the current–voltage curves shown in Figure 5-9, along with the resistive load line. The square marked *B* identifies the dc bias point, and Tables 10 and 11 give the corresponding numerical values.

To analyze the small-signal performance of the MOSFET, we will construct a model similar to the hybrid-pi model for the BJT. To do this, we replace the large-signal circuit of Figure 5-3(b) with the corresponding small-signal circuit given in Figure 5-10. One primary objective at present is to derive the element values of the small-signal circuit, starting with the Level-2 model given in the last section. Then we will analyze the small-signal performance of the MOSFET for the common-source circuit given in Figure 5-8. It will turn out that the small-signal current-gain cutoff frequency of the selected MOSFET is about 0.5 GHz, and that the 3-dB bandwidth of the common-source circuit is about 66 MHz.

The selected dc bias point is in the saturation regime of operation, so that the resistive component of drain current $i_D^R(t)$ is given by Equation 5-30s. The term "resistive component" is used to identify the first term of Equation 5-29, given in detail in

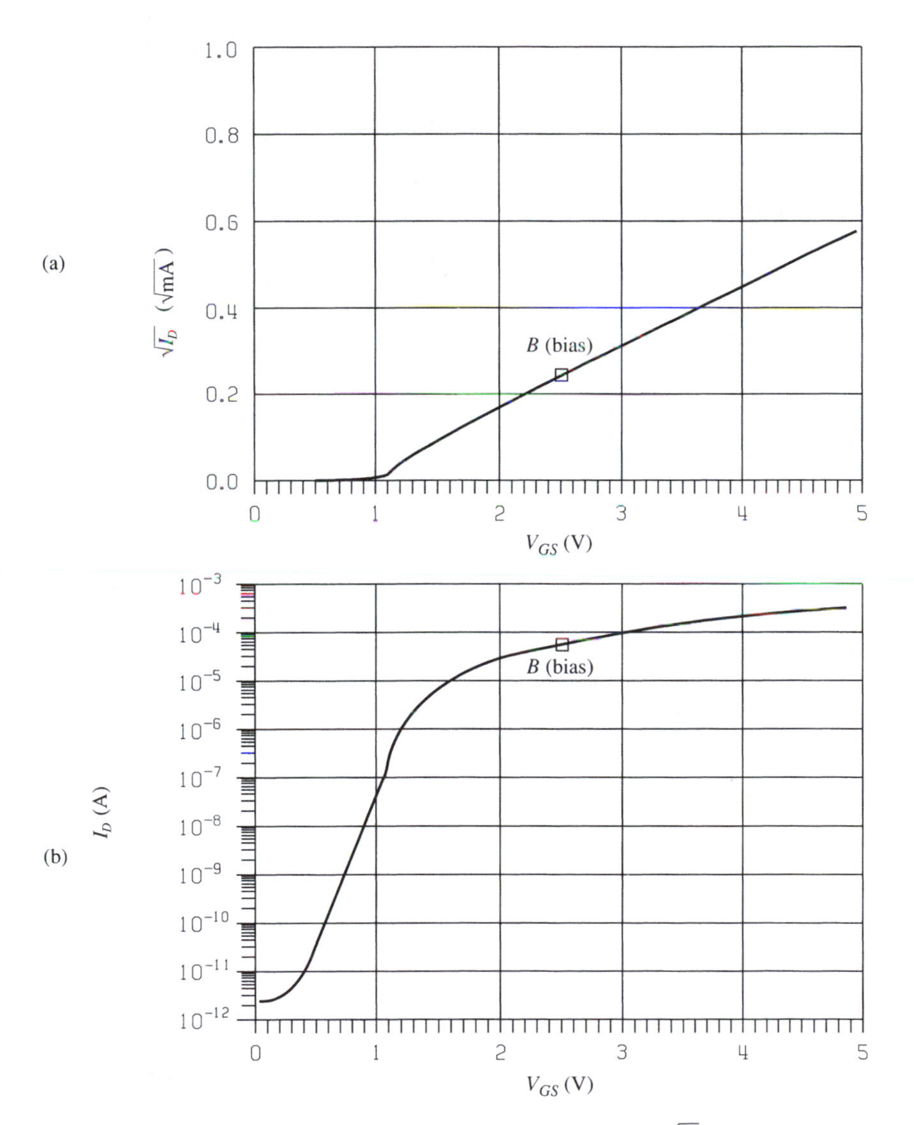

Figure 5–5 Transfer characteristic for $V_{DS} = 2.471$ V. (a) Plot of $\sqrt{I_D}$ versus V_{GS} on a linear scale. (b) Plot of I_D on a logarithmic scale versus V_{GS} on a linear scale.

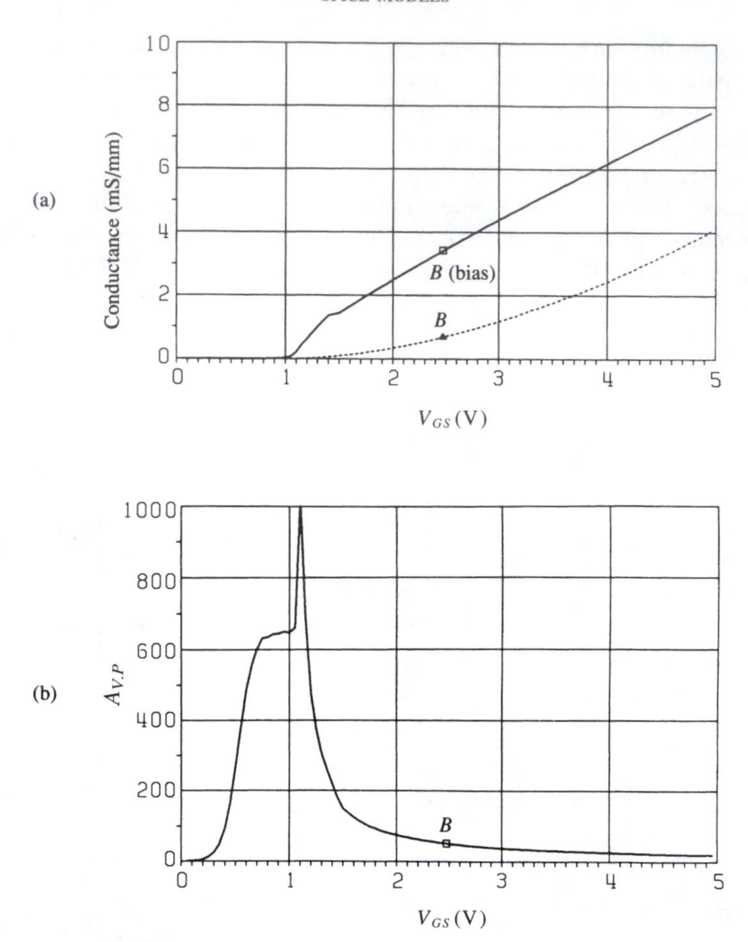

Figure 5–6 MOSFET gain properties as a function of gate–source voltage. Point B is the bias point. (a) Transconductance comprehending parasitic elements $g_{m.P}$ (solid line) and output conductance g_o after multiplication by 10 (dotted line). (b) Small-signal voltage gain $A_{V.P}$ comprehending parasitic elements.

Equation 5-31. Its mathematical form is identical to that for static drain current, Equation 5-2.

Taking advantage of the equivalence just noted, let us revert to the simpler dc symbols to evaluate the intrinsic transconductance of the MOSFET:

$$\begin{aligned} g_m \equiv \frac{\partial I_D^R}{\partial V_{GS}} &= 2K_{\text{EFF}} V_{DS.\text{SAT}} + \frac{I_D^R}{K_{\text{EFF}}} \left(\frac{\partial K_{\text{EFF}}}{\partial V_{GS}} \right) \\ &\quad + 2K_{\text{EFF}} [V_{GS} - V_{FB}^* + 2\psi_B - V_{DS.\text{SAT}} \\ &\quad + \gamma \sqrt{V_{DS.\text{SAT}} - 2\psi_B - V_{GS}}] \left(\frac{\partial V_{DS.\text{SAT}}}{\partial V_{GS}} \right). \end{aligned} \tag{5-37}$$

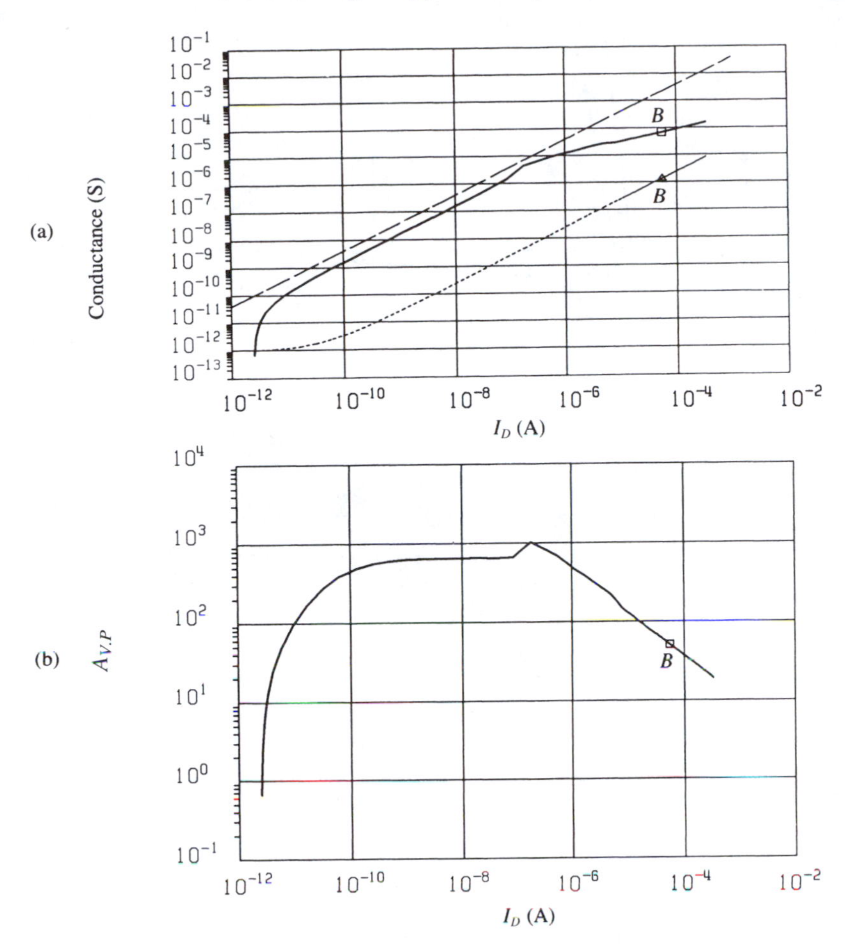

Figure 5–7 Semilog presentation of MOSFET gain properties as a function of drain current (same data as in Figure 5-6). Point B is the bias point. (a) Transconductance comprehending parasitic elements $g_{m.P}$ (solid line) and output conductance g_o after multiplication by 10 (dotted line). Dashed curve shows qI_D/kT limit, seen in the BJT. (b) Small-signal voltage gain $A_{V.P}$ comprehending parasitic elements.

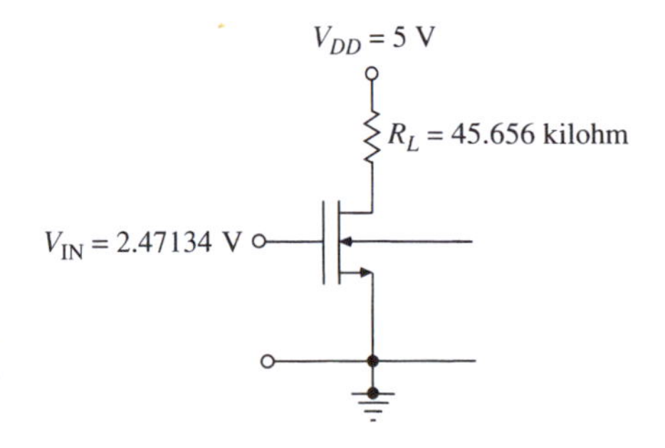

Figure 5–8 Schematic diagram of the resistive-load circuit used for small-signal analysis. The resistor value has been tailored to produce the desired operating point.

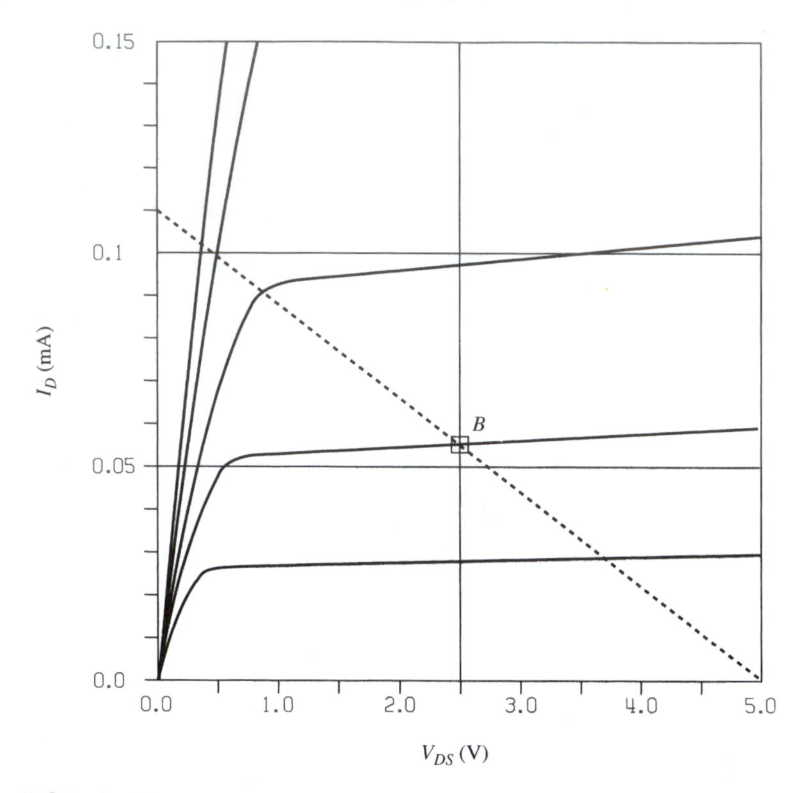

Figure 5-9 MOSFET output characteristics and resistive loadline used for small-signal analysis. The square indicates the bias point.

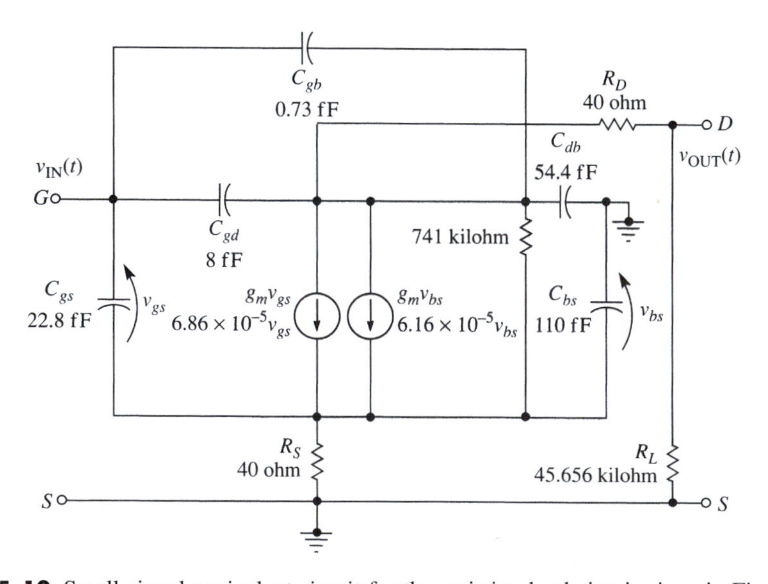

Figure 5-10 Small-signal equivalent circuit for the resistive-load circuit given in Figure 5-8.

In the remainder of this section we will continue to use static symbols for the sake of simplicity. The first partial derivative in Equation 5-37 can be found using Equation 5-22 and is

$$\frac{\partial K_{\text{EFF}}}{\partial V_{GS}} = -\left[\frac{K_{\text{EFF}} U_{\text{EXP}}}{V_{GS} - V_{\text{ON}}}\right] = -2.09 \times 10^{-5} \text{ A/V}^3, \tag{5-38}$$

where we have used Tables 7 and 11 and the Level-2 model to obtain the numerical value. The second partial derivative in Equation 5-37 is found using Equation 5-19, yielding

$$\frac{\partial V_{DS.\text{SAT}}}{\partial V_{GS}} = 1 - \left[1 + \frac{4}{\gamma^2}(V_{DS} - V_{FB} - V_{BS})\right]^{-1/2} = 0.5581. \tag{5-39}$$

Substituting the values just given and values taken from Tables 8 and 11 into Equation 5-37 yields

$$\begin{aligned} g_m &= [7.939 \times 10^{-5} - 1.083 \times 10^{-5} + 3.5 \times 10^{-8}] \text{ A/V} \\ &= 0.069 \text{ mA/V}. \end{aligned} \tag{5-40}$$

The output conductance g_o is given by

$$g_o \equiv \frac{\partial I_D^R}{\partial V_{DS}} = \frac{I_D^R}{K_{\text{EFF}}}\left(\frac{\partial K_{\text{EFF}}}{\partial V_{DS}}\right). \tag{5-41}$$

Using Equation 5-22, we find that

$$\frac{\partial K_{\text{EFF}}}{\partial V_{DS}} = \frac{\lambda K_{\text{EFF}}}{1 - \lambda V_{DS}}, \tag{5-42}$$

so that

$$g_o = \frac{\lambda I_D^R}{1 - \lambda V_{DS}} = 0.001 \text{ mS}. \tag{5-43}$$

In Problem A30 we evaluate the bulk–source transconductance, which is a measure of the effectiveness of the bulk region as a control electrode. The value found there is

$$g_{mbs} = 0.062 \text{ mA/V}, \tag{5-44}$$

which is nearly equal to g_m in value. The remaining values for the small-signal circuit given in Figure 5-10 can be obtained directly from the Level-2 model.

Using standard circuit techniques, we can approximate the model of Figure 5-10 with the simplified model of Figure 5-11. To characterize the performance of the selected MOSFET, we need to evaluate A_V, the small-signal voltage gain, and f_T, the cutoff frequency for current gain. It may at first be puzzling to encounter "current gain"

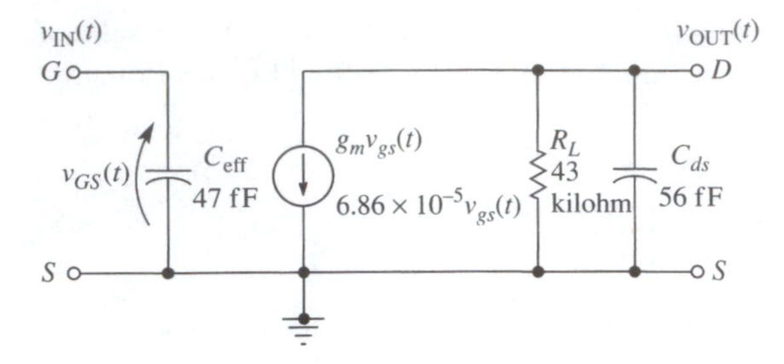

Figure 5-11 Approximate small-signal equivalent circuit derived from Figure 5-10 by using standard circuit analysis.

in the MOSFET case. Static current gain in the ideal device is of course infinite, because the gate insulator is perfect. But in high-frequency operation, capacitive current through the gate terminal is appreciable, so current gain is a finite and useful concept. Consequently, f_T is as valid a figure of merit for the MOSFET as for the BJT, although it does have some limitations.

To characterize the resistive-load circuit associated with the MOSFET in Figure 5-8, we need to know the circuit voltage gain and 3-dB bandwidth ($A_{V,\text{cir}}$ and $f_{3\text{dB}}$, respectively), as well as the intrinsic gain properties of the MOSFET. The intrinsic voltage gain A_V of the MOSFET is given by

$$A_V = g_m/g_o = 69. \tag{5-45}$$

To find f_T, we must first calculate the dynamic current gain A_I. Let us assume a sinusoidal input voltage. Employing the appropriate symbols for ac analysis and referring to Figure 5-11, we obtain the output current:

$$i_d(\omega) = g_m v_{gs}(\omega). \tag{5-46}$$

The input current is given by

$$i_g(\omega) = j\omega[C_{gs} + A_V C_{gd}]v_{gs}(\omega), \tag{5-47}$$

so that

$$A_I(\omega) = \frac{g_m}{j\omega(C_{gs} + A_V C_{gd})}. \tag{5-48}$$

By definition, f_T has the value $2\pi\omega$ when $A_I(\omega) = 1$. Thus,

$$f_T = \frac{g_m}{2\pi(C_{gs} + C_{gd})} = 0.5 \text{ GHz}. \tag{5-49}$$

Hence the selected MOSFET is characterized by an intrinsic voltage gain of 69 and a cutoff frequency of 0.5 GHz.

When the input generator is a pure ac-voltage source, the high-frequency response of the amplifier circuit is determined by the RC time constant of the output circuit.

Thus, the $f_{3\text{dB}}$ of the resistive-load circuit is

$$f_{3\text{dB}} = \frac{g_o + g_L}{2\pi C_{bd}} = 66 \text{ MHz}, \tag{5-50}$$

and the ac voltage gain is

$$A_{V,\text{cir}} = -\frac{g_m}{g_o + g_L} = -3. \tag{5-51}$$

Using SPICE we can calculate the voltage gain as a function of frequency, with the results plotted in Figure 5-12. These calculations and Figure 5-12 show that the resistive-load circuit is characterized by a voltage gain of -3, and a bandwidth of 66 MHz. The circuit rise time and bandwidth are interrelated, so that

$$f_{3\text{dB}} t_{\text{RISE}} = 0.35, \tag{5-52}$$

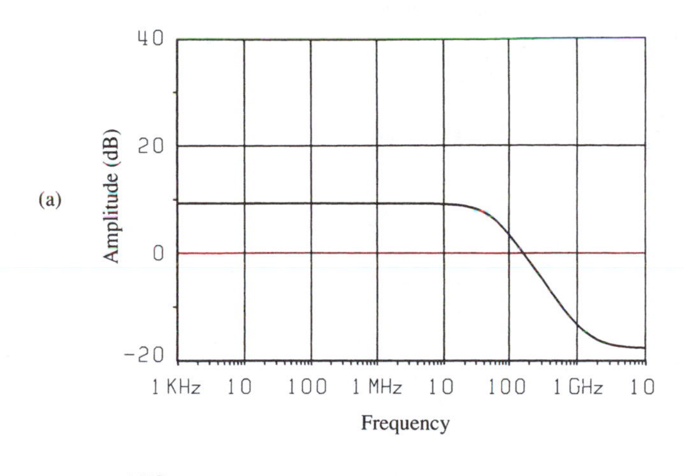

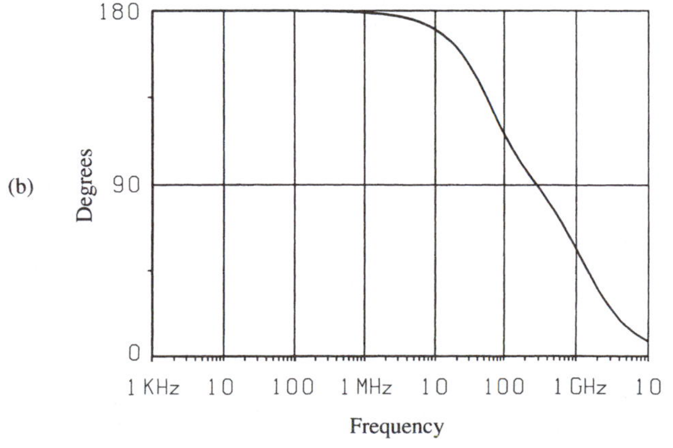

Figure 5–12 SPICE analysis of resistive-load MOSFET amplifier. (a) Frequency dependence of voltage-gain magnitude $A_{V,P}$ in decibels, comprehending parasitic elements. (b) The corresponding phase of $A_{V,P}$ as a function of frequency.

and hence $t_{RISE} = 5.3$ ns. Also, the 10-to-90% rise time is given by 2.2 times the RC time constant of the output circuit, so that

$$t_{RISE} = \frac{2.2C_{bd}}{g_o + g_L} = 5.3 \text{ ns}, \tag{5-53}$$

in agreement with the result just given. The calculated small-signal pulse response of the resistive-load circuit is given in Figure 5-13.

Finally, there is an aspect of SPICE that deserves explanation. SPICE employs certain early physical data that are different from the more up-to-date values used elsewhere in this book, and for understandable reasons, has preferred to adhere to the older values. An example is the too-high value of intrinsic density used in SPICE, $n_i = 1.45 \times 10^{10}/\text{cm}^3$. But because empirical adjustment is an integral part of SPICE modeling, the inaccuracies do not cause problems for circuit simulation.

5-4 Large-Signal Applications of Model

To investigate the large-signal dynamic response of a MOSFET, we consider the two-stage inverter circuit shown in Figure 5-14. The first stage consists of a driver transistor

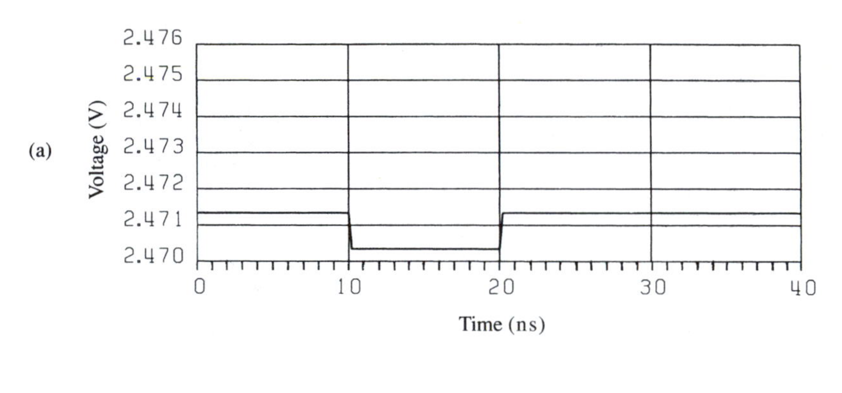

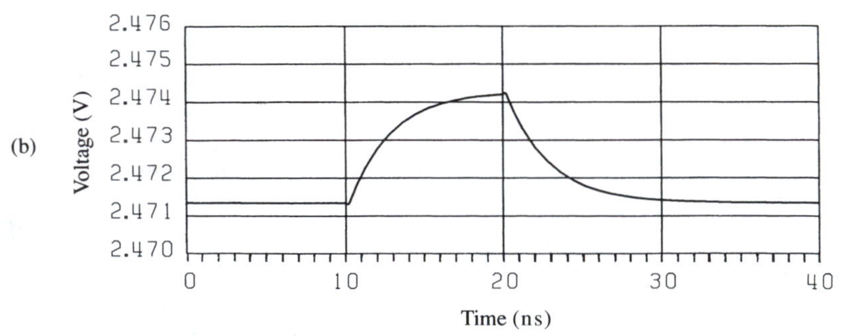

Figure 5-13 SPICE analysis of resistive-load amplifier in Figure 5-8. (a) Input-voltage pulse applied to amplifier. (b) Computer-determined output waveform.

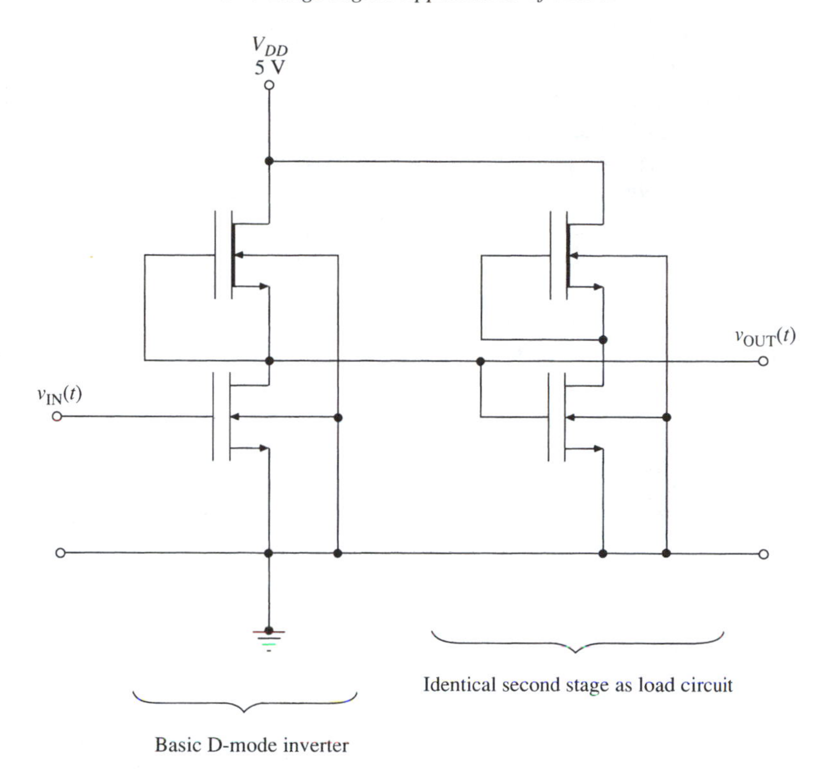

Figure 5-14 Schematic diagram of the two-stage circuit used for large-signal MOSFET analysis. Each stage constitutes an E-D inverter, having an E-mode driver and a D-mode load device. The second inverter serves as a realistic circuit for loading the output node of the first stage.

with a D-mode MOSFET as the load device. The driver is identical to that described in Section 5-3. The load was formed by an additional ion implantation in the channel region; hence we can use the same Level-2 parameters as those for the driver, except that $N_f^* = 1.20 \times 10^{12}/\text{cm}^2$ to include the implanted ions. This value of N_f^* represents an implant dose close to 1.20×10^{12} ions/cm^2 that produces a threshold shift from $+1.03$ V to -5.0 V, which is to say, from E-mode to D-mode properties. The geometry of the load device is different from that of the driver, with $Z = 12 \times 10^{-6}$ m, $L = 4 \times 10^{-6}$ m, $A_D = 320 \times 10^{-12}$ m^2, $A_S = 32 \times 10^{-12}$ m^2, $N_{RS} = 1$, and $N_{RD} = 1$.

The source region of the load and the drain region of the driver are common in the inverter circuit. When a single region does double duty in this manner, it is often incrementally larger in area than one that serves, let us say, as a drain terminal only. At present we will assume a 10% increment and will assign to the source of the load device a value of A_S that is one-tenth its value in the driver device. In this way, the output node of each inverter will have the correct total parasitic junction capacitance; yet we will still have the convenience of having each driver device in Figure 5-14 be identical to the lone driver in Figure 5-8 in all of the latter's properties, including parasitic elements.

The *I–V* characteristics of the driver and load are plotted in Figure 5-15. The solid line represents the load characteristic with the gate, source, and bulk terminals of the load device being common. In the inverter circuit, however, only the gate and source are common, and these are at the same voltage as the drain of the driver. The bulk region is at circuit ground. As a result, the D-mode device has a finite bulk–source voltage V_{BS}. The body effect is surprising in its magnitude, yielding the characteristic shown with a dotted line. Recall that the body effect is incorporated in the model through the parameter γ. Comparing Figures 5-9 and 5-15, we see that the D-mode load approximates a resistive load. However, the D-mode load is small compared to other possible load devices, which is a major advantage in integrated-circuit applications. Points *A*, *B*, and *C* in Figure 5-15 represent the calculated dc bias values listed in the corresponding columns of Table 11.

The dc characteristics of the inverter are best represented using transfer characteristics, as shown in Figure 5-16. Again we have plotted the calculated dc values listed in Table 5-11. The solid line represents the inverter output voltage $v_{OUT}(t)$ as a function of the input voltage $v_{IN}(t)$. The dotted curve is the same function plotted with the

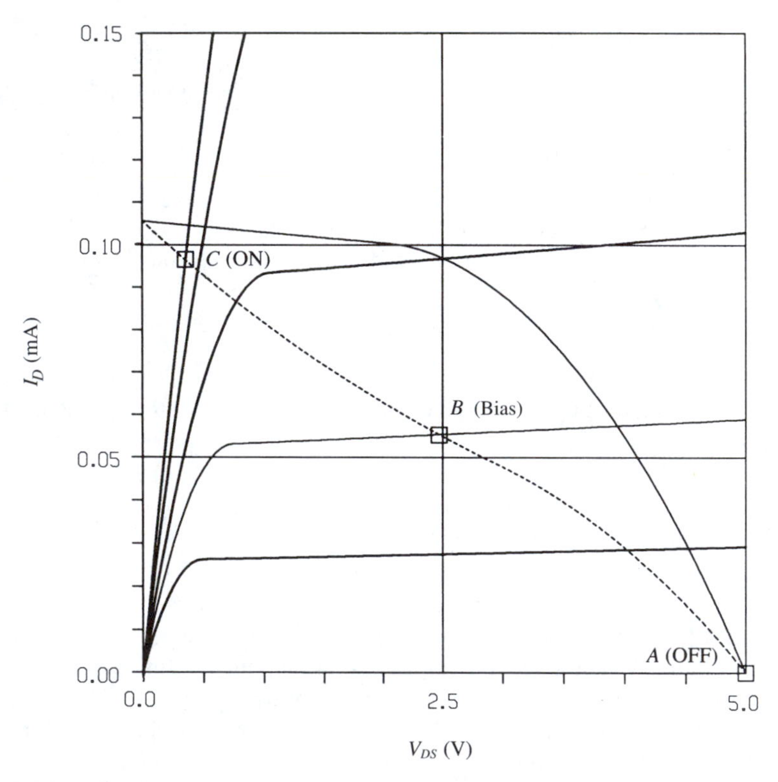

Figure 5-15 MOSFET output characteristics and D-mode loadline. The solid curve corresponds to the case where gate, source, and bulk regions of the load device are common (as would be true with discrete devices). The dotted curve displays the consequences of the body effect on the D-mode load for the present numerical example.

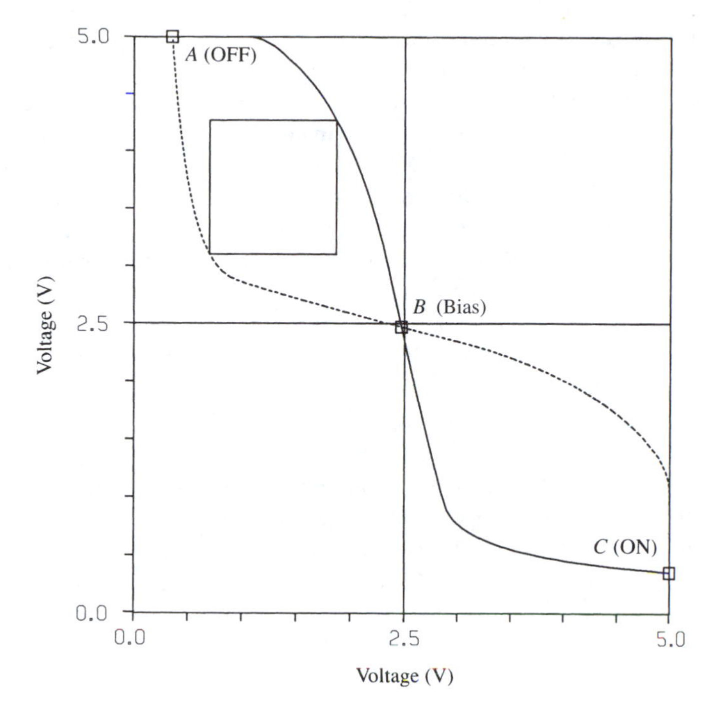

Figure 5–16 The solid curve is the transfer characteristic for the first inverter in Figure 5-14. The dotted curve is obtained by rotating the solid curve 180 degrees about a straight line drawn through the origin and point *B*. The large square illustrates one way of defining noise margin.

axes interchanged. The two curves intersect at the OFF and ON dc bias values, points *A* and *C*, and at the dc bias value, point *B*. The slope of the solid curve at point *B* equals the dc voltage-gain value of about -3 that was calculated in the case of the resistive load.

The *noise margin* in a digital, or large-signal-switching, circuit is of great importance. It constitutes a measure of the noise-voltage magnitude that the circuit can withstand at its input terminal without "flipping," or changing state. Generous noise margins are a key to reliable operation, and hence have a bearing on product yield as well, since testing is a matter of assessing operational reliability. Yield, in turn, has a crucial bearing on product cost. Noise margin is often defined as the length of one side of the largest square that can be fitted between the solid and dotted curves, as illustrated in Figure 5-16. This, however, is an optimistic way of defining noise margin [85].

Now let us treat the dynamic response of the two-stage inverter circuit. To begin, we must examine in more detail than before the capacitive components of the Level-2 model, as given by Equations 5-32 through 5-36. In what follows we will consider only the driver MOSFET in the first inverter circuit. Since the bulk–source voltage is always zero for the driver, the corresponding depletion-layer capacitance C_{BS} approximates its zero-bias value of 110 fF for all input voltages, where 100 fF is contributed by the bottom portion of the implanted region, and 10 fF by the sidewall portion. But because C_{BS} is short-circuited, it can be neglected.

The bulk–drain junction is reverse-biased to its maximum degree when the driver is OFF. Its depletion-layer capacitance rises from 41 to 92 fF as V_{IN} increases from the OFF condition (0.356 V) to the ON condition (5.0 V), as described by Equation 5.42. We saw in Section 5-3 that the small-signal response of the inverter is dominated by C_{BD}. It is not surprising, therefore, to learn that the large-signal response is also dominated by C_{BD}, especially (as before) when the input generator is a voltage source applied directly to the gate–source input port.

The three capacitances associated with the gate are given by Equations 5-32 through 5-34. For a low-impedance input-voltage source, these capacitances do not play a major role because they are short-circuited to ground. But as the impedance of the source increases, they increase in importance. The behaviors of C_{GS}, C_{GD}, and C_{GB} are shown in Figure 5-17 as functions of the gate–source voltage $v_{GS}(t)$. As $v_{GS}(t)$ increases from 0 to 5 V, and the operating point moves along the dc loadline such as the one shown in Figure 5-15, the operation of the MOSFET moves in sequence through the accumulation, depletion, saturation, and curved regimes. In the accumulation regime, C_{GB} is constant and equals the sum of the external gate capacitance and the total oxide capacitance:

$$C_{GB} = C_{OGB}L^* + C_{OX}L^*Z. \tag{5-54}$$

It has a value of 23 fF, with the gate-oxide contribution dominant. In the depletion regime, C_{GB} decreases, let us assume linearly, until it equals the external gate capacitance of 0.73 fF. In a numerical program such as SPICE, it is important that the tran-

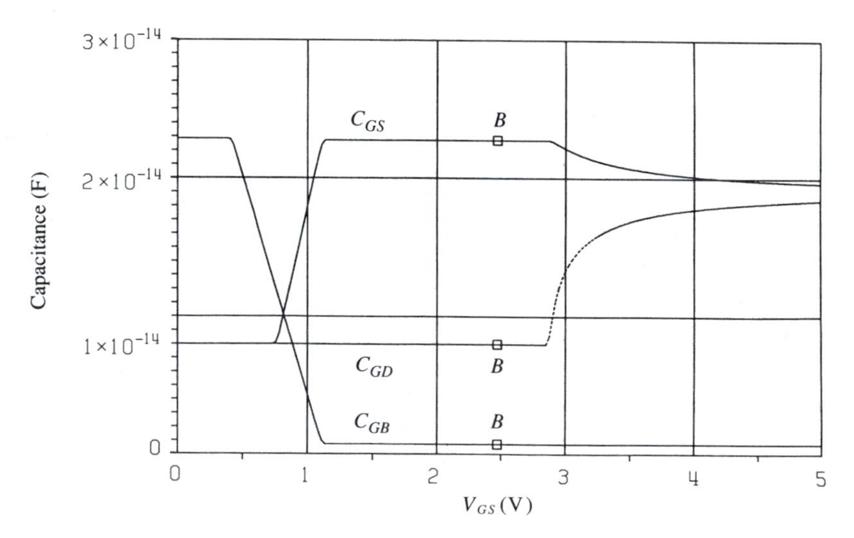

Figure 5-17 Gate-capacitance components as functions of gate–source voltage. The square *B* is the bias point in each case.

sition from one regime to the next be gradual, accounting for the assumption. In the saturation and curved regimes, C_{GB} is constant and equal to this value.

The variation of C_{GS} as V_{GS} increases is as follows. In the accumulation regime, C_{GS} equals the parallel-plate overlap capacitance of 8 fF. In the depletion regime, its value increases linearly until it equals the sum of the overlap capacitance and two-thirds of the gate-oxide capacitance, where the sum equals 19.6 fF. The factor of two-thirds is an empirical approximation that deals with the somewhat complicated combined behavior of the depletion and inversion layers as they develop in the depletion regime—a value based upon analytical calculations and experimental measurements. In the saturation regime, C_{GS} remains constant and equal to 23 fF. Finally, in the curved regime it decreases according to Equation 5-32c, approaching a final value of 19.7 fF.

The corresponding behavior of C_{GD} as $v_{GS}(t)$ increases is as follows. In the accumulation, depletion, and saturation regimes, it is constant and equal to the overlap capacitance of 8 fF. In the curved regime it increases according to Equation 5-33(c), approaching a value of 19.7 fF. In the curved regime, an empirical assignment of gate-oxide capacitance is made once again. Half is placed in parallel with the gate–source junction, and half with the gate–drain junction.

EXERCISE 5–1 Why does this assignment of C_{OX} make sense at low bias levels?

■ With low bias, especially with low V_{DS}, the depletion layer is nearly uniform in thickness from source to drain. Since the device is symmetric with respect to source and drain, an equal division of the gate-oxide capacitance makes sense.

The dynamic behavior of the two-stage inverter circuit is shown in Figures 5-18 and 5-19. The former gives external variables, and the latter, internal variables. Figure 5-18(a) shows the assumed input voltage waveform, where $v_{IN}(t) = v_{GS.P}(t)$, and the corresponding output waveform, where $v_{OUT}(t) = v_{DS.P}(t)$. The input waveform is a voltage pulse that rises linearly from 0.356 V at $t = 10$ ns to 5 V at $t = 20$ ns and falls linearly from 5 V at $t = 91$ ns to 0.356 V at $t = 101$ ns. Figure 5-18(b) through (d) shows the drain current $i_D(t)$, the gate current $i_G(t)$, and the bulk current $i_B(t)$. The gate and bulk currents are significant only during the rise and fall periods of the switching waveform. During these periods, the bulk current dominates because it is required to charge and discharge $C_{BD}(t)$. Recall that this is the capacitance that dominates both the high-frequency and fast-switching behavior of the MOSFET.

Behaviors of the various internal variables are shown in Figure 5-19 for the time interval from 5 to 25 ns. Figure 5-19(a) presents the internal variables $g_m(t)$, $g_{mbs}(t)$, and $g_o(t)$. Note that $g_m(t)$ and $g_{mbs}(t)$ are nearly identical. The resistive portion of the drain current $i_D^R(t)$—as given by Equation 5-30—is shown in Figure 5-19(b). This component has a large peak at $t = 18$ ns because drain current is required to charge the various capacitances. Figure 5-19(c) shows the three capacitive components of gate current corresponding to the three terms of Equation 5-27. Finally, Figure 5-19(d) shows two of the five components of bulk current, specifically the bulk–source and the bulk–drain capacitive currents. The corresponding resistive components can be neglected

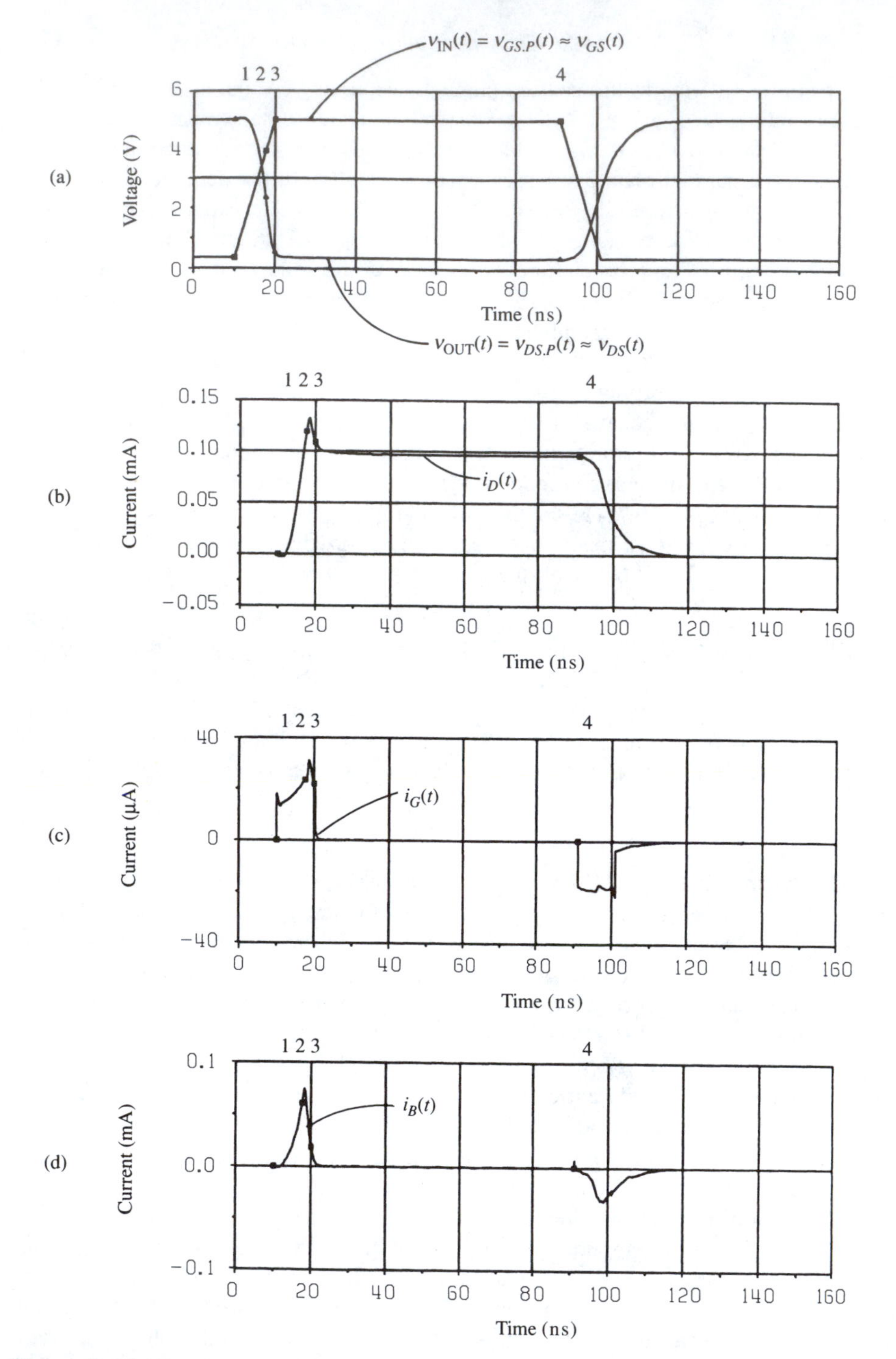

Figure 5–18 Waveforms calculated using SPICE for the terminal variables in the large-signal transient problem, with parasitic effects included. The numbered points correspond to the columns in Table 12. (a) Input and output voltages versus time. (b) Drain current $i_D(t)$ versus time. (c) Gate current $i_G(t)$ versus time. (d) Bulk current $i_B(t)$ versus time.

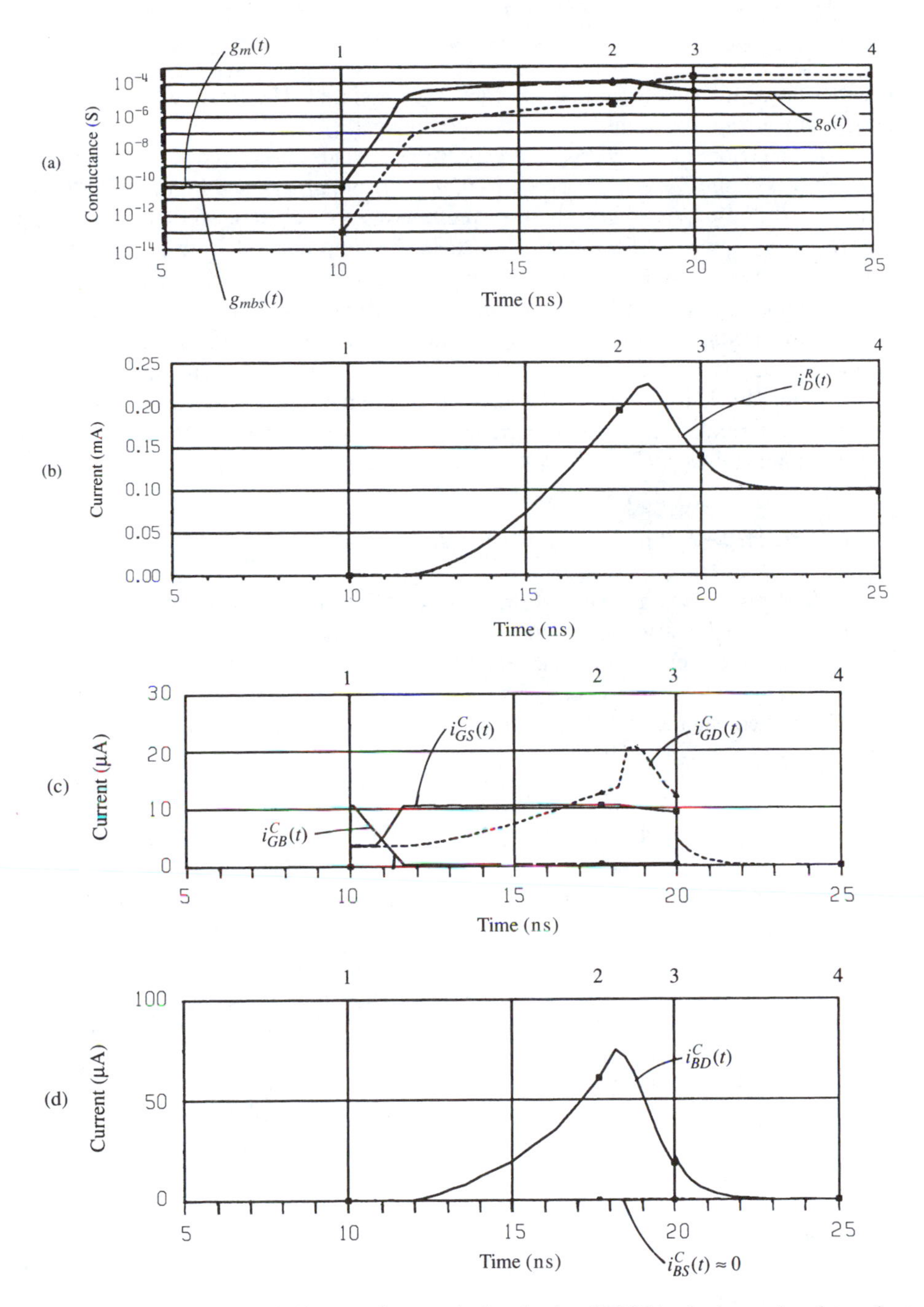

Figure 5–19 Intrinsic-variable waveforms calculated using SPICE in the large-signal transient problem. The numbered points correspond to the columns in Table 12. (a) Gate–source transconductance $g_m(t)$, bulk–source transconductance $g_{mbs}(t)$, and output conductance $g_o(t)$. (b) The resistive component of drain current $i_D^R(t)$ and total (or net) drain current $i_D(t)$. (c) The three components of gate current, $i_{GS}^C(t)$, $i_{GD}^C(t)$, and $i_{GB}^C(t)$. (d) The important capacitive component of bulk–drain current $i_{BD}^C(t)$. The resistive component of bulk–drain current can be neglected because the drain junction is reverse-biased. Also, the bulk–source junction is short-circuited.

because the source and drain junctions are reverse-biased. The final bulk-current component is the negative of the gate-current component $i_{GB}(t)$.

The behaviors of the charge and capacitive variables are given in Figure 5-20(a) through (d). Figure 5-20(a) presents $q_{GS}(t)$, $q_{GD}(t)$, and $q_{GB}(t)$, and Figure 5-20(b) presents $q_{BS}(t)$ and $q_{BD}(t)$. Figure 5-20(c) presents $C_{GS}(t)$, $C_{GD}(t)$, and $C_{GB}(t)$. Finally, Figure 5-20(d) presents $C_{BS}(t)$ and $C_{BD}(t)$. In the time interval from 5 to 25 ns, the MOSFET moves from the accumulation regime through the depletion regime, through the saturation regime, and finally into the curved regime, so that the curves in Figure 5-20(c) are similar in form to the corresponding dc curves given in Figure 5-17. The various capacitive effects cause the dynamic loadline to differ significantly from the dc loadline. This is illustrated in Figure 5-21, where the squares correspond to the values of total time-dependent drain current $i_D(t)$, listed in Table 12. The lower dashed curve is the loadline for MOSFET turn-off.

EXERCISE 5-2 Why is $i_D^R(t)$ much larger than $i_D(t)$ at $t = 17.68$ ns?

■ The resistive component of drain current $i_D^R(t)$ must supply not only the drain current $i_D(t)$, but also the current needed to charge parasitic capacitances. Visualize $i_D^R(t)$ as the current source in an equivalent circuit, wherein its upper node is connected to the load device, carrying $i_D(t)$, and two capacitances, $C_{GD}(t)$ and $C_{BD}(t)$. The current source must "pull" current through all three elements.

The large-signal pulse response of the MOSFET is very complicated, as can be seen from Figures 5-18 and 5-19. SPICE simulation provides an accurate means for predicting the nonlinear switching behavior of a given device. A detailed examination of the SPICE outputs is always a rewarding endeavor.

5-5 Recent MOSFET Models

The SPICE circuit-simulator program was originally developed in the late 1960s at the University of California, Berkeley [86–89]. The initial MOSFET models, Levels 1 to 3, are still important today after more than 30 years of use. These models—called the first-generation models—are discussed in detail in this book because (1) they are widely available, (2) they are based directly on device physics, and (3) they form the basis for understanding the more recent second- and third-generation models. The three generations of MOSFET models are described in detail in Foty [86], and his classifications are summarized in Table 14.

The second-generation models [86, 92–94] were developed in the late 1980s to describe more accurately MOSFETs with submicrometer ("submicron") channel lengths. As noted by Foty, the emphasis was on "mathematical conditioning for faster and more robust [trouble-free] circuit simulation." This emphasis required a large number of model parameters that were difficult and time-consuming to extract from device

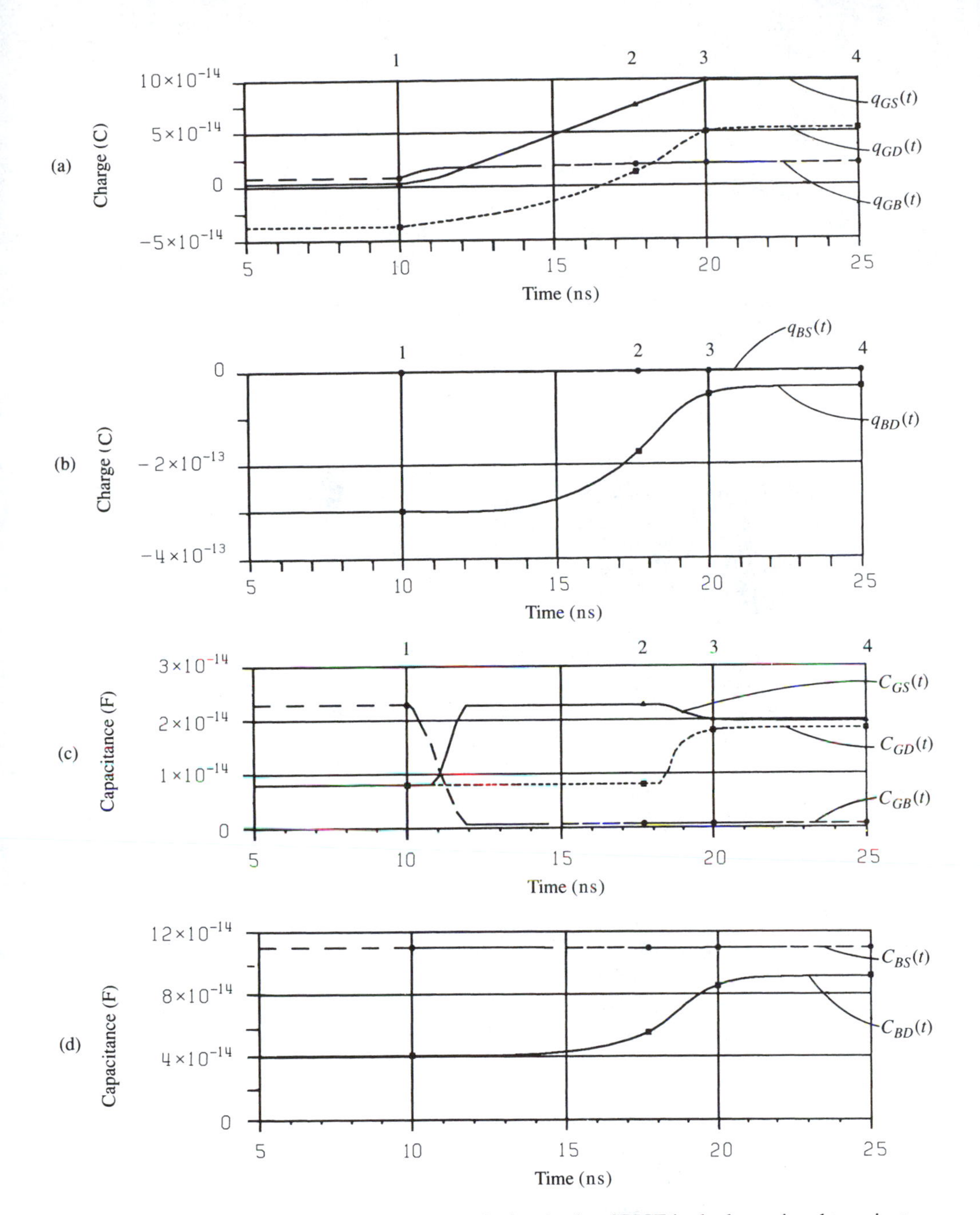

Figure 5–20 Intrinsic-variable waveforms calculated using SPICE in the large-signal transient problem. The numbered points correspond to the columns in Table 5-12. (a) Magnitudes of the charges stored in the three gate capacitances. (b) Magnitudes of the charges stored in two of the three bulk capacitances. The magnitude of the third bulk-capacitance charge is identical to $q_{BG}(t)$ in part (a). (c) Magnitudes of the capacitances associated with the gate terminal. (d) Magnitudes of two of the three capacitances associated with the bulk terminal. The magnitude of the third capacitance is identical to $C_{GB}(t)$ in part (c).

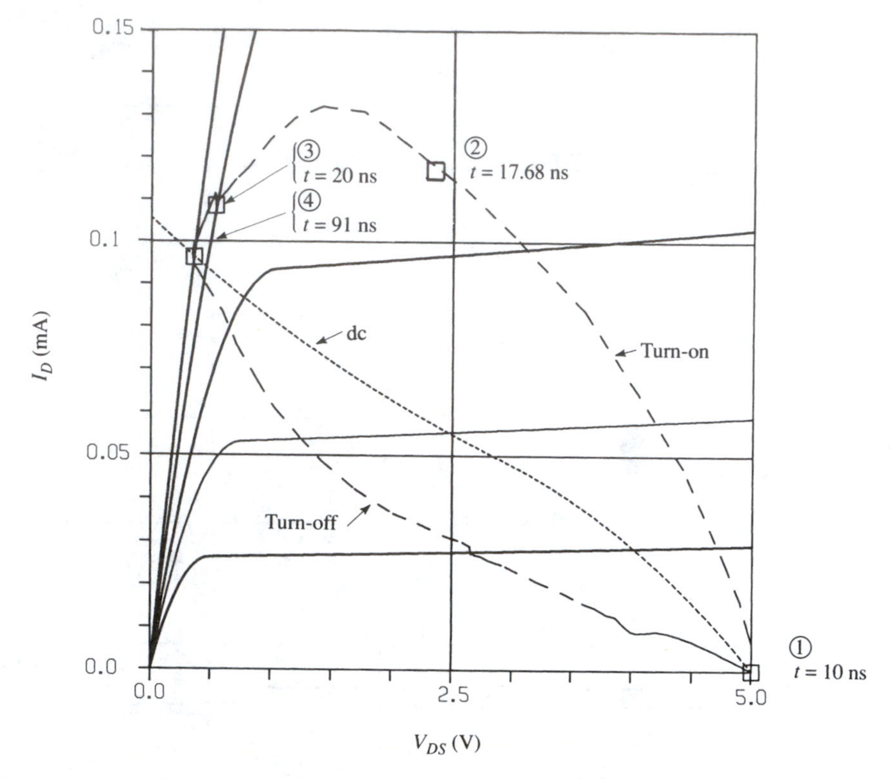

Figure 5–21 Comparing the two dynamic loadlines (dashed) with the static loadline (dotted), superimposed on the MOSFET output characteristics. The four squares correspond to the four columns of Table 12.

measurements. However, the second-generation models have largely displaced the first-generation models for circuits using the shortest-channel MOSFETs.

Third-generation models [86, 92, 97–100] are presently being developed, and they represent a return to more physically based modeling. The number of parameters required has been reduced in comparison with the number needed for the second-generation models, and they are easier to extract. Moreover, considerable emphasis is placed on mathematical efficiency and on dependable convergence behavior during circuit simulation. The models also address some of the problems in analog circuit design, which are described by Tsividis and Suyama [101]. The third-generation models are not well established as yet, and it is too early to determine which model or models will be dominant in the future.

MOSFET–BJT Performance Comparisons

As we have seen, the MOSFET is today's dominant device because it has a favorable combination of properties. However, it has rarely been able to challenge the BJT in raw performance. The topic of performance comparisons emerged in the 1960s when it first became evident that the MOSFET was more than "just another device." The applications considered ranged from integrated circuits [102] to power devices [103]. Increasingly, the performance comparisons have focused on transconductance, which plays such an important part in fixing the circuit and system delays associated with charging parasitic capacitances. And since both devices display marked dependence of transconductance upon output current, the figure of merit g_m/I_{out} has also received considerable attention.

Transconductance comparisons of the BJT and MOSFET have been inhibited by the apples–oranges character of the problem. Operating principles in the two devices are quite dissimilar. But by normalization, we can eliminate a number of the difficulties. Several examinations of MOSFET transconductance properties have been published [54, 78, 79, 104–107], including some that employ normalization [78, 79, 104]. There have also been efforts to make BJT–MOSFET comparisons involving transconductance [102, 105, 108, 109]. Comparisons given here, however, are based upon a recent treatment [110] that goes considerably beyond the previous comparisons with respect to generality.

6–1 Simple-Theory Transconductance Comparison

Following Gummel and Poon [111], we can write collector current as

$$I_C = I_S \exp\left(\frac{qV_{BE}}{kT}\right), \tag{6-1}$$

where I_S is their *intercept current.* It follows, then, that transconductance is

$$g_m = \frac{qI_C}{kT}. \tag{6-2}$$

Using I_S for current normalization, and the thermal voltage for voltage normalization, converts Equation 6-1 to

$$I_{\text{OUT}} = e^{U_{BE}}, \tag{6-3}$$

where I_{OUT} and U_{BE} will be used as symbols for the two normalized quantities. Similarly, using qI_S/kT for transconductance normalization converts Equation 6-3 into

$$G_M = I_{\text{OUT}} = e^{U_{BE}}. \tag{6-4}$$

Thus we have the well-known result that a log–log plot of normalized transconductance versus normalized output current for the BJT is a straight line with a positive slope of unity, as exhibited by the steeper curve in Figure 6-1.

Now turn to the case of the E-mode MOSFET. Equation 1-22 gives drain current in saturation as

$$I_D = K(V_{GS} - V_T)^2. \tag{6-5}$$

Thus MOSFET transconductance is

$$g_m = 2K(V_{GS} - V_T), \tag{6-6}$$

as given before in Equation 1-24. Let us normalize output current and transconductance for the MOSFET, employing the same quantities used before. Using the symbols I_{OUT} and G_M once again for the normalized variables, we first have from Equation 6-5,

$$I_{\text{OUT}} = \frac{K}{I_S}\left(\frac{kT}{q}\right)^2(U_{GS} - U_T)^2, \tag{6-7}$$

where $U_{GS} \equiv (qV_{GS}/kT)$ and $U_T \equiv (qV_T/kT)$ are the normalized input and threshold voltages, respectively. Similarly, from Equation 6-6,

$$G_M = \frac{2K}{I_S}\left(\frac{kT}{q}\right)^2(U_{GS} - U_T). \tag{6-8}$$

Combining Equations 6-7 and 6-8 yields

$$G_M = 2\,\frac{kT}{q}\sqrt{\frac{K}{I_S}}\sqrt{I_{\text{OUT}}}, \tag{6-9}$$

displaying the equally well-known result [112] that transconductance in the MOSFET goes as the square root of drain current. This result for the MOSFET is plotted as the straight line in Figure 6-1 with a positive slope of one-half.

A critical point in Figure 6-1 is the point labeled 1, where the two straight lines intersect. We can provide a measure of interpretation as follows. Recall that anywhere along the line labeled BJT, normalized current and transconductance are identical. Therefore this condition holds at the point of intersection of the two straight lines in Figure 6-1 as well, permitting us to equate the normalized expressions in Equations 6-7 and 6-8, with the result that $U_{GS} - U_T = 2$, or

$$V_{GS} - V_T = 2\,\frac{kT}{q}. \tag{6-10}$$

Thus the intersection point corresponds to an input voltage lying two normalized units above MOSFET threshold. Note, now, that simple MOSFET theory upon which the present analysis is based disallows drain current below threshold, and so Figure 6-1 presents only above-threshold data.

Having thus interpreted point 1 in Figure 6-1, let us hasten to emphasize that the actual transconductance curve for the MOSFET departs appreciably from the simple-theory straight line as one moves toward low-level operation. The phenomenon of "excess" near-threshold conduction causes transconductance to decline as threshold is approached, as indicated by the dashed curve in Figure 6-1. For vanishing oxide thickness, the actual MOSFET curve on its low-current end asymptotically approaches the solid straight line labeled BJT, a MOSFET property that has been noted previously [104, 105, 109]. The assumption of vanishing oxide thickness is an acknowledgment of BJT and MOSFET dissimilarities.

Two other critical points can also be identified on Figure 6-1. These take advantage of the characteristic current exhibited by an E-mode MOSFET, a current value analogous to I_{DSS} in the D-mode device, and one that is displayed on the normalized current axis in Figure 1-10. As Figure 1-10 shows (let us confine ourselves to the N-channel E-mode case for simplicity), the current-to-voltage transfer characteristic is parabolic, having a branch (dashed) that, while not otherwise meaningful, serves to define a drain current equal to that which the device exhibits at $V_{GS} = 2V_T$. Let us define this current as

$$I_D(V_{GS} = 2V_T) \equiv I_2 = KV_T^2. \tag{6-11}$$

The current I_2 defines critical point 2 on the MOSFET curve in Figure 6-1. At this current, then, and at $2V_T$, Equation 6-6 yields

$$g_m = 2KV_T. \tag{6-12}$$

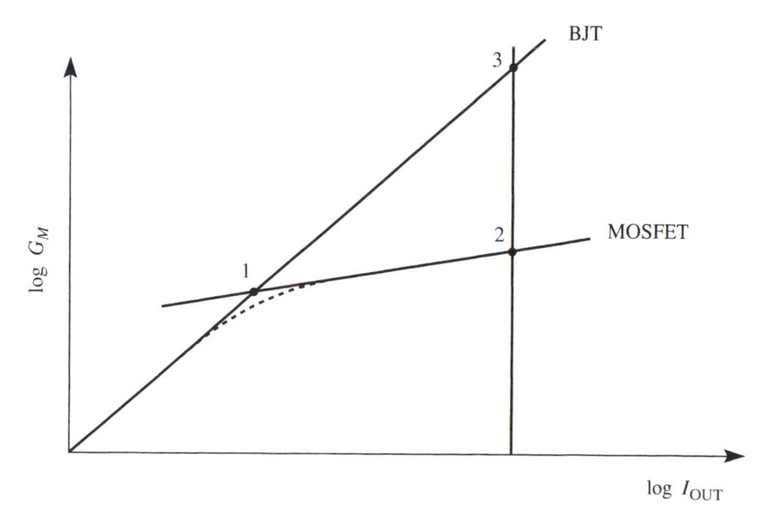

Figure 6–1 Normalized transconductance G_M versus normalized output current I_{OUT} for the BJT and the MOSFET.

It follows that the normalized MOSFET transconductance at this specific input voltage and output current is

$$G_{MM2} = \frac{2KV_T}{qI_S/kT} = \frac{2K}{I_S}\frac{kT}{q}V_T. \tag{6-13}$$

Also, for input-voltage and output-current values at this point we have from Equation 6-11 for the corresponding normalized MOSFET output current,

$$\frac{I_2}{I_S} = \frac{K}{I_S}V_T^2. \tag{6-14}$$

But we have seen in Equation 6-4 that normalized output current, and transconductance as well, for the BJT at any input voltage is $\exp(U_{BE})$. Thus it follows that for the BJT at the current I_2 (point 3 in Figure 6-1), normalized BJT transconductance can be written

$$G_{MB2} = e^{U_{BE}} = \frac{K}{I_S}V_T^2. \tag{6-15}$$

Thus, combining Equations 6-13 and 6-15, we find that the ratio of BJT to MOSFET transconductance at the critical current I_2 is

$$\frac{G_{MB2}}{G_{MM2}} = \frac{\dfrac{K}{I_S}V_T^2}{\dfrac{2K}{I_S}\dfrac{kT}{q}V_T} = \frac{1}{2}\frac{V_T}{kT/q} = \frac{U_T}{2}, \tag{6-16}$$

where U_T is normalized threshold voltage. For $U_T = 10$, then, corresponding approximately to a quarter-volt threshold voltage, the BJT is superior in transconductance by a factor of five at $V_{GS} = 2V_T$, and improves its advantage further as current is increased. Obviously, the BJT advantage also increases as threshold voltage is permitted to increase.

A second comparison between BJT and MOSFET performance is obtained by examining the quotient of transconductance by output current as a function of input voltage. For the BJT, this quotient is a constant and is equal to q/kT,

$$\frac{g_m}{I_{\text{out}}} = \frac{q}{kT}, \tag{6-17}$$

where I_{out} is unnormalized output current. This is a consequence of the exponential dependence of output current on input voltage.

Simple MOSFET theory gives

$$g_m = 2\sqrt{KI_{\text{out}}}. \tag{6-18}$$

Thus,

$$\frac{g_m}{I_{\text{out}}} = 2\sqrt{\frac{K}{I_{\text{out}}}} = \frac{2\sqrt{K}}{(V_{GS} - V_T)\sqrt{K}} = \frac{2}{V_{GS} - V_T}. \tag{6-19}$$

Equating the two expressions for g_m/I_{out} in the BJT and MOSFET, Equations 6-17 and 6-19, one again finds, of course, that they are equal at an input voltage two thermal voltages above threshold, as stated in Equation 6-10. That equation resulted from the equivalent observation that the normalized transconductances from simple theory and at the same current coincide at this particular input-voltage value.

Tsividis also presented curves of g_m/I_{out} for the MOSFET [104], but qualified his curves as being "for extremely small drain–source voltage." That is because he approximated the dependence of I_{out} on V_{GS} by using a first-degree polynomial, and then used the approximate expression in calculating g_m/I_{out}. The subthreshold I–V equation introduced next, however, is valid for larger values of V_{DS}, and we shall now use it to develop a more general subthreshold theory.

6–2 Subthreshold Transconductance Theory

In the subthreshold regime it has been found [113, 114] that

$$I_{\text{out}} = \frac{qn_iD_nL_{Di}Z}{L}\, e^{-3|U_B|/2}(1 - e^{-U_{DS}})\, \frac{e^{W_S}}{\sqrt{W_S - 1}}, \tag{6-20}$$

where we have employed the notation of Section 3-1. It is instructive to use Equation 6-20 to derive a general expression for subthreshold transconductance [71]. The effort is further worthwhile because the result will permit us to determine precisely where the quantity g_m/I_{out} peaks, since in the real device this does occur below threshold. To derive the desired expression for transconductance, we note that

$$g_m \equiv \frac{\partial I_D}{\partial V_G} = \frac{q}{kT}\frac{\partial I_D}{\partial U_G} = \frac{q}{kT}\frac{\partial I_D}{\partial W_S}\frac{\partial W_S}{\partial U_G}, \tag{6-21}$$

where U_G is absolute normalized gate potential with respect to the common reference. Noting that $I_{\text{out}} = I_D$ at present and differentiating Equation 6-20 with respect to W_S yields:

$$\begin{aligned}\frac{\partial I_D}{\partial W_S} &= a\,\frac{e^{W_S}\sqrt{W_S - 1} - (e^{W_S}/2)\sqrt{W_S - 1}}{W_S - 1}\\ &= a\,\frac{e^{W_S}}{\sqrt{W_S - 1}}\left[1 - \frac{1}{2(W_S - 1)}\right]\\ &= I_D\left(\frac{2W_S - 3}{2W_S - 2}\right),\end{aligned} \tag{6-22}$$

where

$$a \equiv \frac{qn_iD_nL_{Di}Z}{L}e^{-3|U_B|/2}(1 - e^{-U_{DS}}). \tag{6-23}$$

The total charge per unit area in the surface region, assuming zero flatband voltage, is given by

$$Q_S = C_{OX}(V_G - \psi_S) = \left(\frac{kT}{q}\right)C_{OX}(U_G - W_S + |U_B|), \tag{6-24}$$

where C_{OX} is the oxide capacitance per unit area. This pair of equations requires some explanation. First, because we have assumed zero flatband voltage, $V_G = \psi_G$, and hence the "mixing" of voltage and potential symbols in the first parenthetic expression is permissible. Second, because we deal here with potential relationships in the MOS capacitor rather than a bulk-silicon sample only, and because the substrate is P-type, it is convenient to choose $(q/kT)\psi_S \equiv W_S - |U_B|$, letting ψ_S and W_S have the same sign. (We shall see in a subsequent diagram that this is indeed the case.)

Next, using Gauss's law and Approximation C from Table 3 for the normalized electric field, we can also write Q_s at the source end of the channel as:

$$\begin{aligned} Q_s &= \left[\frac{kT}{q}\epsilon q n_i e^{|U_B|}\right]^{1/2}\sqrt{2(W_S - 1 + e^{W_S - 2|U_B|})} \\ &= \sqrt{2}\, qn_iL_{Di}e^{|U_B|/2}\sqrt{2(W_S - 1 + e^{W_S-2|U_B|})} \\ &= 2qn_iL_{Di}e^{-|U_B|/2}\sqrt{e^{2|U_B|}(W_S - 1) + e^{W_S}}. \end{aligned} \tag{6-25}$$

(See Section 3-6 and Problem A17.) Although Equation 6-25 gives the areal charge density at the source end of the channel, it is in error by only a few percent along the entire channel. This is because ionic or bulk charge is dominant all the way from source to drain in the subthreshold regime of operation. Setting the last portion of Equation 6-24 equal to the last portion of Equation 6-25 gives us

$$\frac{kT}{q}C_{OX}(U_G - W_S + |U_B|) = 2qn_iL_{Di}e^{-|U_B|/2}\sqrt{e^{2|U_B|}(W_S - 1) + e^{W_S}}, \tag{6-26}$$

or

$$U_G = \frac{q}{kT}\frac{2qn_iL_{Di}e^{-|U_B|/2}}{C_{OX}}\sqrt{e^{2|U_B|}(W_S - 1) + e^{W_S}} + W_S - |U_B|. \tag{6-27}$$

Hence,

$$\frac{\partial U_G}{\partial W_S} = \frac{q}{kT}\frac{qn_iL_{Di}e^{-|U_B|/2}}{C_{OX}}\left[\frac{e^{2|U_B|} + e^{W_S}}{\sqrt{e^{2|U_B|}(W_S - 1) + e^{W_S}}}\right] + 1$$
$$= 1 + \frac{\epsilon}{2C_{OX}L_{Di}}e^{-3|U_B|/2}\left[\frac{e^{2|U_B|} + e^{W_S}}{\sqrt{W_S - 1 + e^{W_S - 2|U_B|}}}\right]. \tag{6-28}$$

EXERCISE 6–1 Explain the change of coefficient in the last step.

■ Multiplying numerator and denominator of the fraction by $\exp(-|U_B|)$ accounts for the exponential portion of the coefficient (as well as for the change in the radical expression). Noting that

$$L_{Di}^2 = \left(\frac{kT}{q}\right)\left(\frac{\epsilon}{2qn_i}\right)$$

shows that

$$\left(\frac{q}{kT}\right) qn_iL_{Di} = \frac{\epsilon}{2L_{Di}}.$$

For subthreshold conditions, $\exp(W_S - 2|U_B|) \ll (W_S - 1)$, so that the exponential term in the denominator can be dropped. Thus,

$$\frac{\partial U_G}{\partial W_S} \approx 1 + \frac{\epsilon}{2C_{OX}L_{Di}}e^{-3|U_B|/2}\left(\frac{e^{2|U_B|} + e^{W_S}}{\sqrt{W_S - 1}}\right). \tag{6-29}$$

Substituting Equations 6-22 and 6-29 into Equation 6-21, we obtain

$$\frac{\partial W_S}{\partial U_G} = \frac{2C_{OX}L_{Di}\sqrt{W_S - 1}}{2C_{OX}L_{Di}\sqrt{W_S - 1} + \epsilon e^{|U_B|/2} + \epsilon e^{W_S - 3|U_B|/2}}$$
$$= \frac{C_{OX}}{C_{OX} + \dfrac{\epsilon}{2L_{Di}}\dfrac{e^{|U_B|/2}}{\sqrt{W_S - 1}} + \dfrac{\epsilon}{L_{Di}}\dfrac{e^{W_S - 3|U_B|/2}}{W_S - 1}}. \tag{6-30}$$

But the last two terms in the denominator are identically C_b and C_i, respectively, at the source end of the channel. (See Problem A17.) Hence, $\partial W_S/\partial U_G = C_{OX}/(C_{OX} + C_b + C_i)$, and

$$g_m = \frac{qI_D}{kT}\left(\frac{2W_S - 3}{2W_S - 2}\right)\frac{C_{OX}}{C_{OX} + C_b + C_i}. \tag{6-31}$$

These expressions for capacitance,

$$C_b = \frac{\epsilon}{2L_{Di}} \frac{e^{|U_B|/2}}{\sqrt{W_S - 1}} \tag{6-32}$$

and

$$C_i = \frac{\epsilon}{2L_{Di}} \frac{e^{W_S - 3|U_B|/2}}{\sqrt{W_S - 1}} \tag{6-33}$$

are consistent with the equations derived by Tsividis [104]. That is, if the denominator of the last factor in Equation 6-28 is replaced by its Taylor's series expansion and the development above is repeated, Equation 6-33 will be identical to his expression for C_i.

Equation 6-31 is a general expression for subthreshold transconductance [71]. Its factor containing W_S is approximately unity. As C_{OX} becomes large, g_m approaches qI_D/kT, which is a consequence of the exponential dependence of drain current on surface potential in the subthreshold regime. The capacitance ratio included in the subthreshold g_m is the result of voltage division in the input loop. The "useful" portion of the applied gate voltage is that which modulates the surface potential, and thus a capacitance ratio appears in the expression for g_m.

6–3 Calculating Maximum MOSFET g_m/I_{out}

Now we are in a position to determine the maximum value of g_m/I_{out} [110]. Dividing Equation 6-31 through by I_{out} gives an expression for g_m/I_{out} that is a function of only one variable, W_S. The maximum in g_m/I_{out} will occur when $C_b + C_i$ is at a minimum, because C_{OX} is constant. The fact that there is a minimum, residing below threshold, in the sum of C_b and C_i is clearly displayed in Figure 3-8.

Neglecting the factor $(2W_S - 3)/(2W_S - 2)$ because it is approximately unity, we can seek the extremum for $C_b + C_i$. Evidently,

$$\frac{dC_b}{dW_S} = \frac{\epsilon}{2L_{Di}} e^{|U_B|/2} \left[\frac{-1}{2(W_S - 1)^{3/2}} \right], \tag{6-34}$$

and

$$\frac{dC_i}{dW_S} = \frac{\epsilon}{2L_{Di}} e^{-3|U_B|/2} \left[\frac{e^{W_S}(W_S - 1)^{1/2} - \frac{1}{2} e^{W_S}(W_S - 1)^{-1/2}}{(W_S - 1)} \right]. \tag{6-35}$$

The minimum in $C_b + C_i$ occurs when

$$\frac{\epsilon}{2L_{Di}} e^{-3|U_B|/2} \left[\frac{-e^{2|U_B|}}{2(W_S - 1)^{3/2}} + \frac{2e^{W_S}(W_S - 1)}{2(W_S - 1)^{3/2}} - \frac{e^{W_S}}{2(W_S - 1)^{3/2}} \right] = 0. \tag{6-36}$$

Thus we find that

$$2e^{W_S}(W_S - 1) - e^{W_S} = e^{2|U_B|}, \tag{6-37}$$

or

$$e^{W_S}(2W_S - 3) = e^{2|U_B|}, \tag{6-38}$$

gives us the value of W_S for which g_m/I_{out} is at its maximum. This does not directly give us the value of V_{GS} for which the maximum occurs, but V_{GS} can be obtained from the U_G-versus-W_S relation, Equation 6-27, if the properties of the device are specified. Note, however, that the maximum in g_m/I_{out} occurs for $W_S < 2|U_B|$ because $(2W_S - 3)$ is greater than unity. Thus the maximum will be below threshold, as we expected. This behavior agrees with the results of Evans and Pullen [109].

Simple MOSFET theory, Equation 6-19, predicts that g_m/I_{out} becomes infinite at the threshold voltage. In reality, however, subthreshold current sets an upper limit for this quotient, as demonstrated above. As oxide capacitance approaches infinity, this limit is q/kT, the same value found in a BJT. Figure 6-2 is a plot of g_m/I_{out} versus V_{in} for the BJT and the MOSFET. Simple theory for the MOSFET in the near-threshold regime is displayed as the rising curve, and the more accurate subthreshold theory is displayed by the lower curve. The transconductance-current quotient of the MOSFET is thus seen to approach that of a BJT at a particular input voltage that is dependent on

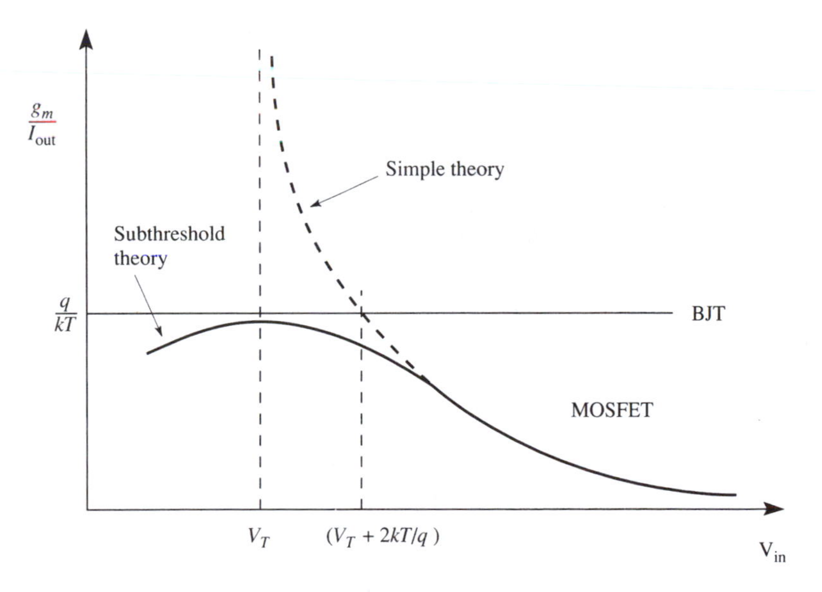

Figure 6–2 Transconductance-current quotient versus input voltage for the BJT and the MOSFET, with results for the MOSFET in the near-threshold regime plotted as the rising curve, and results from more accurate subthreshold theory plotted as the lower curve [110].

the device properties. For any finite value of C_{OX}, g_m/I_{out} will be lower for a MOSFET than for a BJT at all input voltages.

6-4 Transconductance versus Input Voltage

As another method of transconductance comparison, let us examine this property directly as a function of input voltage for the E-mode MOSFET and the BJT. We start with the latter, and again employ the notation of Section 6-1. Recall that Equation 6-4 gives the well-known result that a semilog plot of G_M versus U_{BE} is linear, which in fact is true through many decades of transconductance [102], with the result plotted in Figure 6-3(a).

For the MOSFET case, we have a series of curves, each for a different value of U_T. Specifically chosen in Figure 6-3(b) are $U_T = 10$, 20, 30, and 40. Here we have chosen linear axes, yielding the simplest presentation of the MOSFET curves, but note that the abscissa is coordinated with that of Figure 6-3(a), employing thermal-voltage normalization once more. Figure 6-3(b) makes it clear that reducing threshold voltage U_T (or V_T in unnormalized form) produces a higher transconductance at a given input voltage. To relate the transconductance-versus-voltage curves for the two devices, it is necessary to examine the ratio of the quantities used for normalizing their respective transconductances. The ratio of the BJT quantity to the MOSFET quantity is

$$\text{RATIO} = \frac{qI_S/kT}{(kT/q)K} = \left(\frac{q}{kT}\right)^2 \frac{I_S}{K} = \left(\frac{q}{kT}\right)^2 \left[\frac{q\mu_{nb}\dfrac{kT}{q}\dfrac{n_{0B}}{X_B}A_E}{\dfrac{\mu_{ns}\epsilon_{OX}}{2X_{OX}}\dfrac{Z}{L}}\right], \tag{6-39}$$

where μ_{nb} and μ_{ns} are the bulk and surface mobilities, respectively, n_{0B} is equilibrium minority-electron density in the base region (with uniform doping of the base region assumed), A_E is emitter area, X_B is base thickness, and X_{OX} is oxide thickness, while ϵ_{OX}, Z, and L have their usual MOSFET meanings. Hence,

$$\text{RATIO} = \left(\frac{q}{kT}\right)\left(\frac{q}{\epsilon_{OX}}\right)\left(\frac{\mu_{nb}}{\mu_{ns}}\right)\left(\frac{X_{OX}}{X_B}\right)A_E n_{0B}\left(\frac{L}{Z}\right). \tag{6-40}$$

In view of the fact that LZ is gate area A_G, this becomes

$$\text{RATIO} = \left(\frac{q}{kT}\frac{q}{\epsilon_{OX}}\right)\left(\frac{\mu_{nb}}{\mu_{ns}}\right)\left(\frac{X_{OX}}{X_B}\right)\left(\frac{A_E}{A_G}\right)n_{0B}L^2. \tag{6-41}$$

Interestingly, this form of comparison involves (beside the constant coefficient) a series of three dimensionless factors that directly compare structural and physical features of the two devices, and two additional factors, one for each device. The last two factors display the irreconcilable dissimilarities in the two devices. Let us evaluate the

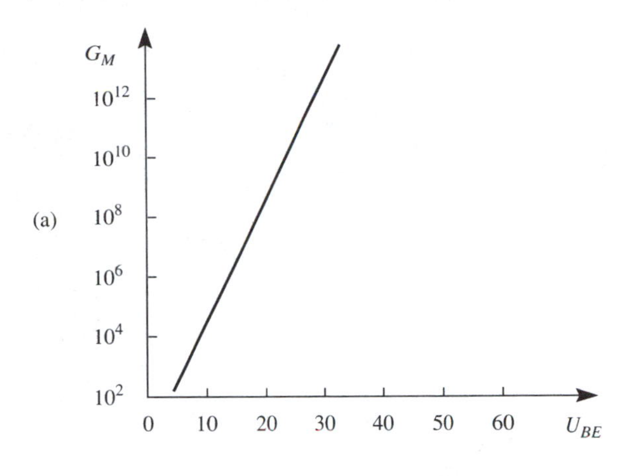

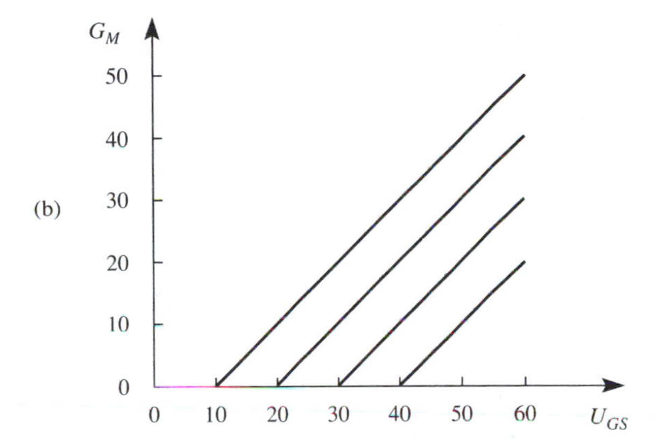

Figure 6–3 Normalized transconductance versus input voltage. (a) Semilog plot for the BJT. (b) Linear plot for the MOSFET, with normalized threshold voltage as a parameter [110].

constant factor and choose reasonable values for the variables, as has been done in Table 13. For the chosen values, we have

$$\text{RATIO} = 1.853 \times 10^{-9}. \tag{6-42}$$

Multiplying the transconductance values in Figure 6-3(a) by this factor yields the modified BJT curve plotted in Figure 6-4. The four MOSFET curves of Figure 6-3(b) have also been plotted in semilog form in Figure 6-4. The BJT curve can readily be adjusted for other values of the variables, and variable ratios just given.

6–5 Physics of Subthreshold Transconductance

Certain details of MOSFET operation, especially subthreshold operation, have been vigorously debated in the literature. Following the arguments used in a recent exami-

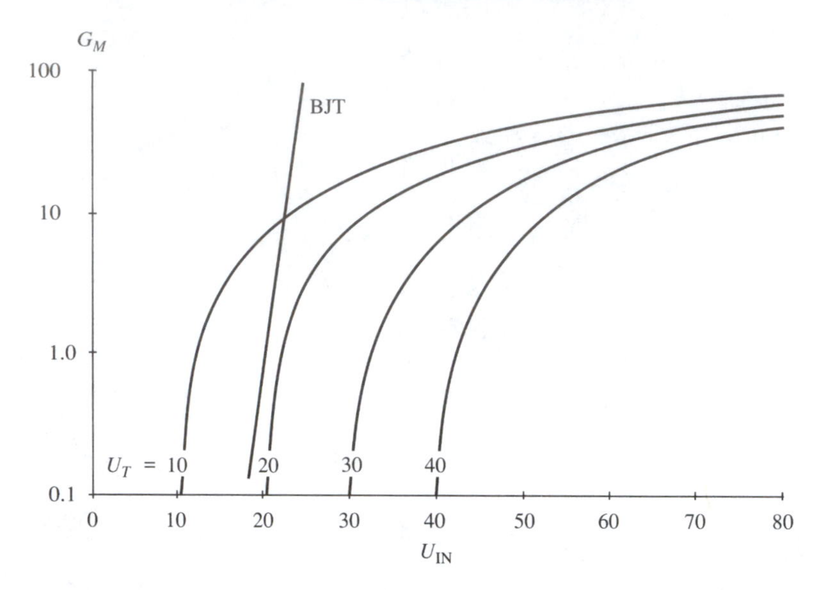

Figure 6–4 Normalized transconductance versus input voltage for the specific BJT–MOSFET variable ratios and variables listed in Table 13 [110].

nation of this matter [71], let us review the debate. Early among the authors who examined subthreshold properties of the long-channel MOSFET were Evans and Pullen [109], who in 1966 measured the transconductance-current quotient, finding that it peaked in the subthreshold regime at a value under *q/kT.* The subthreshold regime of operation was treated analytically by Barron in 1972 [113]. Starting from the Pao–Sah double-integral formulation [54], he derived an approximate-analytic expression for drain current by making mathematical approximations appropriate to subthreshold conditions. Subsequent papers by Van Overstraeten, Declerck, and Broux [114] and by Troutman [115] dealt with the problem similarly (Pao–Sah formulation with approximations) but recognized in addition that subthreshold drain current is primarily a diffusion current. Still later, drain-current equations were derived by Brews [78] and by Fichtner and Pötzl [116] by treating the source and drain junctions at the channel ends much like the emitter and collector junctions of a BJT. Brews based his analysis on the charge-sheet approximation (employing a modified depletion approximation and a channel of zero thickness), while Fichtner and Pötzl introduced an effective channel thickness.

The Pao–Sah paper did not specifically address the subthreshold regime, but is very relevant because it was the first to consider the contribution to channel current (and hence to drain current) made by the diffusion-transport mechanism, and in so doing, produced a quantitative model so accurate that it is the standard by which other models are judged. Because diffusion is essentially the only transport mechanism at very small current values, as just noted, their work led directly to valid subthreshold analysis.

The 1982 paper by Tsividis [104] critically examined the regime of moderate inversion, lying between the strong-inversion and subthreshold regimes, and presented normalized curves of transconductance and of transconductance-current quotient throughout all three regimes. Yet another interesting contribution to the understanding of the subthreshold MOSFET was made by Johnson in 1973 [105]. Noting the Evans–Pullen observation of a transconductance-current quotient approaching q/kT, he invoked a BJT-like model for the subthreshold MOSFET, since a strict q/kT quotient exists over many current decades of BJT operation [102]. He did not note the pure-diffusion character of the subthreshold channel current, which would have strengthened the comparison, but he did call specific attention to the barrier between the source region and the source end of the channel, accurately describing it as a quasi "high–low" junction. Other authors have called attention to this potential barrier, although not as explicitly. In particular, Sah and Pao [106] in 1966 had remarked on the "injection" of carriers from the source region into the channel. Johnson went further, introducing a capacitance into the MOSFET model that he likened to the diffusion capacitance associated with the BJT emitter junction and base region. He suggested that this capacitance was a result of joint properties of the channel and channel-source barrier and, linking it to the aforementioned transconductance behavior, labeled it the "control capacitance." The theory of subthreshold transconductance in the MOSFET developed in Sections 6-2 and 6-3 shows that the control-capacitance concept is unnecessary, and the discussion that follows shows that the concept is in fact flawed. As an accompaniment we now include a detailed description of the source-channel barrier region.

Consider the source end of the channel region. Let it be a plane N^+P step junction that is normal to the oxide–silicon interface. The properties of such a junction are well known now from discussions in Appendixes F and G and Section 3-3, properties summarized in Figure 6-5. The space-charge layer consists of ionic charge on the high side, while on the low side it is a combination of ionic charge and carriers in an inversion layer—electrons in the present case, illustrated in Figure 6-5(a). Given an equilibrium electron density n_0 on the N^+ side, we know that the equilibrium electron density at the junction is n_0/e, as depicted in Figure 6-5(b), consistent with a potential drop of kT/q on the high side [8]. With forward bias, this density at the junction increases, and both ionic-charge layers shrink. (See Section 3-3).

In the MOSFET operating above threshold, the inversion layer formed by the field plate merges with the inversion layer of the source-region N^+P junction. At this point, two-dimensional considerations are no longer avoidable. For a first-order treatment, let us use an orthogonal merging of two one-dimensional solutions [71, 117] (also see Section 4-1), confining ourselves to the long-channel-MOSFET case in order to give this approximation the greatest meaning. We have already made reference to the two merged inversion layers. But ionic space-charge regions in the P-type substrate also merge. In fact, the channel region of the MOSFET resides totally in a region that is completely depleted when the device is in above-threshold operation, and all depletion-layer boundaries are spaced well away from the region of interest. Even in subthreshold operation, the incipient inversion layer resides in a well-depleted region. As a final simplification in the present problem, let us assume that the source and substrate regions are electrically common.

Areal charge densities in the channel inversion layer remain far below those in the source-junction inversion layer, even in normal operation of the MOSFET, and as a consequence, the potential barrier at the source end of the channel has appreciable mag-

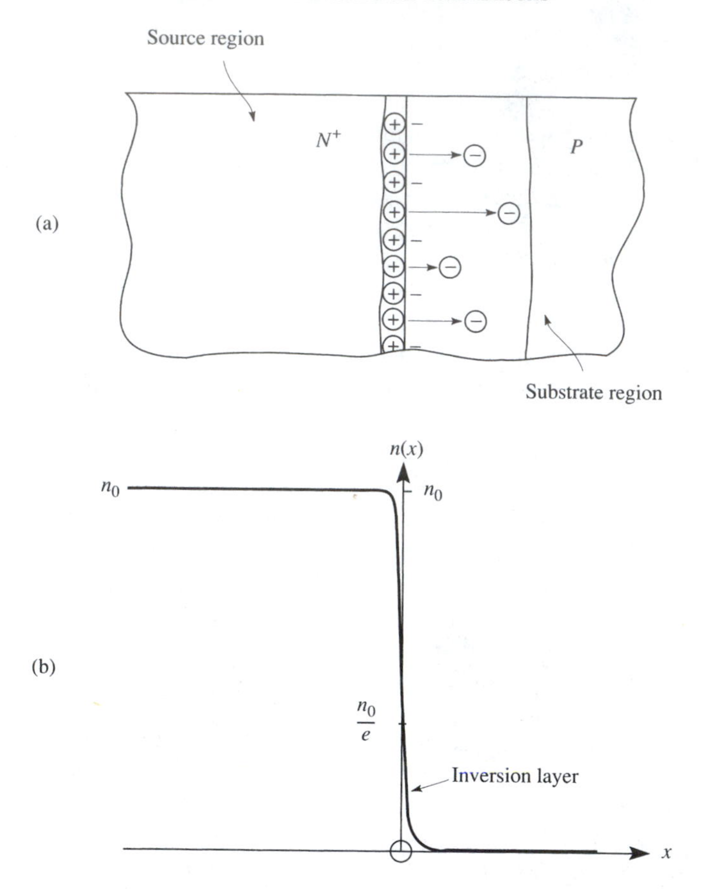

Figure 6–5 Space-charge layer of an N^+P step junction [71]. (a) Physical representation, showing that charge on the high side is ionic, while on the low side it is a combination of ionic and inversion-layer charge. (b) Linear representation of the electron distribution through the space-charge region.

nitude. In the subthreshold case, the barrier is even larger. Let us examine this case further. Figure 6-6 offers a magnified illustration of the orthogonal and merging inversion layers (idealized in one-dimensional terms). The notation y_t designates the threshold plane in the N^+P-junction's inversion layer—the plane at which $n(y) = p_0$, where p_0 is the equilibrium majority-carrier density in the substrate. Since we have assumed subthreshold conditions, there is no analogous plane labeled x_t. But we do have intersecting "intrinsic planes" designated y_i and x_i, along which $n = p$.

Before proceeding, let us make a quantitative examination and comparison of the two inversion layers depicted in Figure 6-6. The junction inversion layer, in addition to being "stronger" than the channel inversion layer, is only slightly altered by device operation, while the latter experiences orders-of-magnitude change in normal device operation. With no drain–source bias, the source junction and the inversion layer remain at equilibrium. A positive voltage increment on the gate causes a strengthening of the channel inversion layer, in turn causing a local change in the source junction near the oxide–silicon interface, a change that "creates a new" high–low junction. We shall examine this situation in more detail shortly, but for now need only point out the response of the junction profile depicted in Figure 6-5(a) to such a positive gate-voltage increment; the potential drop on the left-hand side shrinks somewhat, and electron density at the metallurgical junction rises. But note that possible change in the former has a "hard ceiling" of 26 mV, and the corresponding change in the latter cannot be more than a factor of *e*. By contrast, surface potential in the channel region (the analog of junction potential in the source junction) changes by more than a volt when the device is switched from "off" to strongly "on." And for every volt of change in surface potential, electron density at the oxide–silicon interface changes by nearly seventeen orders of magnitude! Therefore it is a good approximation to treat the source-junction depletion layer as static in examining its interaction with the channel.

Using the depletion-approximation replacement [53] and its extensions, we can easily determine dimensions for the inversion layers that are depicted to scale in Figure 6-6. Let us assume net substrate doping of $1 \times 10^{15}/\text{cm}^3$. For this case, the inversion-

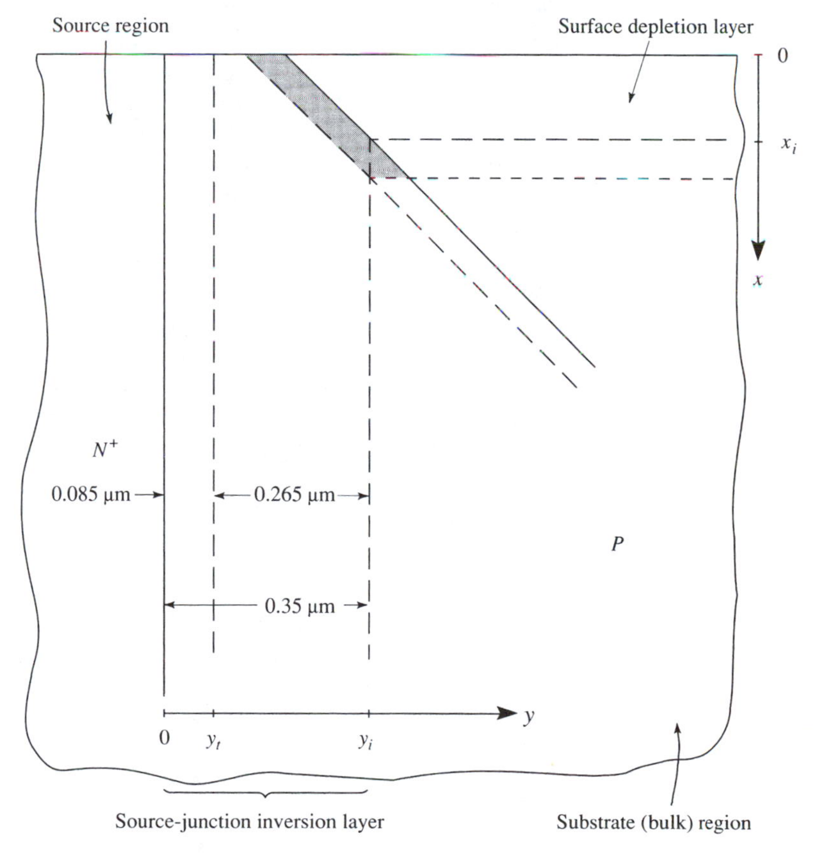

Figure 6–6 Orthogonal and merging depletion and inversion layers at the source end of a MOSFET channel [71].

layer thickness from threshold plane to junction or surface saturates at about $0.66L_D$ [71], where L_D is the general Debye length, which in the present example gives us a thickness of 0.085 μm. In addition, the distance from the threshold plane to the intrinsic plane is 0.265 μm, giving an overall thickness for the junction inversion layer of 0.35 μm, measured from the intrinsic plane. One would not ordinarily measure inversion-layer thickness from the intrinsic plane, but since we are interested in subthreshold conditions, it is a relevant dimension.

The subthreshold behavior of the source junction in the region where it intersects the channel is that of a "variable" one-sided junction at equilibrium, meaning this: absent source–drain voltage, as noted earlier, this portion of the junction remains at equilibrium even in the face of channel-strength variations. The size of the junction's static potential barrier at any x position is given by the Boltzmann relation, employing the source-region electron density n_0 and the electron density $n(x)$ at that position in the channel. Altering channel strength alters $n(x)$ and thus the barrier potential at that point, which is equivalent to altering doping ratio in the equivalent *PN* junction (keeping the high side fixed). But with the onset of weak inversion in the channel, a qualitative change takes place in the affected region of the source junction. The portion of the junction wherein $n > p$ now exhibits the properties of a high–low junction, which are markedly different from those of a *PN* junction [118]. The forward characteristic in the *PN* case exhibits an "offset voltage," while that of the high–low junction does not, a fact that makes the latter an extremely useful ohmic contact. The high–low junction exhibits high conductance, even at very small voltage. By contrast, in the *PN* case conductance is very low, even up to a forward voltage of some 0.4 V.

Noting the potential barrier at the source end of the channel, Johnson [105] reasoned correctly that variations in V_{GS} would lead to a modulation of this barrier, a modulation that will occur, as we have seen, in spite of the external short-circuiting of source and substrate. Specifically, a positive voltage increment on the gate would produce a reduction in the height of the barrier. This relationship between a barrier increment and a charge increment resembles to a limited degree, once drain–source bias is applied, the BJT phenomenon that is characterized as *diffusion capacitance*, in which an increment of forward bias on the emitter junction raises minority-carrier density at the emitter boundary of the base region. It has been shown that, assuming continuous reverse bias on the collector junction and a uniformly doped base region, the result of such an emitter-bias increment is to cause the linear base-region profile to rotate about a point on the x axis that is defined by the depletion-approximation boundary of the collector junction on the base side [119]. The resulting wedge-shaped increment in base-region minority carriers is a stored-charge increment of one sign. An equal and opposite charge increment consists of majority carriers drawn from the base contact in one dielectric relaxation time, which we shall assume to be very short. Quasineutrality is preserved in the base region and the two diffusion-capacitance charge increments are stored in the base region in balanced fashion.

Carrier behavior in the subthreshold MOSFET diverges to a significant degree from that in the BJT, however. Briefly, at a given depth below the surface, the electron profile from source to drain is linear and transport occurs only by diffusion, a point on which there is general agreement. The source–drain carrier gradient has its maximum value at the surface, and declines monotonically to zero as distance from the surface increases. Volumetric electron-density profiles at several x positions in the subthreshold channel are shown in Figure 6-7. The crucial difference of this case from the BJT case is that an *elevation* in channel-electron density at a given point is accompanied by a further *depression* in hole density. In other words, we deal here with *surplus* carriers, characterized by conditions of quasiequilibrium nonneutrality, rather than the *excess* carriers of the BJT-base case, characterized by quasineutral nonequilibrium. The fatal flaw, then, in the control-capacitance concept examined in this way, is its lack of an increment of holes to accompany the increment of electrons.

Having dispensed with Johnson's control capacitance we are obligated to explain the transconductance behavior of a MOSFET from the present point of view. A cross-sectional view of the MOSFET under analysis is shown in Figure 6-8(a). Note that the spatial origin of the y axis is located at the depletion-layer boundary for consistency with the notation used by previous authors. The potential profile at the surface in the source–drain direction for this device is plotted in Figure 6-8(b), and the corresponding diagram using normalized potential notation is given in Figure 6-8(c). Note that W_S in Figure 6-8(c) and ψ_S in Figure 6-8(b) indeed have the same sign, a point made in Section 6-2 in explanation of Equation 6-24.

Thus we have offered a physical picture of subthreshold conditions and phenomena in the long-channel MOSFET, underlying the analytic calculation of subthreshold transconductance in Sections 6-2 and 6-3. Combining these physical and analytical pictures yields an understanding that can be summarized in these terms: The transconductance of the long-channel MOSFET approaches qI_D/kT as oxide capacitance becomes large and as drain current simultaneously becomes small. This behavior is a consequence of three features of subthreshold operation that were integrated for the

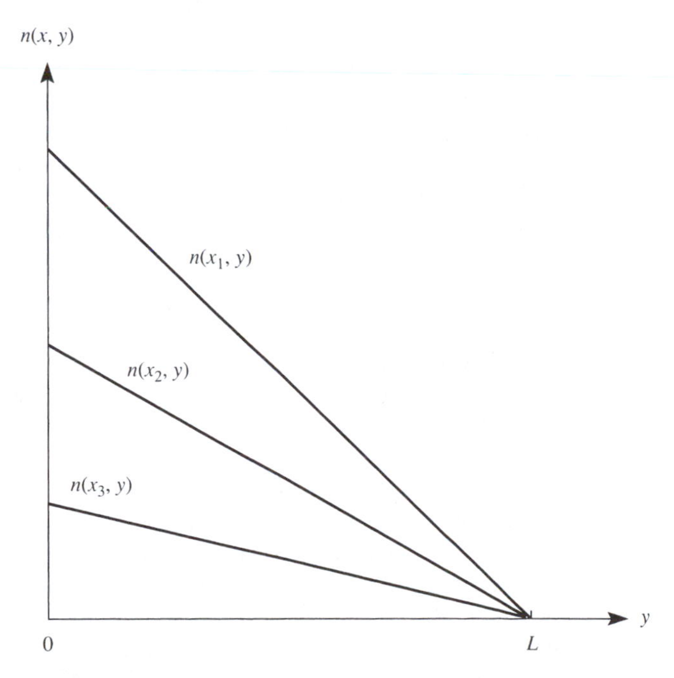

Figure 6–7 Volumetric electron-density profiles at several positions in the subthreshold channel of a MOSFET.

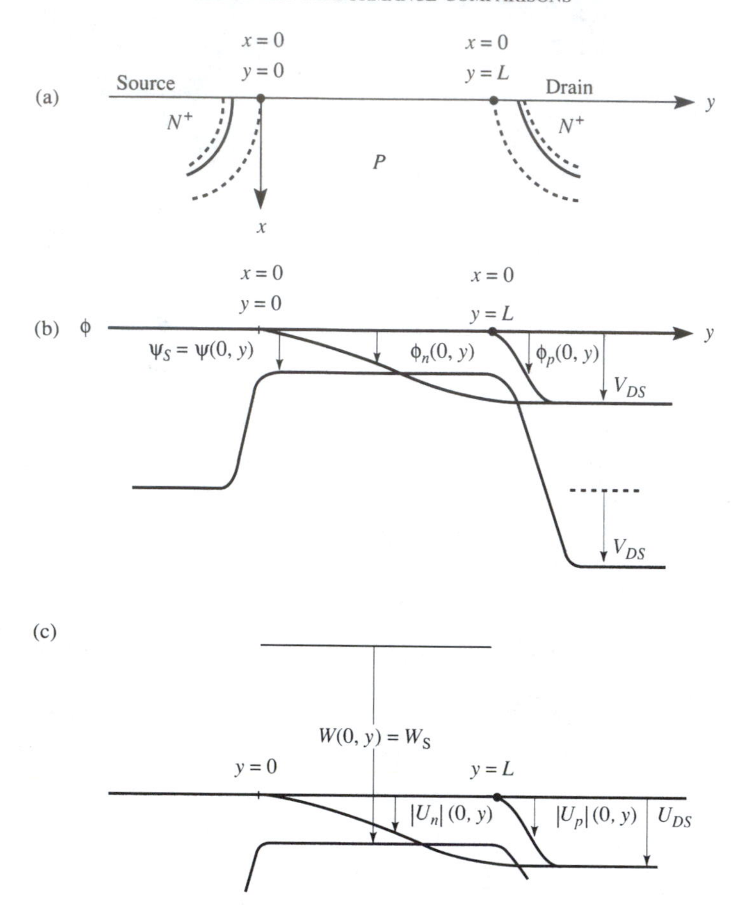

Figure 6–8 Physical and potential-profile representation of the device under analysis. (a) Cross section of *N*-channel MOSFET. (b) Profiles at the surface in the *y* direction of electrostatic potential $\psi(0, y)$, and quasi Fermi levels $\phi_n(0, y)$ and $\phi_p(0, y)$. (c) Quantities of part (b) represented in normalized form.

first time in Reference 71, but each of which had been noted at least once before. These factors are (1) the essentially linear source–drain carrier profile, declining in slope in each successive channel lamella as one moves farther from the surface, these profiles associated with purely diffusive transport; (2) carrier-density values at the source end of each lamella that are exponentially related to source-end surface potential; and (3) a channel–source barrier that exhibits the high-conductance properties of a high–low junction. These properties bear some similarity to those of a BJT (which exhibits a strict qI_C/kT transconductance over many decades) with the primary exception being that the emitter junction is a *PN* junction rather than a high–low junction, accounting for the BJT's offset voltage, absent in the MOSFET. Further, surface potential in the MOSFET is controlled by "reaching through" the gate oxide, introducing a capacitor ratio into the transconductance expression, an expression developed in Section 6-2 as Equation 6-31.

Summary

At the heart of the MOSFET is the MOS (metal–oxide–semiconductor) capacitor, a three-layer structure with a metallic field plate on top and silicon (in the most important case) on the bottom. The intervening dielectric layer is usually SiO_2 grown thermally on the silicon. Doped regions, opposite to the substrate in type, are positioned at opposite edges of the capacitor and become two terminals of the MOSFET, with the field plate or gate being a third. The gate is the input or control electrode. In early devices it was usually made of aluminum, while today it is usually heavily doped polycrystalline silicon. Sufficient bias of the polarity that pushes majority carriers away from the oxide–silicon interface creates an inversion layer in the silicon at the interface. This is the depletion-inversion polarity. The inversion layer, or channel, permits conduction between the two doped regions. Carriers flow from the region termed the *source* to the region termed the *drain*, with the latter having the same bias polarity as the gate. Voltages in the MOSFET are usually referred to that of the source terminal.

Majority carriers in the source and drain regions are the same as those of the channel, and these single-type flowing carriers are primarily responsible for MOSFET properties. For this reason a field-effect transistor (FET) is termed *unipolar*, in contrast to the bipolar junction transistor (BJT), where carriers of one type flow into the base region, and opposite-type carriers flow through the collector junction. FETs in general can be described as voltage-controlled resistors. The MOSFET is inferior to the BJT in a number of respects. But its overall combination of properties is extremely advantageous, causing the MOSFET to be today's dominant solid-state device. MOS technology requires a degree of control of conditions in the oxide, and especially at the oxide–silicon interface, that vastly exceeds that required in BJT technology. For this reason, the MOSFET arrived about a decade later than the BJT, even though it was one of the earliest solid-state amplifiers (or switches) to be conceived in principle.

The *N*-channel MOSFET has an advantage over the *P*-channel device because electron mobility exceeds hole mobility. Inversion-layer mobilities for both, however, are typically one-half to one-third their bulk values. (The substrate region is often termed *bulk* for brevity and convenience.) In the *N*-channel case the bulk region is *P*-type, and we assume it to be uniformly doped (apart from the heavily and oppositely doped source and drain regions, of course). When gate–source bias voltage is adjusted to yield hole density at the oxide–silicon interface equaling that deep in the bulk region, the flat-band condition exists, and the associated critical gate–source bias is termed the *flat-band voltage*. Variations in gate–source voltage do not disturb quasiequilibrium conditions in the substrate, so the Fermi level is constant or "flat," irrespective of band bending. When the surface potential (the mid-gap potential at the interface) equals the

Fermi potential (or level), we have a condition called the *threshold of weak inversion*. Doubling the band bending, or causing the surface potential to be equal in magnitude and opposite in sign to the bulk potential, brings us to the threshold of strong inversion. The corresponding critical gate–source voltage is the threshold voltage. Further increases in gate voltage cause very little further thickening of the depletion layer that preceded inversion, because it is now "easier" for MOS-capacitor lines of force to terminate on inversion-layer electrons (in the *N*-channel example) right at the interface than on additional negative acceptor ions at the more remote depletion-layer boundary. Band-bending character also changes qualitatively just beyond threshold, with a sharp increase in curvature, because channel carriers form a thin, dense layer.

Rudimentary MOSFET analysis ignores depletion-layer charge, assuming that all charge on the silicon plate of the MOS capacitor is inversion-layer charge. It produces "square-law" (parabolic) characteristic curves in the regime of drain current versus drain–source voltage nearest the origin of the output plane. Physical reasoning predicts constant current beyond the maximum (or vertex) of each parabola, nicely confirmed experimentally in the long-channel MOSFET. This is the important saturation regime of operation, wherein drain current is independent of drain–source voltage, and proportional to the square of the gate–source voltage that is in excess of the threshold. This is the most important square-law property of the MOSFET.

The MOSFET just described is enhancement-mode (E-mode) in the sense that gate–source voltage above the threshold voltage must be applied to produce significant conductance from source to drain. By doping a thin region near the oxide–silicon interface with impurities of the same type as those in the source and drain regions, one creates a channel that exists even with the device at equilibrium. Gate–source voltage of the depletion-inversion polarity will further increase drain–source conductance. But gate–source voltage of the opposite polarity causes a depletion layer that forms at the interface to diminish channel conductance. This is a depletion-mode (D-mode) device. The gate–source voltage that fully depletes the doped channel is known as the *pinch-off voltage*. Channel doping in such a device is usually accomplished by ion implantation. The D-mode MOSFET is also a square-law device.

The transconductance of the MOSFET, or the rate of change of drain current with respect to gate–source voltage, is a linear function of the latter variable. Transconductance in the MOSFET is a function of the dimensional and structural properties of the device. The BJT, by contrast, has transconductance that to first order is independent of the dimensional and structural properties of the device. It is proportional to output current, and hence exponentially dependent upon input voltage. As the linear-exponential comparison suggests, BJT transconductance vastly exceeds MOSFET transconductance through wide operating ranges. Transconductances in the two devices have their closest approach in the regimes of very low output current and input voltage.

An E-mode MOSFET with gate and drain common is a diode exhibiting a concave-up (square-law, again) current–voltage characteristic. Used as a load with another E-mode device as driver, it yields a linear voltage amplifier; the nonlinearities of the two devices precisely cancel. The same inverter can be used for voltage switching in a logic circuit. The diode-connected device operates always in the saturation regime, and is hence known as a *saturated load*. The inverter is small in area and therefore economical, but relatively slow in switching. A linear (resistive) load yields switching that is over an order of magnitude faster, but such a load is awkward technologically. A D-mode load combined with an E-mode driver (known as an E–D inverter) is an advantageous combination, small in area, also much faster than the saturated-load inverter. One of the fastest MOS inverters uses *N*-channel and *P*-channel E-mode devices with gates common and drains common, the latter node serving as the output node. Such a complementary MOS (CMOS) inverter is both fastest in switching and lowest in power dissipation at low switching rates, because it requires no "standby" current. In either state of the switch, either the *P*-channel device is OFF, or the *N*-channel device is OFF.

Switching delay in MOS circuits is fixed largely by the time required to charge parasitic capacitances in the circuit. Thus, to optimize an integrated circuit, the sizes of its various MOSFETs (and hence their transconductances) must be adjusted according to the capacitive loads they have to "drive," so that rough parity of switching time throughout the circuit is achieved. This kind of optimization is best handled by computer, and has been so accomplished since the mid-1960s. The practice with bipolar integrated circuits has been very different because the BJT has an enormous transconductance "surplus" that to first order is geometry-independent.

The silicon portion of the MOS capacitor poses a problem that is analytically identical to the problem posed by one side of a step junction at equilibrium. Each change of bias on the capacitor, therefore, produces a condition similar to that in one region of a step junction, with doping on the other side altered. Electric field is discontinuous at the oxide–silicon interface because of the threefold ratio of dielectric constants, silicon to oxide. Treating the oxide portion as a conventional parallel-plate capacitor, and the silicon portion by means of the depletion approximation, one can produce a serviceable model of the MOS capacitor. The potential profile in the oxide portion is linear, and in the silicon portion, parabolic. The depletion-approximation replacement (DAR) provides more accurate analysis of the silicon portion.

A real MOS capacitor embodies a number of parasitic properties. If the silicon and gate materials exhibit work-function differences, then the potential (or energy) interval from the Fermi level on a given side to the conduction-band edge in the oxide will be different from that on the other side. The difference of these two potentials we term barrier-height difference. It contributes a term to threshold voltage. Another parasitic effect involves electronic states located at the surface of the silicon single crystal, into which electrons can flow when energetically favorable, that is, when band bending places these states below the Fermi level they tend to fill, and vice versa. The resulting charge in these states constitutes a highly undesirable voltage-dependent variable in the MOS capacitor. Many years of effort have devised ways to reduce the density of such "surface states" to negligible values. The carrier charges residing in such states are termed interface trapped charge. When such states are present at high densities, they act somewhat like an electrostatic shield, because of their ability to take on and give up charge. This shielding action frustrated early field-effect experiments.

Another parasitic charge component is a quite stable positive charge near the oxide–silicon interface, believed to be associated with unreacted (unoxidized) silicon. The density of such charge correlates with silicon surface-atom density, which is a function of crystal orientation. It also correlates with the last portion of the silicon-oxidation process, because the oxidation reaction occurs at the interface and not at the oxide surface. This oxide *fixed charge*, as it is known, is in contrast to the oxide mobile charge that plagued early MOS technologists. Alkali ions were mainly responsible in the latter case because of their comparatively high mobility in silicon dioxide. The cure was "cleaning up" the process, notably by avoiding the filament evaporation of a metal for the gate. The final parasitic-charge category is oxide trapped charge. It occurs when an electron passes part way through the oxide-layer conduction band and becomes trapped, or "lodged." In an analogous way, a hole can pass part way through the oxide-layer valence band and become trapped. This can result when certain device actions and phenomena (such as avalanche breakdown) create energetic carriers near the interface. All these parasitic effects, when present, play a part in determining threshold voltage. Their combined effects were responsible for the early popularity of *P*-channel MOSFETs; only after control of the oxide–silicon system had advanced appreciably could the benefits of *N*-channel devices be exploited.

MOS-capacitor modeling and characterization are easier than step-junction modeling and characterization in the sense that the silicon remains in quasiequilibrium in the face of bias changes. The problem is more difficult, however, in the sense that the oxide capacitance is permanently in series with the semiconductor phenomena to be modeled. The approximate-analytic method and its DAR culmination are useful for accurate capacitor modeling, especially because the MOS capacitor poses a quasiequilibrium problem.

There are several simple and useful equivalent-circuit models for the MOS capacitor, depending upon the regime of operation. In accumulation, the oxide capacitance suffices. With modest depletion, two capacitances are in series, the oxide and depletion capacitances. With the inversion layer present, as well as an arrangement for "feeding" carriers to it, an inversion-layer capacitance must be added to the previous model, in parallel with the depletion capacitance. With heavy inversion, we revert to the lone oxide capacitor.

The depletion and inversion capacitances are equal to each other precisely at the threshold of strong inversion. This relationship has been demonstrated and explained experimentally, theoretically, and by means of a physical model. The MOS capacitor is free of diffusion capacitance because of its insulating layer. In the junction, by contrast, diffusion capacitance and depletion capacitance are in parallel, mutually shunted by a conductance in the general case. Inversion capacitance exists in a grossly asymmetric *PN* junction, but it is miniscule and negative.

Observing capacitance as a function of voltage is a key method for characterizing real devices, and was the preeminent method employed for analyzing and finally understanding the oxide–silicon system. Interpretation of *C–V* curves is carried out in terms of the equivalent-circuit models just described. Substrate doping, conductivity type, and oxide thickness are primary variables. The parasitic charges also affect the result, in distinctive ways in some cases. Although the formation of an inversion layer inhibits further expansion of the depletion layer under quasistatic conditions, it is possible to cause a transient depletion-layer expansion by employing a voltage pulse, revealing a deep-depletion characteristic. Avalanche breakdown in the depletion layer imposes a limit on deep depletion.

Transition from MOS-capacitor analysis to MOSFET analysis means a shift from a useful and accurate one-dimensional model to one that is at best two-dimensional.

For the long-channel device, a useful treatment uses the orthogonal superposition of two one-dimensional analyses.

Integrated-circuit application of any MOSFET model requires inclusion of the body effect. This term signifies that reverse biasing the bulk–source junction, a common integrated-circuit situation, means that the gate must exert a higher degree of band bending to achieve the threshold condition in the channel region than it does when that junction is not reverse-biased. In other words, the threshold voltage of such a device is increased.

Other major contributions to long-channel MOSFET modeling were made by Pao and Sah, who considered carrier diffusion in the channel and carried out numerical analyses in two orthogonal directions to achieve an extremely accurate model. Also, Baccarani et al. and Brews developed a charge-sheet model. It is relatively simple, and quite accurate as well, because of countervailing approximations.

The initial rudimentary MOSFET analysis becomes the Level-1 model in the SPICE program. Taking account of ionic charge, neglected in the first case, yields the Level-2 SPICE model. Instead of a square-law result, the ionic-charge (or Ihantola–Moll) model yields a three-halves-power result, which is awkward to treat numerically because of computation time. For this reason and others, an empirically adjusted Level-3 SPICE model was generated. It yields accurate results while avoiding the earlier shortcomings. The Level-1 model in empirically adjusted (but still square-law) form is also useful. Unadjusted, it predicts too high a current, because it assumes all charge in the silicon consists of carriers in the channel.

SPICE considers four regimes of operation. Moving upward along a loadline, we encounter the accumulation, depletion, saturation, and curved regimes. The subthreshold regime includes portions of the first two of these. The MOSFET makes its closest approach to the remarkable BJT transconductance property in the subthreshold regime. This fact is exploited in circuitry for low-power applications, especially CMOS circuitry.

In small-signal SPICE analysis, one finds that the bulk region is approximately as effective as a control electrode as is the gate itself. The concept of current gain has meaning for the MOSFET because significant current enters the input terminal with a high-frequency signal applied and passes through parasitic capacitances. Consequently, the frequency for unity current gain f_T also has meaning. For the device analyzed here, it amounted to 0.5 GHz.

Digital, or large-signal, applications of MOSFETs are by far the most important. The E–D combination (E-mode driver, D-mode load) is particularly advantageous and is analyzed here. Body effect degrades the current-regulating ability of a D-mode load, and it can end up approximating a resistor. Nonetheless, the D-mode load outperforms a saturated load by more than a factor of ten in switching speed. In addition to speed, noise margin is an important feature of a digital circuit. It is a measure of the noise voltage at the input port that the circuit can absorb without changing state. Noise margin is intimately related to product reliability, yield, and economy.

Parasitic capacitances in the MOSFET that are junction-related are voltage-dependent. SPICE models this voltage dependence; it is capable of predicting the waveforms of capacitive current as well as resistive current. The gate–drain capacitance (Miller capacitance) is of particular importance in both digital and linear applications.

Because of parasitic capacitance, the dynamic turn-on loadline lies well above the static loadline in a pulse experiment, and the turn-off loadline, well below.

A simple-theory comparison of the MOSFET and the BJT with respect to transconductance versus output current shows an intersection close to but above threshold. The two curves do not actually intersect, but have their closest approach just below threshold. When input voltage is used as the independent variable, the BJT displays a single characteristic curve, while the MOSFET displays a family of such curves with lower slope, having threshold voltage as a parameter. One particular comparison of normalized transconductances, BJT to MOSFET, produces an intriguing product of ratios—bulk to surface mobility, oxide to base thickness, and emitter to gate area.

When subthreshold physics is examined, the source–channel junction is seen to be an N^+N junction after the source end of the channel passes the threshold of weak inversion. This transition from NP-like to N^+N-like properties is significant, because there is an offset voltage in the conducting I–V characteristic of the former, but none in the latter. This source-end property, the electron diffusion from source to drain along linear electron profiles, and electron density at the silicon source-end surface that is exponentially dependent upon gate voltage, all combine to produce a subthreshold transconductance that approximates the BJT-like value of qI_{out}/kT.

Appendixes A–G

Appendix A
Ohm's Law Sign Convention and Double-Subscript Notation

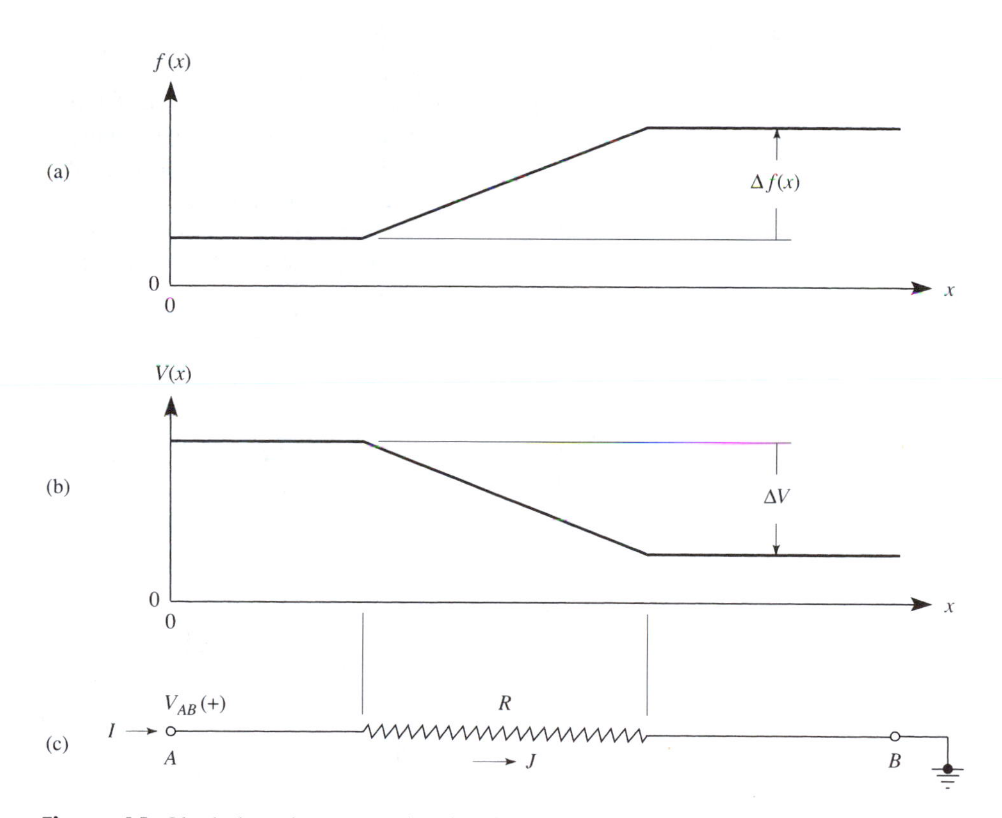

Figure A1 Ohm's-law sign convention for algebraic signs. (a) The usual mathematical convention takes an *increase* in a function such as $f(x)$ to be a positive increment. (b) The Ohm's-law convention takes a *decrease* in vcltage, here $V(x)$, or a voltage drop to be positive. (c) This choice is made to have algebraic-sign consistency in $\Delta V = IR$, since an inward-flowing terminal current was arbitrarily chosen to be positive. In addition, construction of this diagram makes the current density J within the resistor positive as well, by reference to the positive-x axis.

In Figure A1(a) we note that an increase in a function is normally considered a positive increment. The Ohm's-law convention reverses this choice. A voltage *drop*, as illustrated in Figure A1(b), is taken to be a positive voltage increment ΔV. Under such conditions, a conventional current I flows *into* terminal A of the resistor illustrated in Figure A1(c). Conventional current *into* a terminal is taken as positive, and this then yields sign consistency in Ohm's law, $\Delta V = IR$.

At this point it is convenient to adopt the double-subscript notation that eliminates any remaining sign ambiguity. Consistent with the Ohm's-law convention we can write

$$V_{AB} \equiv V_A - V_B. \tag{A1}$$

Thus a positive value of V_{AB} in Figure A1(c) ensures that a positive current I will flow *into* terminal A from outside the resistor.

Appendix B
Resistivity and Its Use in Ohm's Law

The magnitude of the current density J that accompanies a given electric field E depends upon carrier availability (density) and also upon the responsiveness of the carriers to the field. These two factors determine the resistance of a given material to the passage of a current or, more precisely, its *resistivity*. To define this quantity, let us shift attention to a bar of uniform material, with perfect (nonresistive) electrical contacts made to its ends, as shown in Figure B1. The fact that the length L of the sam-

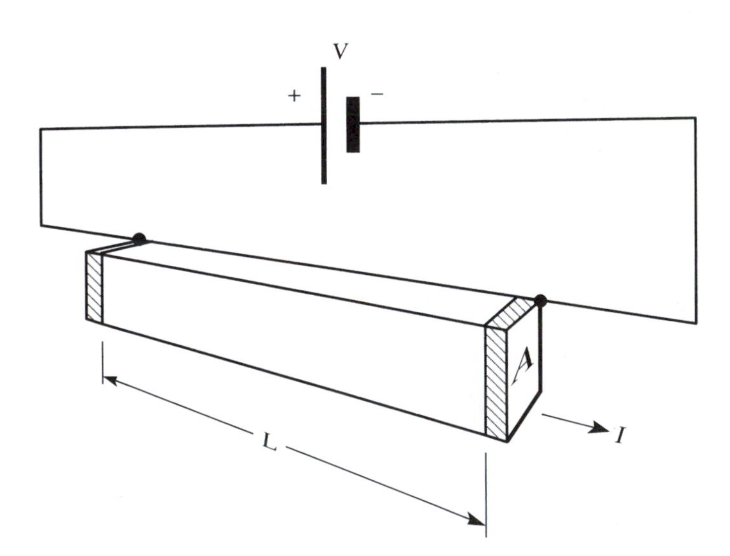

Figure B1 Uniform macroscopic sample of length L and cross-sectional area A for defining resistivity ρ. The current I flows one-dimensionally, and the end-to-end voltage drop in the sample is V.

ple is large compared to the sample's cross-sectional dimensions does not upset the (much desired) one-dimensional character of this problem, because current is confined to the interior of the sample. That is, there are no "fringing effects" in this physical situation. It is self-evident that when this sample is used as a resistor by passing a current through it, we will find its resistance to be directly proportional to its length L and inversely proportional to its cross-sectional area A. That is, doubling L places two resistors in series; doubling A places two in parallel. Therefore, introducing a constant of proportionality ρ, on a purely intuitive basis we can write

$$R = \rho\left(\frac{L}{A}\right), \tag{B1}$$

where ρ is known as *resistivity*. Solving this expression for ρ yields RA/L, so that the dimensions of resistivity evidently are

$$\frac{[\text{ohm}][\text{cm}^{\cancel{2}}]}{[\cancel{\text{cm}}]} = [\text{ohm}\cdot\text{cm}]. \tag{B2}$$

Notice that ρ is measured in ohms *times* centimeters (and not ohms per centimeter).

The inverse of resistivity is *conductivity*, denoted by σ.

$$\sigma \equiv \frac{1}{\rho}. \tag{B3}$$

Obviously the dimensions of conductivity are 1/ohm·cm, or S/cm, where S denotes siemens, the unit for conductance.

Armed with the units of resistivity ρ, we next examine the dimensions of the product of ρ and current density J:

$$[\text{ohm}\cdot\cancel{\text{cm}}]\frac{[\text{A}]}{[\text{cm}^{\cancel{2}}]} = \frac{[\text{V}]}{[\text{cm}]}. \tag{B4}$$

From this we infer that the passage of a current having density J through a uniform material of resistivity ρ is accompanied by the presence of an electric field E. Or,

$$E = J\rho. \tag{B5}$$

This, in fact, is a form of Ohm's law, as can be seen by yet another dimensional argument. Writing the unit equation corresponding to Equation B5 and multiplying both sides by cm yields

$$[\text{V}] = [\text{A}][\text{ohm}], \tag{B6}$$

the unit equation corresponding to $V = IR$, Ohm's law. An important feature of Ohm's law written as in Equation B5 is that all three of the quantities involved can be defined at a mathematical point.

Appendix C
Dielectric Permittivity, Polarization, and Displacement

The parallel-plate capacitor that is so useful for discussing electric field (and to a lesser degree, resistivity) has further basic contributions to make, and is intrinsically important as well. In general, a capacitor can be defined as a two-terminal device that is capable of charge storage when a voltage difference is applied from terminal to terminal. Specifically, the applied voltage V causes a charge $+Q$ to be stored on one plate, and $-Q$ on the other. Insofar as the storage is associated with an internal field, these charges will always be equal in magnitude. (In fact one must go to considerable trouble to store unequal charges on the two plates of such a capacitor, introducing significant electric field outside the device.) The next point to realize is that "stored charge" refers to the charge on *one* plate. Let us arbitrarily take the negative plate as voltage reference, and then focus on the positive voltage applied to the other plate and its corresponding positive charge. Because the charge stored is proportional to the voltage applied, the capacitor is a linear device, and its governing law is

$$Q = CV. \tag{C1}$$

The constant of proportionality between these two variables is of course *capacitance*, for which the unit is the *farad* or, equivalently, 1 coulomb per volt. Now we can be more specific with regard to the plate whose charge is to be considered here. For algebraic-sign consistency in Equation C1, we must employ the charge Q on the *same* plate to which the voltage V is being applied, with the other plate taken as voltage reference.

In the capacitors we considered initially in Section 2-1, the space between the plates was devoid of atoms, molecules, and charges. In short, a *vacuum* existed between the plates. Somewhat more loosely we often use the term *air-dielectric capacitor* to denote the same device, because the presence of air molecules (mainly nitrogen and oxygen) between the plates has little effect on the device's properties through wide ranges of temperature, pressure, and applied voltage. The classical term denoting vacuum—or an approximation to it in the electrostatic context—is *free space*. (If the workers of a century ago had shared the modern picture of interplanetary and even interstellar space as a low-density "soup" composed of many species, some neutral and some charged, they probably would have preferred the term *vacuum*.)

In Exercise 2-1 we found it convenient to divide both sides of Equation C1 by plate area A, yielding charge per unit area (C/cm^2) on the left-hand side of the resulting equation. Let us write a unit equation corresponding to Equation C1 as thus modified:

$$\frac{[\mathrm{C}]}{[\mathrm{cm}^2]} = \frac{[\mathrm{F}]}{[\mathrm{cm}]}\,\frac{[\mathrm{V}]}{[\mathrm{cm}]}. \tag{C2}$$

Note that we have chosen to divide each factor on the right by cm, with the result that the second factor on the right corresponds dimensionally to electric field E. And the first factor on the right, farads per centimeter, corresponds dimensionally to *permit-*

tivity, previously encountered in Exercise 2-1. Specifically, for the air-dielectric case it is the permittivity of free space ϵ_0 that is appropriate, so that charge per unit area on one of the plates is given by

$$\frac{Q}{A} = \epsilon_0 E. \tag{C3}$$

Therefore, permittivity can be regarded as a factor that, by multiplication, converts electric field at a particular place (since field can vary from place to place) into charge per unit area on the plates of a capacitor introduced locally to produce the same electric field.

Permittivity has the same meaning when the electric field under scrutiny is located in the practically important class of *dielectric* media rather than in a vacuum. This adjective usually implies *nonconducting* or *insulating*. That is, a dielectric material of ideal properties contains no free carriers that will move continuously in the presence of an electric field. With a dielectric material filling the space between capacitor plates, such continuous charge motion would constitute a "leakage current," normally considered to be parasitic in a capacitor.

Even with continuous charge motion ruled out, however, there is a second kind of charge motion possible in a dielectric medium that makes it useful in a capacitor, because the resulting effect is an enhancement of the capacitor's ability to store charge. This second kind of motion is termed *polarization*. In the presence of an electric field in a *polarizable* medium, constituent positive and negative charges on an atomic or molecular scale undergo a slight separation. Figure C1(a) shows the two kinds of charges schematically under the condition of no electric field. In Figure C1(b) the same material is shown equally schematically after the local charge separation, or *polarization*, has taken place, in response to the presence of the electric field E indicated in the diagram. Notice that if one chooses an arbitrary volume inside the polarized material that has large dimensions (compared to atomic dimensions), then no net charge is found inside the arbitrary volume. But at the outer faces normal to the polarizing field, there is a sheet of net charge, negative on the left and positive on the right in Figure C1(b). It is this thin sheet of net charge on either side that accounts for the storage-enhancing effect of placing a polarizable medium in a capacitor.

To see how this works, consider the air-dielectric capacitor depicted in a charged state in Figure C2(a). Let the applied voltage be V. In Figure C2(b), then, we show the effect of placing a particular polarizable (dielectric) material between the plates. The applied voltage is still V, and the electric field E is unaltered, as the unaltered density of lines of force indicates. However, charge storage on the plates of the capacitor has doubled in the example arbitrarily chosen. This is a consequence of the polarization and the resulting monolayers of additional charge existing just outside the conducting plates and inside the contiguous dielectric medium. In Figure C2(b) we have omitted the complex representation of Figure C1(b) to improve clarity, but we intend that the same kind of situation is being presented there. Further, caution is needed because Figure C2 offers a two-dimensional representation of a three-dimensional problem.

In this example, charge storage has been doubled by the presence of the polarizable medium, so according to the description of permittivity already given, we can say

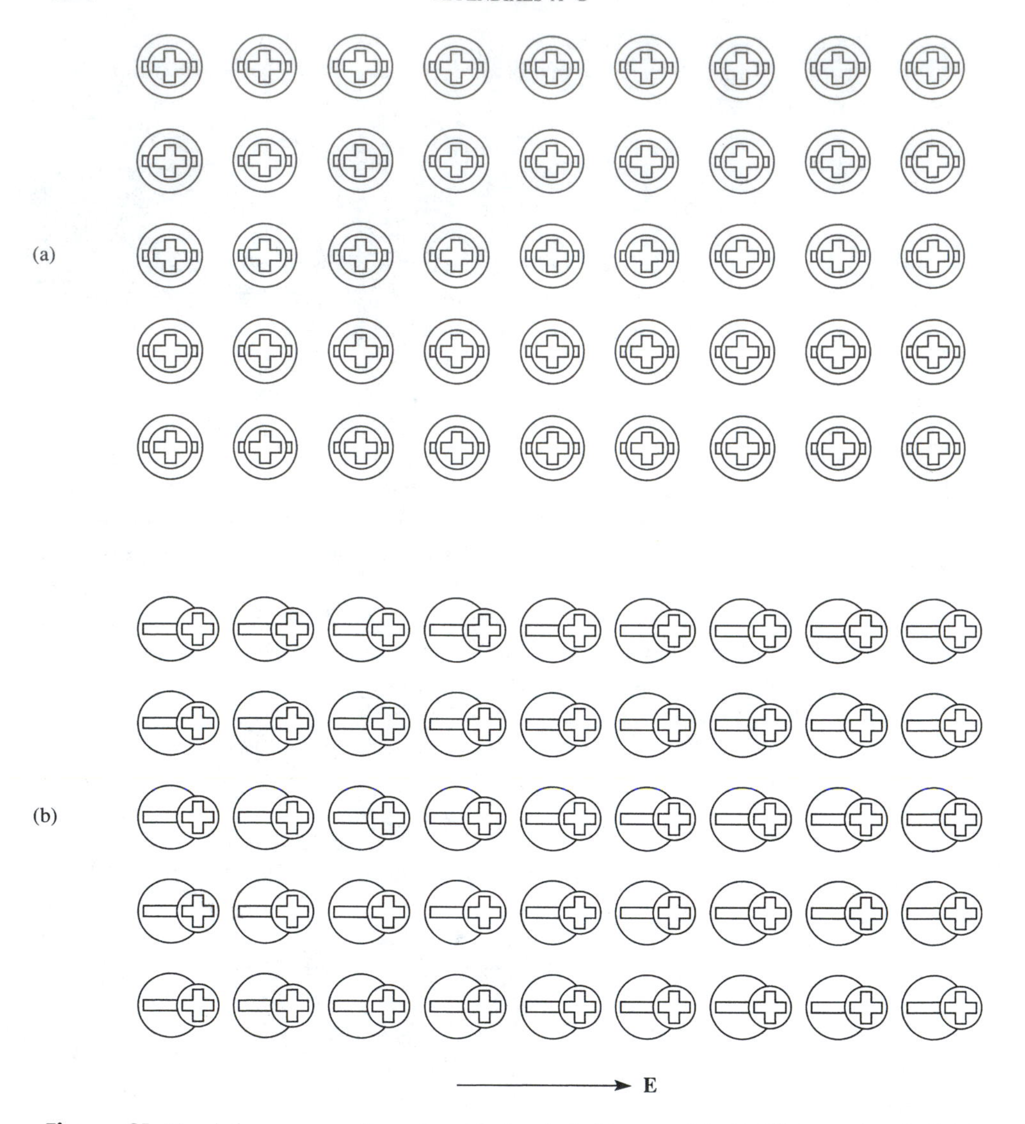

Figure C1 Heuristic representation of an ideal polarizable material. (a) With no electric field present, positive and negative charge populations neutralize each other throughout the sample on an atomic or molecular scale. (b) In the presence of an electric field, charge separation, or *polarization*, occurs. Net charge density in an arbitrarily chosen interior volume is still zero, but sheets of opposite net charge appear on the faces of the volume that are normal to field direction.

that the permittivity of this material is twice that of free space. In other words, the *relative dielectric permittivity* of this material is

$$\kappa \equiv \frac{\epsilon}{\epsilon_0} = 2 \tag{C4}$$

in this case. (The value $\kappa = 2$ was chosen in a completely arbitrary way for convenience in drawing Figure C2.) The everyday term for κ is *dielectric constant.* Charges in the metal, free to come and go from the adjacent circuitry, are sometimes termed *free* charges, and those that make up the charge sheet in the dielectric material, *bound* charges.

Refer again to Figure C2. The act of introducing a dielectric material with a dielectric constant of 2 caused the charge storage to double, but did not alter the electric field E. The doubling of charge storage can also be expressed by saying that the product of permittivity and electric field has doubled, remembering that the dimensions of this product are C/cm^2, an areal charge density. To identify this important quantity, the term *displacement vector* was chosen by Maxwell [C1], for which the symbol is $\mathbf{D}$. In the simple situation being treated here, the displacement vector can be written as

$$\mathbf{D} \equiv \epsilon \mathbf{E}. \tag{C5}$$

The term *displacement* (or the term *electric displacement* that is sometimes used) calls forth a picture of the local charge separation that constitutes polarization, the phenomenon represented in Figure C1(b). But it is important to realize that because of the way it is defined, Equation C5, the areal charge density represented by dis-

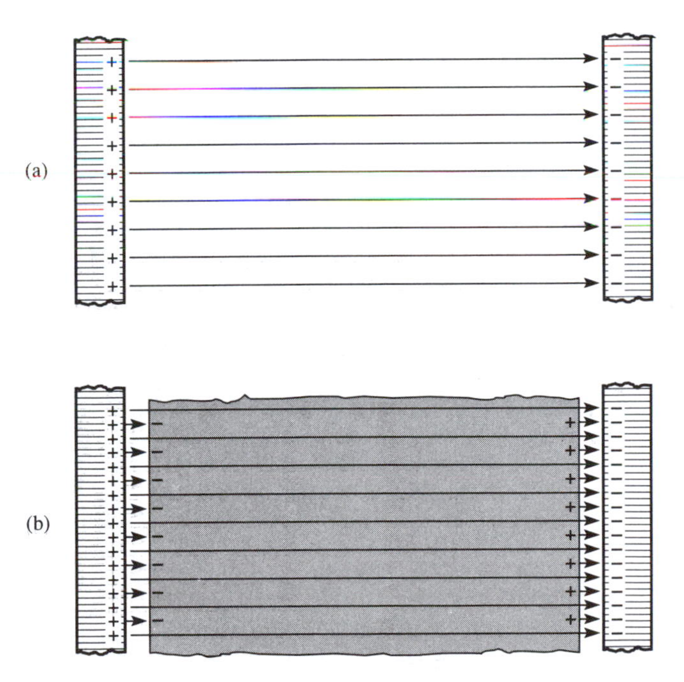

Figure C2 The ability of a polarizable (dielectric) material to enhance charge storage in a capacitor. (a) Charge storage in the plates of an air-dielectric parallel-plate capacitor. (b) Doubled charge storage that results when a slab of material having a dielectric constant $\kappa = 2$ is introduced into the capacitor of part (a) with the same applied voltage.

placement magnitude is the *sum* of the charge stored with the dielectric material absent *and* the additional charge stored with the dielectric material present. That is, in Figure C2(a) we have charge storage in an air-dielectric capacitor, and in Figure C2(b) we have that charge plus the added charge contributed by the dielectric material's presence. The definition of D is based upon the *total* charge per unit area. (Here we have dropped the vector symbol for displacement, just as one frequently

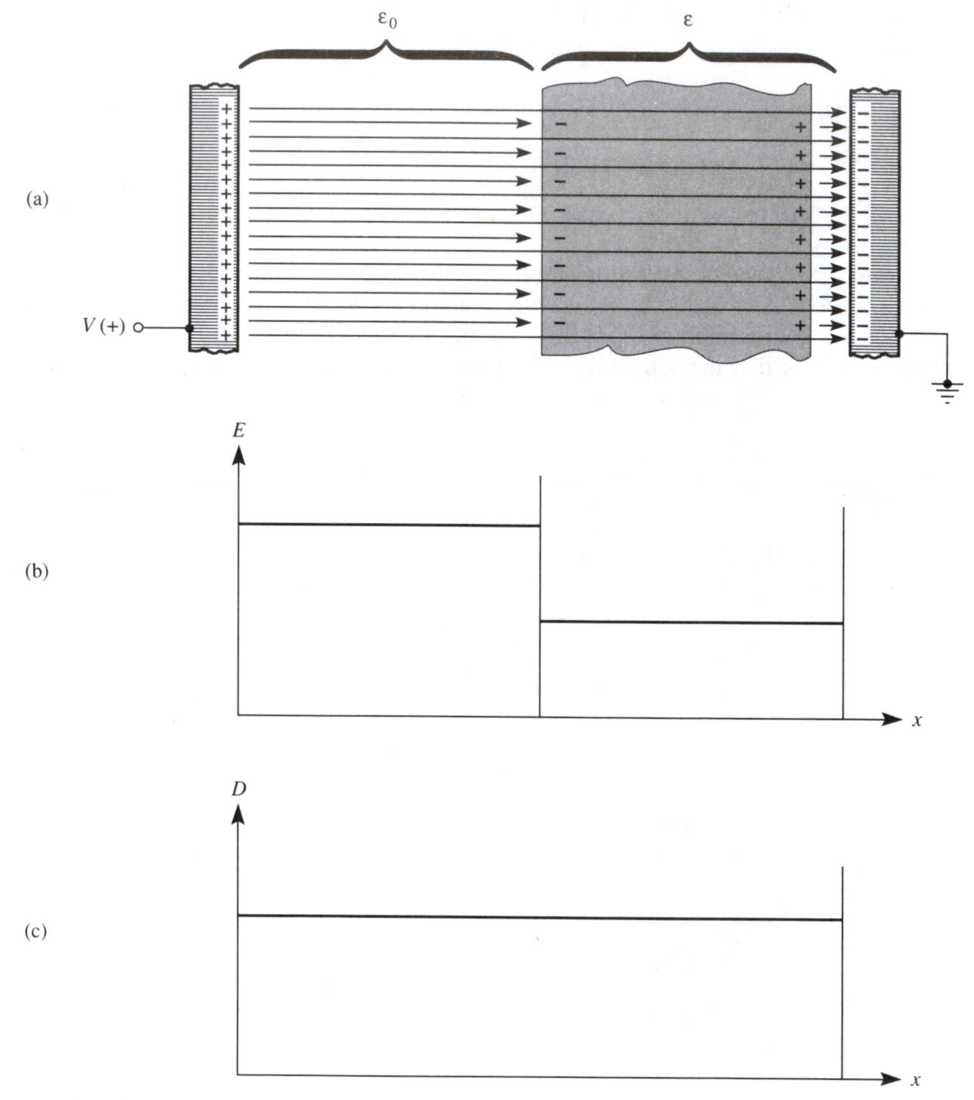

Figure C3 Capacitor having a layered dielectric region. (a) Representation of a charged parallel-plate capacitor with $\kappa = 1$ in the left layer and $\kappa = 2$ in the right. (b) Electric-field profile in the charged capacitor, as inferred from the densities of lines of force. (c) Electric-displacement profile as inferred from charge per unit area on the capacitor plates.

does for electric field.) Therefore, *displacement* is defined and has meaning even in the air-dielectric case.

A number of complications can alter the simple picture just outlined, complications that are treated at length in standard texts on electricity and magnetism [C2–C4]. One complication is that it is possible to have nonuniform polarization in some materials, with the result that net charge can exist within the volume of the material, and not just at the external faces [as in Figure C1(b)]. When this situation exists, the problem is obviously more complicated, because net charge within the volume would affect capacitor-plate charge density, and hence the magnitude of D. We shall avoid this situation by confining ourselves to *homogeneous* materials. A second complication is that for reasons involving the detailed structure of the dielectric material on an atomic or molecular scale, the charge separation may occur along an axis that is not aligned with the electric field. Such a material is *nonisotropic*. We confine ourselves to *isotropic* materials, those that are "the same through all angles."

When a capacitor's dielectric region is not homogeneous, we no longer have the field strength given simply by V/d. A nonhomogeneous case of substantial relevance to solid-state electronics involves a layered dielectric region, for which a simple example is depicted in Figure C3(a). In this case, the left half of the region between the plates is free space, while the right half is a dielectric material with $\kappa = 2$ once more. By inspection of the lines-of-force density, it becomes evident that the electric-field "profile" has the form depicted in Figure C3(b). Displacement, however, is constant throughout the region between the plates. It is evident that both plates have the same areal charge density, which is an interpretation we have placed upon displacement magnitude.

A more complicated case is one in which the **E** and **D** vectors are not normal to the interface between regions of differing permittivity. In such a case, the electric-field lines of force are refracted. The laws governing this situation are simple. The tangential component of the vector **E** is continuous through the interface, and the normal component of the vector **D** is continuous through the interface. Such problems, of course, are at least two-dimensional, and fortunately these are rare for us.

Finally, it is important to note especially the sheet of charge that is present at the median plane of Figure C3(a). A fully analogous charge sheet exists at the oxide-silicon interface of a charged MOS capacitor because of the differing dielectric constants of oxide and silicon. In the cause of simple representation, however, this complication has been omitted from Figures 2-4(b), 2-5, and subsequent similar diagrams.

References

C1. J. C. Maxwell, *Treatise on Electricity and Magnetism*, Cambridge University Press, 1873.

C2. E. M. Pugh and E. W. Pugh, *Principles of Electricity and Magnetism*, Addison-Wesley, Reading, MA, 1960.

C3. G. P. Harnwell, *Principles of Electricity and Electromagnetism*, McGraw-Hill, New York, 1938.

C4. J. D. Jackson, *Classical Electrodynamics*, Wiley, New York, 1962.

Appendix D
Band Diagrams in Terms of Electrostatic Potential

It is often convenient to convert from electron-energy (ξ) band representations to electrostatic-potential (ψ) representations. This is because the semiconductor structures that these diagrams describe will ultimately be used in circuits and will have to be related to circuit voltages.

Suppose we move an electron at the valence-band edge to an isolated state somewhere in the energy gap. The electron of course experiences an increment in total energy $\Delta\xi$ as a result. The agency that imparts this energy increment to the electron must do work on it, and so the corresponding increment in electrostatic potential is given by

$$\Delta\psi = \frac{\text{work}}{Q} = \frac{\Delta\xi}{-q}. \tag{D1}$$

Recall now that the origin or reference is an arbitrary matter for both ψ and ξ. That is, we can replace $\Delta\xi$ by $\xi - \xi_{\text{REF}}$, and $\Delta\psi$ by $\psi - \psi_{\text{REF}}$; but in order for $\Delta\xi$ and $\Delta\psi$ to differ by a constant factor, as in Equation D1, it is necessary for ξ_{REF} and ψ_{REF} to be chosen in a consistent manner. In other words, it is necessary to have $\psi_{\text{REF}} = 0$ when $\xi_{\text{REF}} = 0$. Having made this arrangement, we can then convert from familiar and important energy-value designations such as ξ_C, ξ_V, ξ_F, and ξ_I to the corresponding potential-value designations ψ_C, ψ_V, ψ_F, and ψ_I. Thus the potential ψ corresponding to an arbitrary energy ξ can be written, with the aid of Equation D1, as

$$\psi = -\frac{\xi}{q}. \tag{D2}$$

The band diagrams in terms of ξ and ψ will look the same except for labeling and one additional important feature. The fact that the electronic charge is negative, $-q$ in Equation D1, introduces an awkward necessity. We must either let electrostatic potential increase *downward* in the second case, or else flip the band diagram over, placing the valence band at the top. We usually choose the first option, so that the two band diagrams will look the same. Figure D1 presents the results for an arbitrarily chosen *N*-type sample. Notice that Fermi level has been chosen as reference in both cases. This is frequently a convenient choice because it simplifies important algebraic expressions. Notice also that the ξ and ψ axes are oppositely directed.

Consider now the effects of net-doping changes on these band pictures. Regard the Fermi level as fixed in position because it has been chosen as reference. For the intrinsic case, the intrinsic level coincides with the Fermi level because the band-symmetry approximation will be in use from this point forward. The conduction-band edge lies half the gap above it; the valence-band edge lies half the gap below it. With increases in net *N*-type doping, all three of these features move downward together; with changes in the *P*-type direction, all three move upward together. Heavy-doping effects cause a bandgap reduction ("narrowing") that distorts the simple picture just sketched, but through wide doping ranges, this effect is small enough to ignore. Any one of the three features could be taken as the indicator of electron energy in the first band-diagram case, and as the indicator of electrostatic potential in the second, because

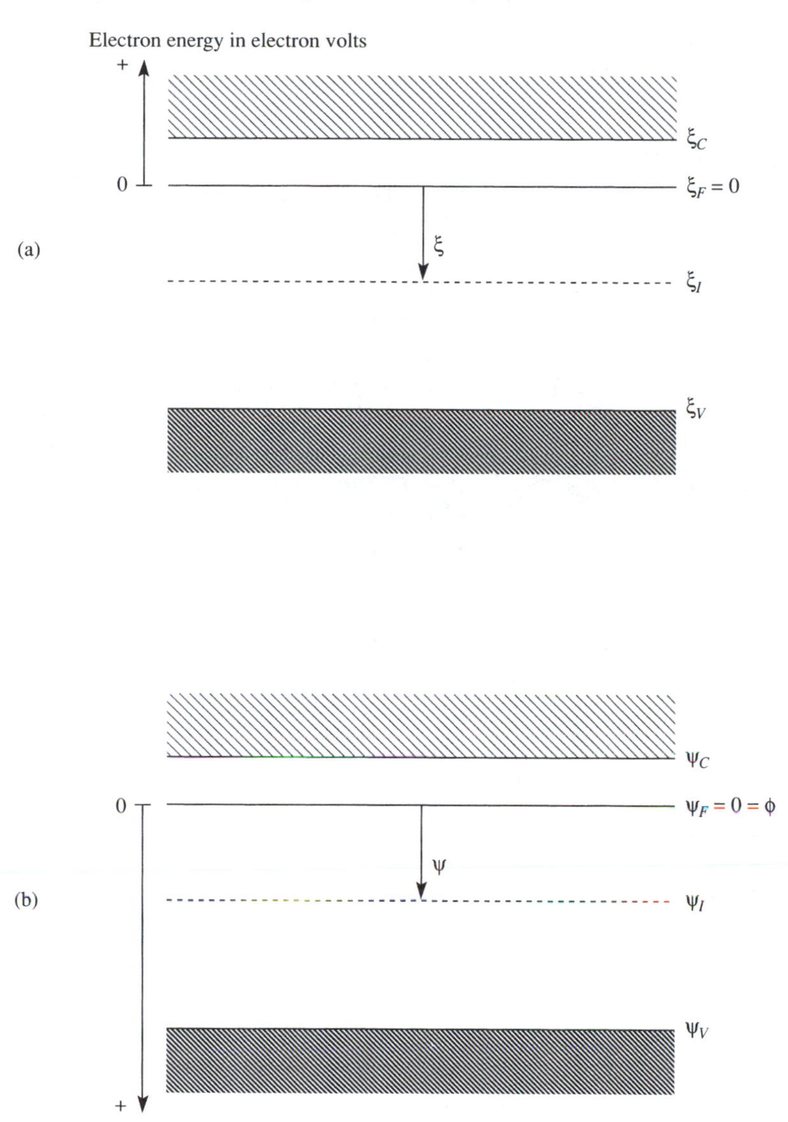

Figure D1 Equivalent band diagrams for an *N*-type sample in terms of (a) electron energy and (b) electrostatic potential. In both cases Fermi level has been taken as reference. Because both diagrams have the same orientation (conduction band above), their ordinate axes are oppositely directed, a consequence of the electron's negative charge.

both of these variables already involve an arbitrary constant. But by choosing the intrinsic level to represent energy or potential, we can take advantage of symmetry by being "even-handed" with respect to holes and electrons. With these choices, it is evident in Figure D1 that ξ in an *N*-type sample is negative, while ψ in an *N*-type sample is positive. Thus the intrinsic level in Figure D1(b) is the *potential* ψ of the sample. Note also in Figure D1(b) that the symbol ϕ (phi) is shown as equivalent to

ψ_F. It is a longstanding custom to let ϕ represent Fermi level in terms of electrostatic potential because of the phonic aid to memory.

Appendix E
Carrier Densities in Terms of Electrostatic Potential

The equivalent-density-of-states approximation places all the conduction-band states right at the conduction-band edge, and considers their volumetric density to be N_C. The valence band is handled similarly. The Boltzmann probability function is a pure-exponential approximation to either "tail" of the Fermi–Dirac probability function that may be used when the Fermi level is a few times kT/q away from the nearest band edge, as is the case in Figure D1(b). The product of these approximate formulations in the conduction-band case yields a simple and useful expression for the volumetric density of conduction electrons as

$$n = N_C e^{q(\psi_C - \phi)/kT}, \tag{E1}$$

where the conversion from ψ_F to ϕ has also been made. Now add and subtract ψ ($= \psi_I$) inside the parentheses in the numerator of the exponent of Equation E1,

$$n = N_C e^{q(\psi_C - \psi + \psi - \phi)/kT} = N_C e^{-q(\psi - \psi_C)/kT} e^{q\psi/kT}, \tag{E2}$$

where we have once again let $\phi = 0$, as was done in Appendix D. But since $\psi - \psi_C$, a positive electrostatic-potential increment, is just half the energy gap, it is evident that the coefficient of $\exp(q\psi/kT)$ on the right-hand side of Equation E2 is equal to n_i, the intrinsic carrier density present in each band when band symmetry is assumed and when the Fermi level is at gap center. For its value, see the front-matter table, Properties of Silicon. Hence,

$$n = n_i e^{q\psi/kT}. \tag{E3}$$

Proceeding in a completely parallel fashion, we find that

$$p = n_i e^{-q\psi/kT}. \tag{E4}$$

These results again emphasize the utility of the constant n_i.

Appendix F
Approximate-Analytic Model for the Step Junction

A closed-form analytic expression can be obtained for electric field as a function of electrostatic potential in a step junction at equilibrium. But the subsequent integration needed to obtain electrostatic potential as a function of position can be carried out only for special and asymptotic cases. As a result, starting in the 1950s, computer methods have been used extensively to complete the task. While these solutions are precise, they fail to provide the physical insight that is afforded by an analytic treatment, even an approximate-analytic treatment. The depletion approximation is an example of the latter case. As we have seen, it is primitive in formulation, but still very useful. Its popularity five decades after its introduction [F1] confirms that even a coarse analytic solution is a useful supplement to an exact numeric solution. Recent work has provided another option, however, an approximate-analytic treatment that is almost as simple as the depletion approximation and almost as accurate as a computer solution.

The first workers to address the problem in this way were specifically analyzing semiconductor surface regions at equilibrium [F2–F4], but their results are equally applicable to the equilibrium step-junction problem that is our immediate concern. Following them, we must first formulate the appropriate differential equation, and then integrate it to obtain electric field as a function of electrostatic potential.

What is needed is an equation in potential and its spatial derivatives that comprehends *space charge* as well. In short, Poisson's equation is appropriate, just as it is for initiating the depletion-approximation solution. The four basic approximations are explicitly retained here—Boltzmann statistics, equivalent densities of states, band symmetry, and complete ionization. From the last one it follows that in the neutral end regions of the junction sample, well away from the metallurgical junction,

$$N_D - N_A = n_0 - p_0. \tag{F1}$$

The symbols n_0 and p_0 are as usual the constant, neutral-equilibrium values of carrier density found in a uniformly doped sample. Thus Equation F1 sets net impurity density equal to the difference of the majority- and minority-carrier densities, a relationship that must hold in a neutral sample. Near the metallurgical junction, however, the sample is *not* neutral, and the density of positive space charge can be written

$$\rho_v = q\{[p_0(x) - n_0(x)] - [N_D - N_A]\}. \tag{F2}$$

The functional notation in the first term on the right-hand side is a reminder that these carrier densities, while equilibrium values, are spatially varying because of a departure from uniform doping in the sample of interest. The departure here is of course the *PN* junction itself. Combining Equations F1 and F2 yields

$$\rho_v = q\{[p_0(x) - n_0(x)] - [p_0 - n_0]\}, \tag{F3}$$

the expression needed for substitution in the right-hand side of Poisson's equation, yielding:

$$\frac{d^2\psi_n(x)}{dx^2} = -\frac{q}{\epsilon}\{[p_0(x) - n_0(x)] - [p_0 - n_0]\}. \tag{F4}$$

Once again using the convenient normalized voltage $U_0 \equiv q\psi_0/kT$, and invoking Equations E3 and E4, which are based on Boltzmann statistics, we have for the second term in brackets,

$$p_0 - n_0 = n_i(e^{-U_0} - e^{U_0}). \tag{F5}$$

But recalling that $\sinh \lambda \equiv [\exp(\lambda) - \exp(-\lambda)]/2$, we can rewrite Equation F5 as

$$p_0 - n_0 = -2n_i \sinh U_0. \tag{F6}$$

For the first term in brackets on the right-hand side of Equation F4 it is convenient to simplify notation by dropping both subscript and functional notation, letting $p_0(x) \equiv p$, and $n_0(x) \equiv n$; similarly, let $\psi_0(x) \equiv \psi$, and $U_0(x) \equiv U$, but remember that these quantities are spatially varying equilibrium values. With this change,

$$p - n = -2n_i \sinh U. \tag{F7}$$

Since

$$\frac{d^2U}{dx^2} = \frac{q}{kT}\frac{d^2\psi}{dx^2} \tag{F8}$$

from the definition of normalized potential, Equation F4 (Poisson's equation), can be rewritten as

$$\frac{d^2U}{dx^2} = \frac{2qn_i}{\epsilon}\frac{q}{kT}(\sinh U - \sinh U_0). \tag{F9}$$

With the definition

$$\left(\frac{\epsilon}{2qn_i}\frac{kT}{q}\right)^{1/2} \equiv L_{Di}, \tag{F10}$$

Equation F9 finally becomes

$$\frac{d^2U}{dx^2} = \frac{1}{L_{Di}^2}(\sinh U - \sinh U_0). \tag{F11}$$

This expression is often identified as the *Poisson–Boltzmann equation*, a felicitous description in view of its derivation from Equation F4, Poisson's equation, and Equation F5, which is based upon Boltzmann statistics. The quantity L_{Di} is the *intrinsic Debye length*, a special case of the general Debye length, a characteristic length we take up next.

There is a tendency for a fixed charge in a semiconductor sample to produce an enhanced density of opposite-type mobile charge in its vicinity. As a result, lines of force emanating from the fixed charge are terminated in the surrounding cloud, so that the fixed charge is "invisible" when "viewed" from an appropriate distance. This phenomenon is named *screening*, and it determines the distance at which a fixed charge can be "felt" in such a medium (to use another anthropomorphic term). When mobile charges of both signs are present, they both participate in the screening phenomenon; in the neighborhood of the fixed charge, density is depleted for like-sign mobile charges and enhanced for opposite-sign mobile charges. This problem was first addressed in the context of electrolytes by Debye and Hückel [F5, F6]. Hence the characteristic length associated with the screening phenomenon is known as the *Debye–Hückel length*, usually shortened simply to *Debye length.*

Let us move now to a one-dimensional problem. Think, for example, of a sheet of positive charge buried in a block of N-type silicon. Electron density on both sides of the sheet will exceed n_0, and can be assumed to decline exponentially with departure from the sheet. The constant with length dimensions that enters into the exponent can be described as a *characteristic screening distance*, or one-dimensional Debye length. We shall confine our attention to one-dimensional cases. (For a two-dimensional or three-dimensional problem, the term *Debye radius* is sometimes encountered.)

For a sample in which both holes and electrons have significant density, it is the *general Debye length* that is relevant:

$$L_D = \left[\frac{\epsilon}{q(n_0 + p_0)}\frac{kT}{q}\right]^{1/2}. \tag{F12}$$

This expression can be rewritten as

$$L_D = \left(\frac{\epsilon}{2qn_i}\frac{kT}{q}\frac{1}{\cosh U_0}\right)^{1/2} \tag{F13}$$

by recalling that $\cosh \lambda \equiv [\exp(\lambda) + \exp(-\lambda)]/2$. Comparing this result with Equation F10 makes it evident that

$$L_{Di} = L_D \sqrt{\cosh U_0}, \tag{F14}$$

a relationship that will be used below. Evidently the conversion from L_D to L_{Di} can be made easily by noting in Equation F12 that $(n_0 + p_0) = 2n_i$ in the intrinsic case, or in Equation F13 that $\cosh U_0 = 1$ in the intrinsic case. To repeat, the quantity L_{Di} is the *intrinsic Debye length.*

A further and more important fact is that for an extrinsic sample, either n_0 or p_0 will drop out of Equation F12. Hence, the *extrinsic Debye length* is

$$L_{De} = \left(\frac{\epsilon}{qn_{0m}}\frac{kT}{q}\right)^{1/2} \approx \left(\frac{\epsilon}{q|N_D - N_A|}\frac{kT}{q}\right)^{1/2}, \tag{F15}$$

where n_{0m} stands for majority-carrier density, n_0 or p_0 as appropriate.

The relationships among L_D, L_{Di}, and L_{De} are shown for silicon in Figure F1. While we have not derived expressions for L_D and L_{De}, we showed that the quantity L_{Di} is "natural" in the sense that it appears spontaneously when the Poisson–Boltzmann equation, Equation F11, is derived. Notice in Figure F1 that the approximation $L_D \approx L_{De}$ is excellent for all but near-intrinsic conditions. For this reason, many authors ignore the distinction. However, when Debye length is used for normalization—our primary application of it—it is convenient to maintain the distinction. A physically meaningful length used as a normalizing constant is often described as a *scaling length*, for the reason that a well-chosen constant can simplify a problem dramatically.

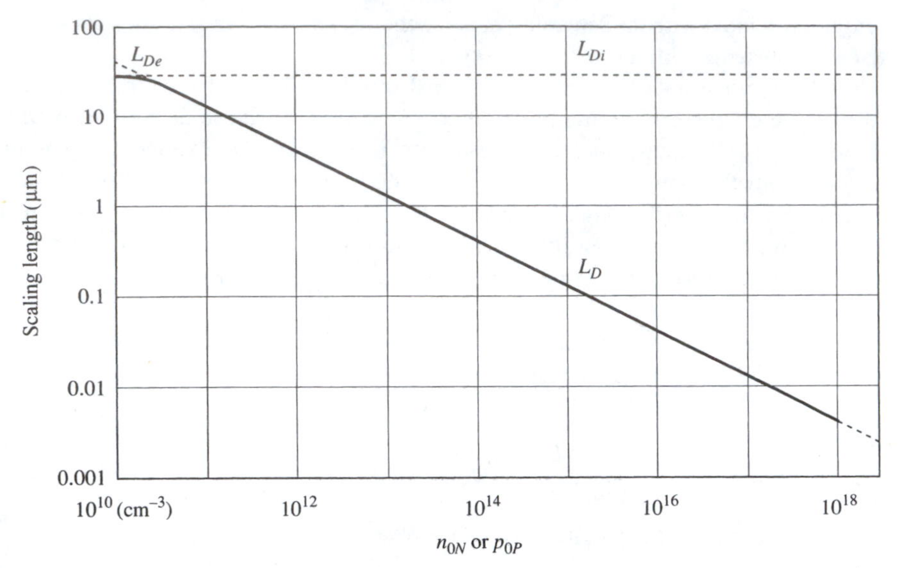

Figure F1 Three scaling lengths: the general (L_D), extrinsic (L_{De}), and intrinsic (L_{Di}) Debye lengths at 24.8°C as a function of majority-carrier density.

EXERCISE F–1 Carry out a full normalization of Poisson's equation. Assume the case of an extrinsic sample, and start with Poisson's equation in the form

$$\frac{dE}{dx} = \frac{qN}{\epsilon},$$

where $N \equiv |N_D - N_A|$. Use the thermal voltage and the extrinsic Debye length for normalizing potential and distance, respectively.

■ Multiply both sides of the given equation by L_{De} to normalize distance:

$$\frac{dE}{d(x/L_{De})} = L_{De}\frac{qN}{\epsilon}.$$

By definition,

$$E = -\frac{d\psi}{dx} = -\frac{kT}{q}\frac{dU}{dx} = -\frac{1}{L_{De}}\frac{kT}{q}\frac{dU}{d(x/L_{De})},$$

where potential is normalized in the second step and distance in the third. Substituting this result into the equation immediately preceding it yields

$$-\frac{d}{d(x/L_{De})}\left[\frac{dU}{d(x/L_{De})}\right]\frac{1}{L_{De}}\frac{kT}{q} = L_{De}\frac{qN}{\epsilon}.$$

Thus

$$\frac{d^2U}{d(x/L_{De})^2} = -L_{De}^2 \frac{qN}{\epsilon} \frac{q}{kT}.$$

But in view of Equation F15,

$$\frac{d^2U}{d(x/L_{De})^2} = -1.$$

This result illustrates the power of normalization, and one of several interpretations of Debye-length significance. In normalized form, Poisson's equation informs us that the field-profile slope is unity in a region of constant space-charge density. The electric field rises one normalized unit in each normalized unit of distance. This situation exists near a step junction, where the depletion of majority carriers leads to a constant space-charge density of magnitude qN.

Another interpretation of Debye length is that it is the distance over which carrier density changes by a factor of e in a screening situation, a point made previously. Yet another, and nonobvious, meaning is the distance that a majority carrier can travel in one dielectric-relaxation time t_D when it moves in an uninterrupted path at its thermal velocity v_t [F7]. This link to dielectric-relaxation time is significant. We noted that t_D depends on permittivity ϵ and resistivity ρ, and the latter is closely related to carrier density. Debye length at a fixed temperature also depends on these two factors, permittivity and mobile-charge density.

Just as there are an infinite number of possible step junctions, there would seem to be an infinite number of accompanying field profiles. But several researchers noticed that redundancy among the possible solutions is vastly reduced when either the general or the extrinsic Debye length is employed for spatial normalization [F8–F10]. In each case, curiously, the discovery was made independently. The fact that observation was not made by the large number of additional workers who addressed the step-junction problem is partly explained by the fact that in the Poisson–Boltzmann equation (repeated below for convenient reference) it is the *intrinsic* Debye length that in essence volunteers to serve for normalization:

$$\frac{d^2U}{dx^2} = \frac{1}{L_{Di}^2}(\sinh U - \sinh U_0). \tag{F16}$$

It is desirable to define sample geometry explicitly, as has been done in Figure F2. This choice, and the notation and approach that will be used, follow prior practice [F10]. Note that normalized potential U (a spatially varying but equilibrium quantity) increases downward, and the Fermi level is used as reference. The junction position is x_J, and the potential there is U_J. Focus on the right-hand side of the sample. Our object is to integrate Equation F16 from the end region where $U = U_{02}$ to an arbitrary

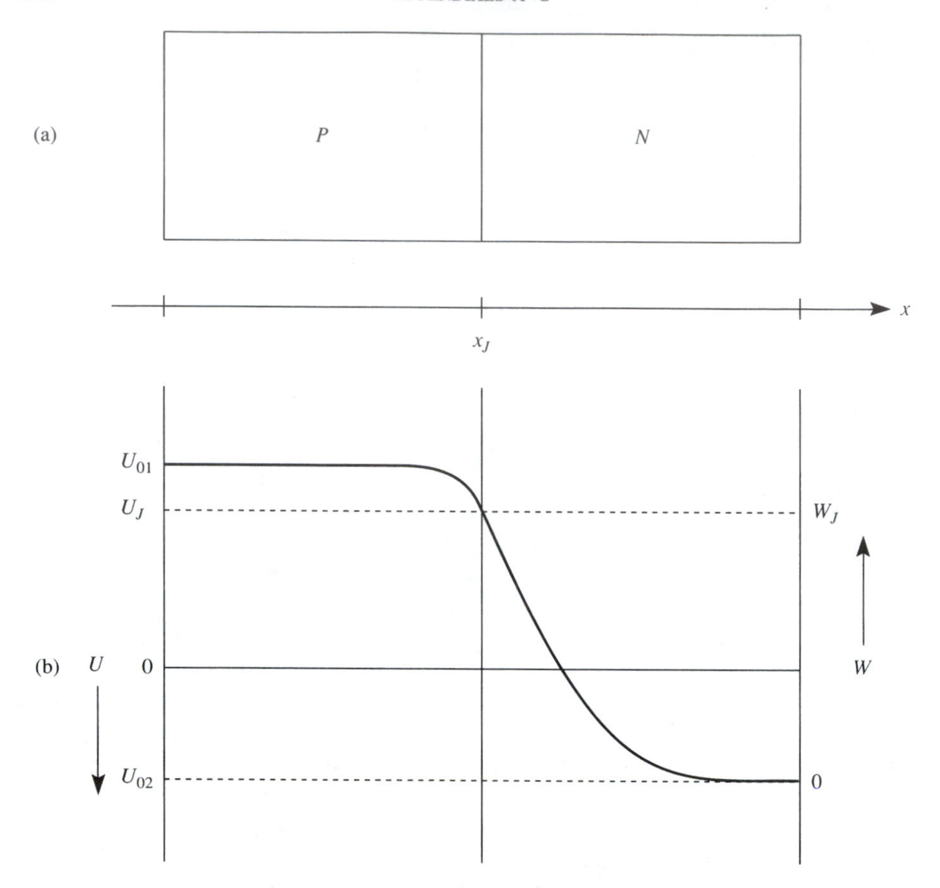

Figure F2 Step-junction sample of the specific orientation for explaining the approximate-analytic solution. (a) Physical representation. (b) Normalized potential as a function of position.

position x where the potential is U. The left-hand side of Equation F16 is rendered easily integrable by means of a straightforward manipulation:

$$\int \frac{d^2U}{dx^2}\, dU = \int \frac{d}{dx}\left(\frac{dU}{dx}\right) dU = \int \left(\frac{dU}{dx}\right) d\left(\frac{dU}{dx}\right). \tag{F17}$$

Substituting this revised expression for the left-hand side of Equation F16 makes it clear that consistent limits for integration are from zero to the variable value (dU/dx) on the left-hand side, and from U_{02} to the corresponding variable value U on the right-hand side. Choosing U' as a dummy variable places Equation F16 in shape for integration:

$$\int_0^{dU/dx} \left(\frac{dU'}{dx}\right) d\left(\frac{dU'}{dx}\right) = \frac{1}{L_{Di}^2} \int_{U_{02}}^{U} (\sinh U' - \sinh U_{02})\, dU'. \tag{F18}$$

Integrating, then, gives

$$\frac{1}{2}\left(\frac{dU}{dx}\right)^2 = \frac{1}{L_{Di}^2}[(U_{02} - U)\sinh U_{02} - (\cosh U_{02} - \cosh U)], \tag{F19}$$

and extracting the square root of this result yields an expression for partly normalized electric field as a function of normalized potential:

$$\frac{dU}{dx} = \pm\frac{\sqrt{2}}{L_{Di}}[(U_{02} - U)\sinh U_{02} - (\cosh U_{02} - \cosh U)]^{1/2}. \tag{F20}$$

The positive sign is appropriate for the situation shown in Figure F2, where dU/dx and x were made consistent in sign by choosing to have $U_{01} < U_{02}$.

Now, use Equation F14 to eliminate L_{Di} from Equation F20 and to introduce the general Debye length L_D, yielding

$$\frac{dU}{d(x/L_D)} = \pm\sqrt{2}\left[\frac{(U_{02} - U)\sinh U_{02} - (\cosh U_{02} - \cosh U)}{\cosh U_{02}}\right]^{1/2}. \tag{F21}$$

Next, convert from hyperbolic-function notation to exponential notation:

$$\frac{dU}{d(x/L_D)} = \pm\sqrt{2}\left[\frac{(U_{02} - U)(e^{U_{02}} - e^{-U_{02}}) - (e^{U_{02}} + e^{-U_{02}} - e^{U} - e^{-U})}{e^{U_{02}} + e^{-U_{02}}}\right]^{1/2}. \tag{F22}$$

Now it is convenient to introduce a new normalized potential,

$$W \equiv U_{02} - U. \tag{F23}$$

The effect of this information is displayed in Figure F2. First, the potential origin is switched from the Fermi level to the equilibrium potential U_{02} in the end region, often called the *bulk potential.* The term "bulk" is meant to convey a region well away from disturbances (such as a junction) so that neutral-equilibrium conditions exist therein. Second, W and U are opposite in sign. Substituting the quantity $U = U_{02} - W$ into Equation F22 yields the result sought,

$$\frac{dW}{d(x/L_D)} = -\sqrt{2}\left[\frac{e^{U_{02}}(e^{-W} + W - 1) + e^{-U_{02}}(e^{W} - W - 1)}{e^{U_{02}} + e^{-U_{02}}}\right]^{1/2}, \tag{F24}$$

an exact, closed-form expression for normalized electric field as a function of normalized electrostatic potential W. The negative sign is appropriate for the conditions of Figure F2.

More useful, to be sure, would be a relationship between electric field and position, but computer intervention is necessary to get there, as will be described below.

The result of that procedure is shown in Figure F3 [F11]. Observe the extensive region wherein normalized electric field versus normalized position exhibits a slope of unity magnitude. This is the deeply depleted region discussed above, and the field profile can legitimately be described as a "universal" curve, applicable in part to any junction. The term *general solution* has also been used to emphasize its universality [F10].

An extrapolation of the linear part of the curve to the spatial axis defines a spatial origin that will be exploited below. The origin falls in the vicinity of the space-charge-layer boundary—the region of transition from neutrality to deep depletion. The utility of this particular origin choice can be illustrated by writing Poisson's equation in fully normalized form. It can be written, with separated variables, as

$$d\left[\frac{dU}{d(x/L_D)}\right] = -d(x/L_D). \tag{F25}$$

(For increased generality, L_D has been substituted for L_{De}.) Obviously this expression can be integrated with ease to yield normalized electric field versus normalized posi-

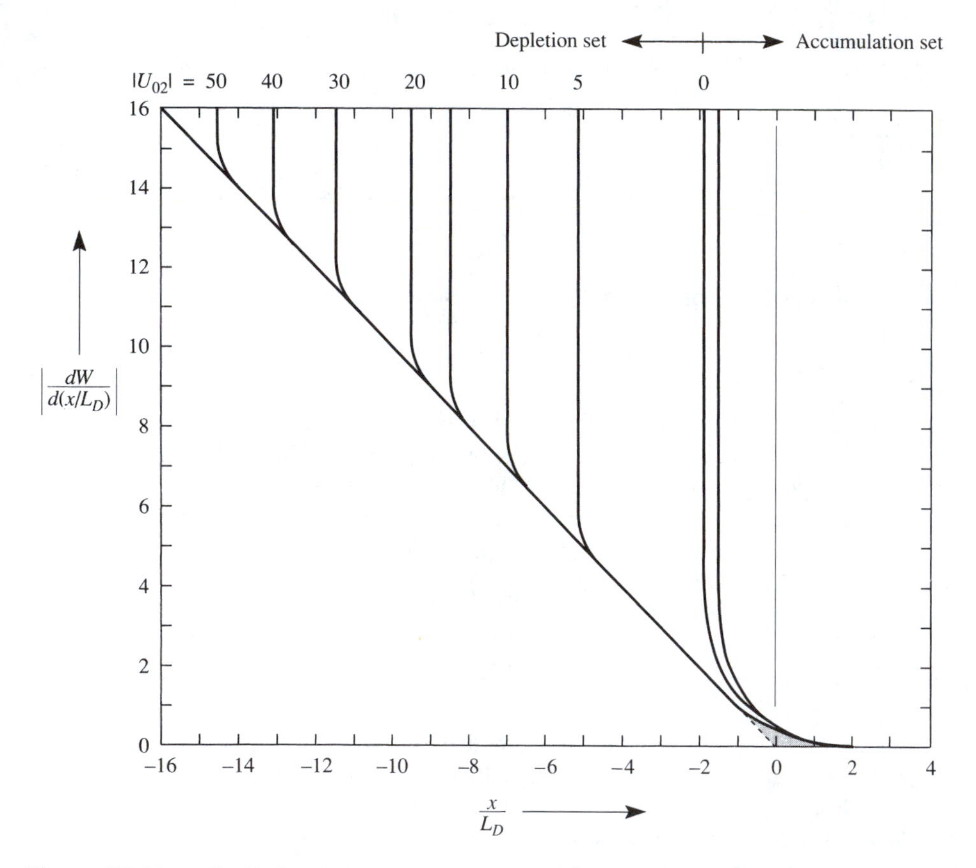

Figure F3 Normalized electric-field magnitude as a function of normalized distance, showing spatial-origin definition in terms of linear-profile extrapolation [after F11].

tion. If arbitrary limits are used, the result will include a constant of integration. But with the origin choice just made we have coordinated limits, and extrapolated electric field vanishes at the positional zero. Under these conditions, the result of integration is

$$\frac{dU}{d(x/L_D)} = -\frac{x}{L_D}, \tag{F26}$$

which is valid in the region of deep depletion where the field profile is linear. The negative sign is consistent with the case shown in Figure F3. For *P*-type material on the right-hand side of the junction (instead of the choice made in Figure F2), the sign would of course be positive.

This issue of algebraic sign is easily resolved by physical reasoning. More serious issues exist in Equation F24, however, where U_{02} and W each can be either positive or negative. Therefore it has been shown to be worthwhile to rewrite Equation F24 in a way that eliminates sign ambiguity [F12]:

$$\left|\frac{dW}{d(x/L_D)}\right| = \sqrt{2}\left[\frac{e^{|U_{02}|}(e^{-W} + W - 1) + e^{-|U_{02}|}(e^{W} - W - 1)}{e^{|U_{02}|} + e^{-|U_{02}|}}\right]^{1/2}. \tag{F27}$$

In this form, the electric-field expression describes depletion conditions (consistent with Figure F3) and is valid for U_{02} positive or negative (an *N*-type or *P*-type region on the right-hand side of the junction), with the further proviso that W should be taken as positive always. Absolute-value signs have been placed on the normalized-field symbol, however, as a reminder that algebraic sign for the overall expression must be established, to repeat, by physical reasoning.

A prominent feature of Figure F3 is the family of near-vertical curves that proceed smoothly upward from all portions of the primary curve. These are a consequence of the phenomenon of *inversion* that will be addressed below. In addition, the dashed lines shown near the spatial origin in Figure F4 will also be explained.

The next aim is to determine normalized potential W as a function of normalized position x/L_D. Hence the first step is to separate variables in Equation F27, yielding

$$|d(x/L_D)| = \frac{1}{\sqrt{2}}\left[\frac{e^{|U_{02}|} + e^{-|U_{02}|}}{e^{|U_{02}|}(e^{-W} + W - 1) + e^{-|U_{02}|}(e^{W} - W - 1)}\right]^{1/2} dW. \tag{F28}$$

It has been determined numerically that the value of W at the spatial origin in Figure F3 is

$$W(0) = 0.59944 \approx 0.6. \tag{F29}$$

Thus consistent limits are from zero to an arbitrary value of x/L_D, and from $W(0)$ to an arbitrary (and positive) value of W. Carrying out this numerical integration yields the result shown in Figure F4. In the spatial interval where the field profile is linear, the potential profile is of course purely parabolic. For now we will concentrate on this portion of the solution and again defer consideration of the steeply rising portions.

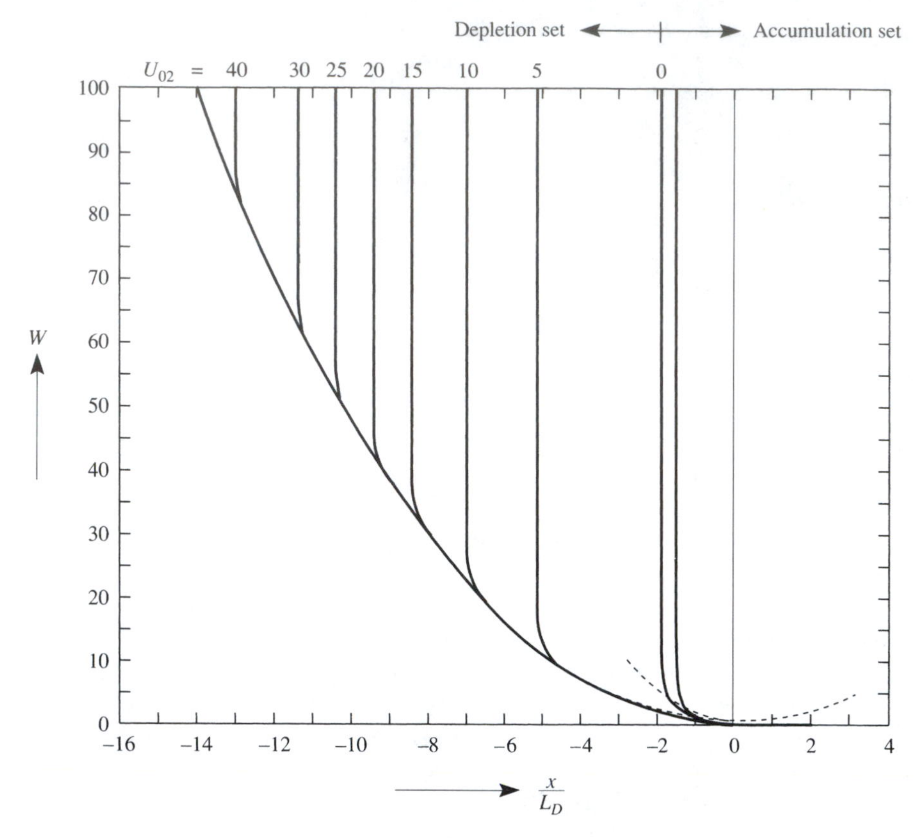

Figure F4 Normalized potential as a function of normalized distance for a step junction with various values of normalized bulk potential U_{02} on the right-hand side [F13].

To apply these findings, let us choose the simplest case first, namely, the symmetric step junction. As a result, in Figure F2 the junction potential $U_J = 0$, or $W_J = U_{02}$. Of primary interest is the carrier profile $n_0(x)$, which for convenience will once again be labeled simply as n. Knowledge of the density profile comes immediately from the potential profile. From the definition of W, Equation F23, we have

$$W \equiv U_{02} - U = \ln \frac{n_n}{n_i} - \ln \frac{n}{n_i} = \ln \frac{n_n}{n}. \tag{F30}$$

Consequently,

$$\frac{n}{n_0} = e^{-W}. \tag{F31}$$

Plotting this function with the aid of the numerically determined function $W(x/L_D)$ yields the result shown in Figure F5. The single curve is valid for all possible symmetric step junctions. The position of this curve in relation to the metallurgical junction, however, is dependent on doping, and the label x_J shows junction position for several net-doping values.

On the right-hand side of Figure F5 it is indicated just where the sharp depletion-layer boundary defined by the depletion approximation will fall; in this comparison of exact and depletion-approximation results, the two have been adjusted for equal contact potential. The depletion-approximation boundary is near, but not precisely at $x = 0$, although it tends toward $x = 0$ as $|N_D - N_A| = N$ approaches infinity.

The same exact solution can be applied to the unsymmetric step junction, as shown in Figure F6. This time the right-hand side of the junction is taken as having a constant and "typical" doping of $10^{16}/\text{cm}^3$. When the left-hand side is also $10^{16}/\text{cm}^3$, we have a symmetric junction and a position x_J corresponding to a doping ratio of unity. The correctness of the x_J position in this case can be verified by referring back to Figure F5 for the case $N = 10^{16}/\text{cm}^3$. Next let the left-hand side doping be decreased by successive factors of ten. Once again the x_J position advances toward the depletion-layer boundary. Recall that potential drop on the high side of a grossly asymmetric junction assumes a limiting value of kT/q, or one normalized unit. But when $W = 1$, Equation F31 yields

$$\frac{n}{n_0} = e^{-1} = \frac{1}{e}, \tag{F32}$$

so that

$$n = \frac{n_0}{e}. \tag{F33}$$

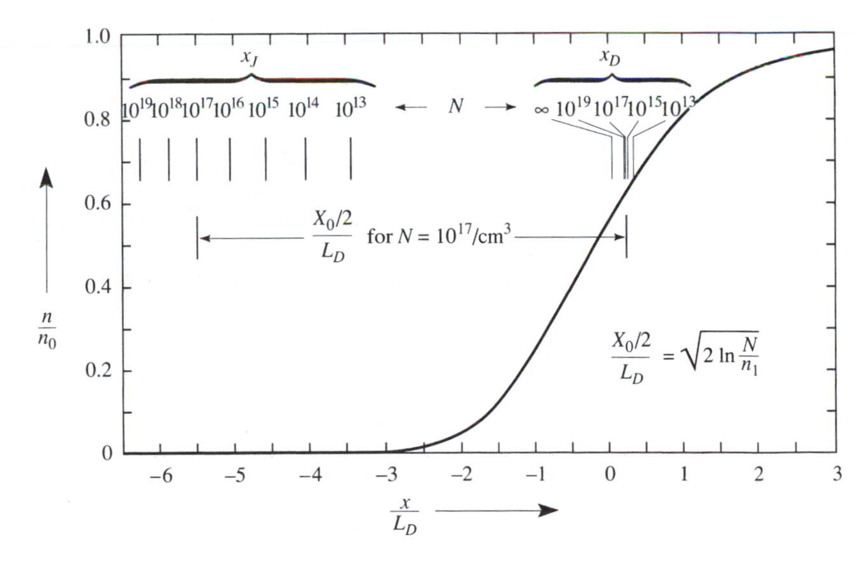

Figure F5 Relative majority-carrier density as a function of position in a symmetric step junction. Depletion-approximation boundary position x_D and junction position x_J are shown for several values of net doping N [F13].

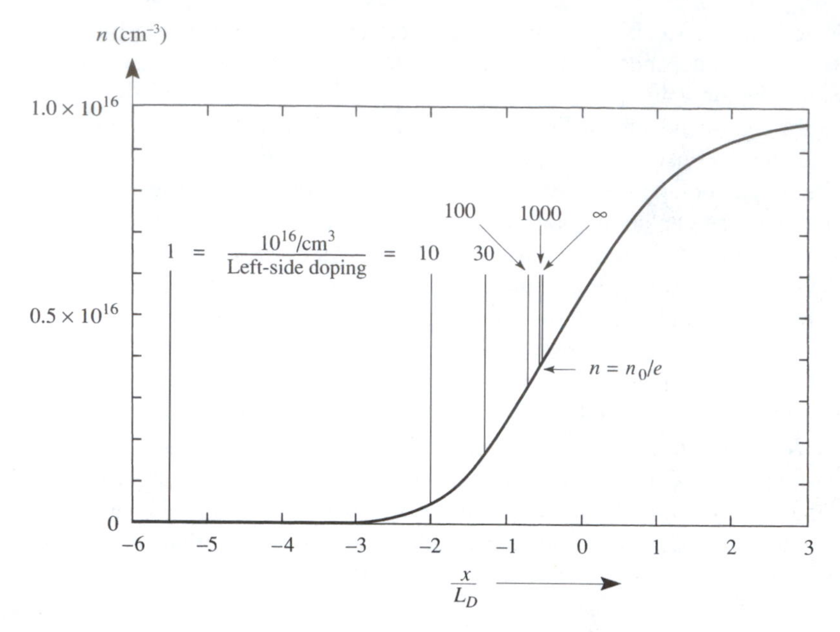

Figure F6 Majority-carrier density as a function of normalized position for asymmetric step junctions, showing junction position x_J for various doping ratios.

This point is indicated in Figure F6, corresponding to the indicated doping ratio of infinity.

The universality of the curves in Figures F3 through F6 suggests the writing of approximate-analytic expressions to describe them. This is a task that was greatly simplified by defining the spatial origin that appears in Figures F3 through F6. The expressions that result have been collectively named a *depletion-approximation replacement* [F13], sometimes abbreviated DAR.

To start, examine the potential profile in Figure F4 and note the dashed lines near the spatial origin. The one projecting toward the right is the right-hand branch of the parabolic function that accurately gives $W(x/L_D)$ in the deep-depletion regime toward the left. The vertex of this parabola has the coordinates (0, 1), rather than (0, 0) as one might expect. That is, the vertex is shifted upward on the W axis, falling one unit above the origin of the diagram. The parabola so positioned is an accurate asymptote in the deep-depletion regime toward the left, and the function can be written

$$W = \frac{1}{2}\left(-\frac{x}{L_D}\right)^2 + 1. \tag{F34}$$

EXERCISE F–2 Examine the field profile in Figure F3 and make an observation that stems from the fact that $W = 1$ in Equation F34 when $x = 0$.

■ Near the spatial origin, the field profile deviates slightly from the linear asymptote, producing the familiar "skirt." The present observation tells us, recalling that $\Delta\psi = -\int E dx$, that the small shaded area shown there amounts to one normalized unit, or kT/q.

The dashed curve projecting toward the left in Figure F4 is asymptotically valid also, and applies in the region of near neutrality, toward the right (or where $W \ll 1$). It is an exponential function, and can be written

$$W = \exp\left(-\frac{x}{L_D} - 0.41209\right). \tag{F35}$$

It is evident that the spatial origin could have been chosen in a manner that eliminates the second term in the exponent, but at the expense of complicating Equation F34. We did not take that option, however, because of the useful and clear meaning of having extrapolated field vanish at the origin, as in Figure F3.

For a clearer comparison of the two asymptotes, it is helpful to expand the scale and examine them only near the spatial origin. Further, as shown in Figure F7, a semilog presentation displays the exponential asymptote in linear fashion. Both asymptotic curves are plotted as solid lines, and are individually valid outside the boundaries of the rectangle sketched inside the diagram. The numerical solution that smoothly joins them at opposite corners of the rectangle is also shown with a solid curve. But then it becomes evident that a further approximation is possible. The exponential expression plotted as a dashed line nearly matches the intervening numerical solution (inside the rectangle), and hence can serve as a kind of "bridge" approximation between the asymptotes. This third equation is

$$W = 0.582 \exp\left(\frac{-0.9x}{L_D}\right). \tag{F36}$$

Similar treatment can be accorded to electric field and other important variables in the problem [F11]. Further, a much more detailed and extended treatment of this subject matter is available in the literature [F14].

For step junctions that are symmetric or only modestly asymmetric, the three approximate-analytic expressions, Equation F33 through F36, are sufficient. They are to be applied to one side of the junction at a time, producing an overall approximate solution. The only information lacking at this point in order to tie the two together is the value of junction potential U_J, but it can be readily calculated [F10]:

$$U_J = \frac{\cosh U_{02} - \cosh U_{01} + U_{01} \sinh U_{01} - U_{02} \sinh U_{02}}{\sinh U_{01} - \sinh U_{02}}. \tag{F37}$$

But as asymmetry increases, it becomes necessary to deal with the additional phenomenon that we address next.

Let us first take the extreme case of a grossly asymmetric step junction exhibiting the limiting high-side potential drop of kT/q. Suppose net doping in the P-type re-

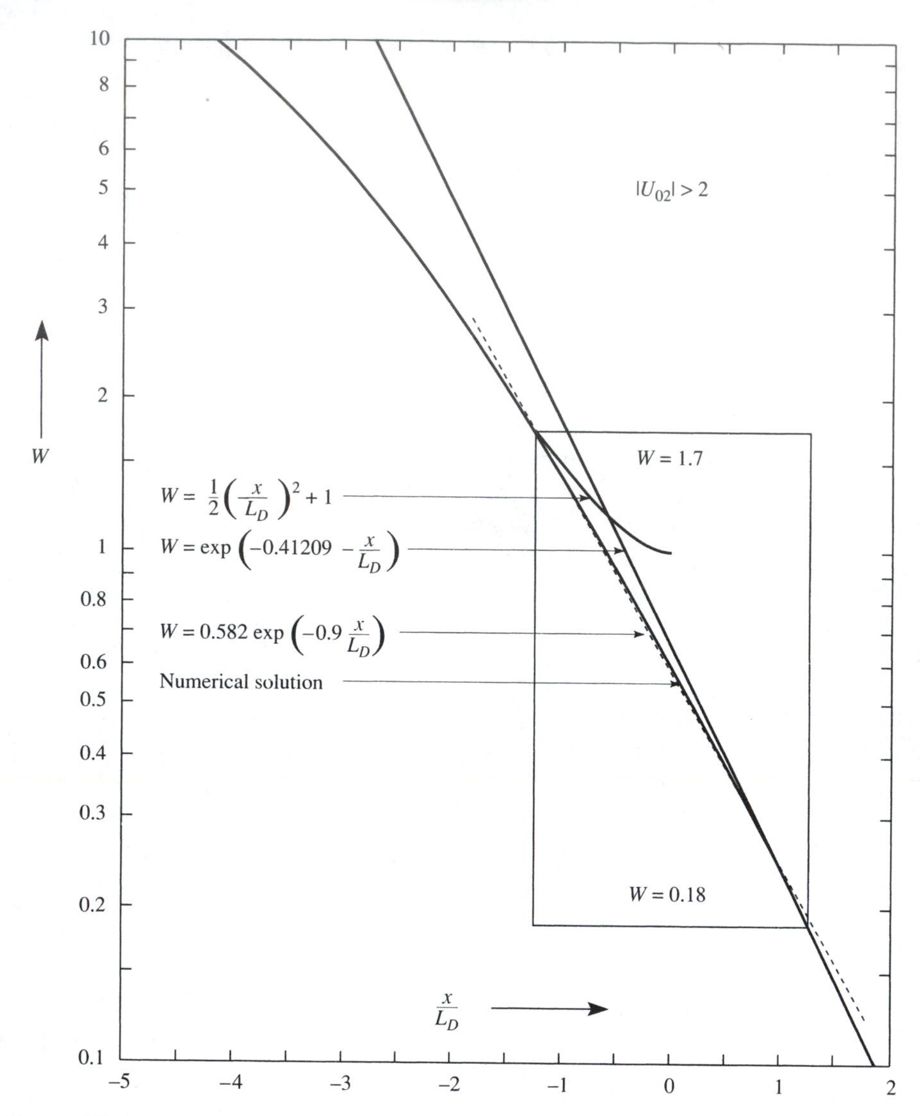

Figure F7 Semilog plot of normalized potential as function of normalized position showing three approximate-analytic functions. Useful and nonuseful (extrapolated) portions of asymptotic functions are shown with solid lines, as is numerical solution. Approximate "bridge" expression for use inside the rectangle is shown with a dashed line [F13].

gion on the left is $10^{13}/\text{cm}^3$, and in the N-type region on the right is $10^{16}/\text{cm}^3$. Insofar as the N-type region is concerned, this situation is depicted in Figure F6, where the indicator for a doping ratio of 1000 falls roughly at the point where $n = n_0/e$. In other words, in the example chosen, the electron density right at the metallurgical junction *on the P-type side* is $n_0(\approx 0) \approx 0.37 \times 10^{16}/\text{cm}^3$, because $n_0(x)$ must be continuous.

The density profile then declines in some manner toward the left, eventually reaching its minority-carrier equilibrium density of $n_{0P} = (n_i^2/p_{0P}) = 10^7/\text{cm}^3$. Thus we have here a situation not treated before. There are electrons present in a P-type region *at equilibrium* at densities orders of magnitude larger than $p_{0P} = 10^{13}/\text{cm}$. That is, we deal with a region having *P-type* net doping in which *electrons* make a vastly greater contribution to conductivity than do holes.

EXERCISE F–3 What is the hole density where $n_0(\sim 0) \approx 0.37 \times 10^{16}/\text{cm}^3$?

■ Because equilibrium obtains,

$$p_0(\sim 0) = \frac{n_i^2}{n_0(\sim 0)} = 2.7 \times 10^4/\text{cm}^3.$$

This situation is often described by saying that conductivity type is *inverted* in this region, and so the layer of electrons in the P-type region is termed an *inversion layer.* Also, the term *surplus carriers* has been introduced to distinguish such "extra" carriers clearly from excess carriers [F12].

EXERCISE F–4 What is the distinction between surplus carriers and excess carriers?

■ Surplus carriers exist in a region of gross space charge. They constitute a case of *equilibrium nonneutrality*. Excess carriers, in the simplest possible case constitute a case of *nonequilibrium neutrality.*

An asymmetric junction is shown in Figure F8 that is similar to the one just considered [F14]. A doping ratio of 100 has been chosen this time, as well as lower and (unrealistic) doping values, because so doing simplifies construction and scaling in the diagram. This can be seen in the fact that normalized scaling of distance on the two sides differs by a factor of ten, because $L_D \propto 1/\sqrt{N}$. The location of the inversion layer is displayed at the left of the diagram. Notice that $n_0(x)$ in the inversion layer does not follow the universal curve, but lies above it.

EXERCISE F–5 Why must the inversion layer profile lie above the "universal" curve?

■ The latter curve is valid when electric field continues to rise linearly toward the left. But in the present case, field peaks at the metallurgical junction and then *declines* toward the left. Declining field means declining drift, which in turn requires declining diffusion in this equilibrium situation, and that in turn requires declining density gradient, which finally brings us to an electron profile that must now be higher.

The situation here is closely related to that in the high–low junction, depicted (with reverse orientation) in Figure F9. There, too, electrons have poured over to the lightly doped side. These surplus carriers do not, however, constitute an inversion layer, but rather an *accumulation layer*, because they are in a same-type region. This fact makes an important difference in the two situations, specifically in the contact potential.

EXERCISE F–6 Compare $\Delta\psi_0$ for a *PN* junction and a high-low junction in which the respective net dopings are $10^{18}/\text{cm}^3$ and $10^{14}/\text{cm}^3$.

■ For the *PN* junction

$$\Delta\psi_0 = \frac{kT}{q} \ln \frac{N_D N_A}{} = 0.71 \text{ V},$$

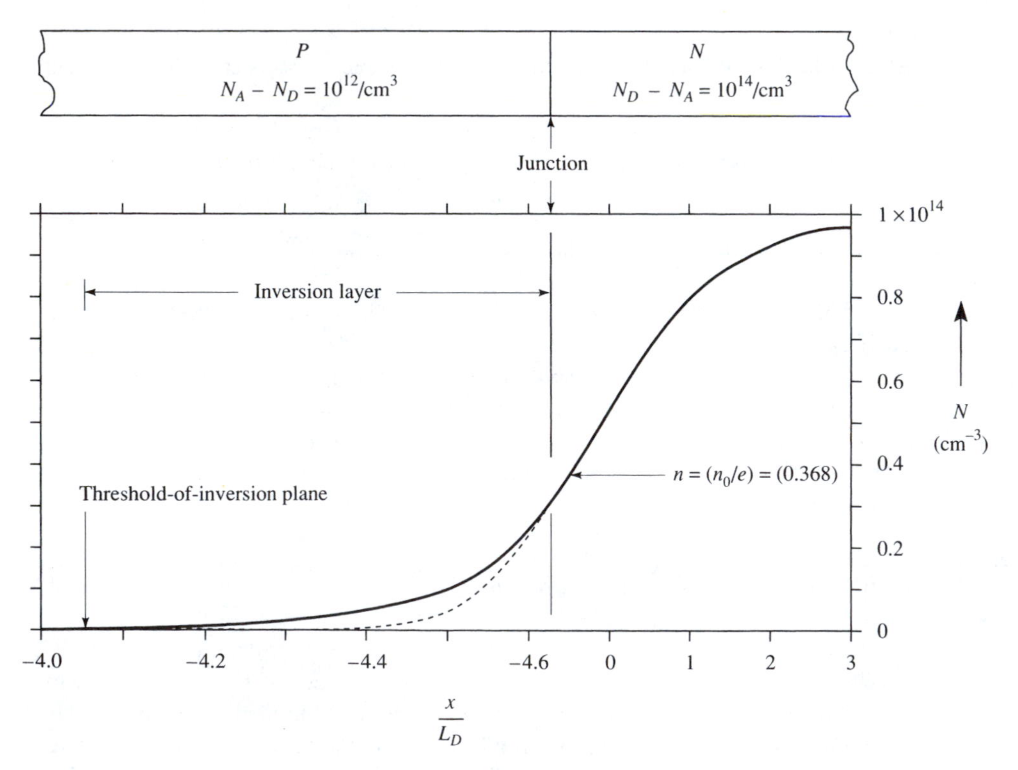

Figure F8 Majority-carrier density as a function of normalized position for an asymmetric step function possessing an inversion layer [F14].

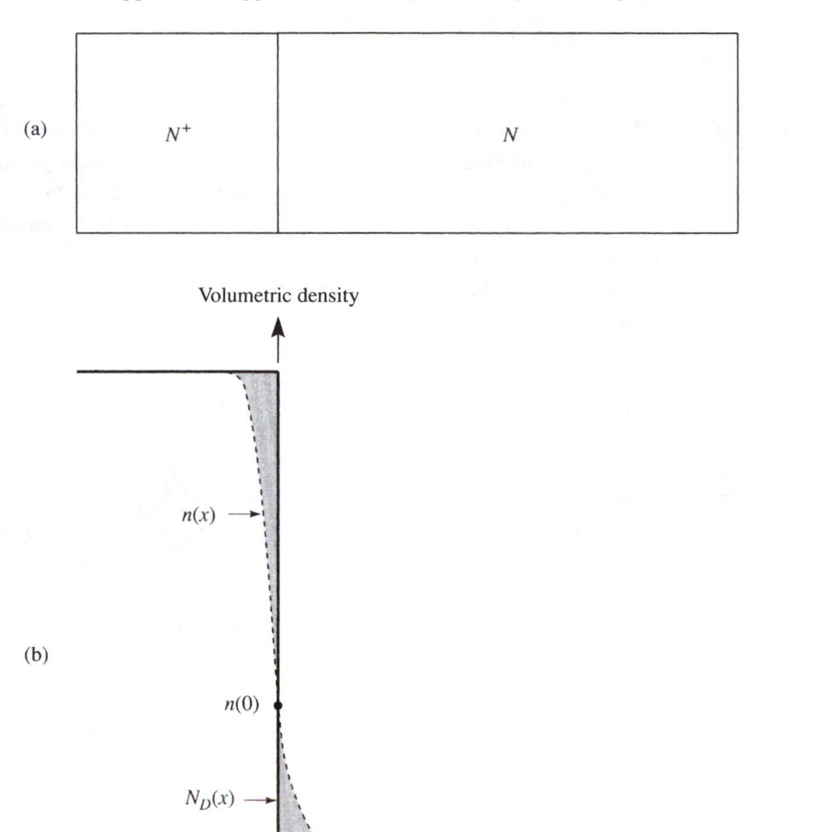

Figure F9 N^+N high–low junction. (a) Physical representation. (b) Net-doping profile $N_D(x)$ and electron profile $n(x)$.

where N_D and N_A are net-doping values on the two sides. For the high-low junction

$$\Delta\psi_0 = \frac{kT}{q} \ln \frac{N_H}{N_L} = 0.24 \text{ V},$$

where N_H and N_L are the net-doping values on the high and low sides, respectively.

EXERCISE F–7 In each case just given, the high-side potential drop will be about one thermal voltage, and the low side has a thin, dense layer of carriers concentrated adjacent to the metallurgical junction. What accounts for the difference in $\Delta\psi_0$ values?

■ In the *PN* junction, the low side exhibits a depletion layer that is *very* thick compared to the inversion layer. *Most* of the charge on the P-type side is in the inversion layer, and only a small fraction is in the ionic (depletion) layer. But charge placement has a major effect on voltage, with the long lines of force reaching out to the ionic charges leading to a large value for the field integral. In the high-low junction, all the low-side charge is in the accumulation layer and there is *no* depletion layer on the lightly doped side, so the contact potential is smaller—almost three times smaller in this example.

Now refer to Figure F10, which represents a one-sided *PN* junction with a low-side doping of $10^{15}/\text{cm}^3$. It can be seen that the electron density in the *P*-type region at equilibrium rises gradually as the junction is approached. But when it reaches the point at which

$$n_{0P}(x) = p_{0P}, \tag{F38}$$

it exhibits sharply increased slope. Equation F38 defines a condition known as the *threshold of strong inversion.* The corresponding position is sometimes termed the

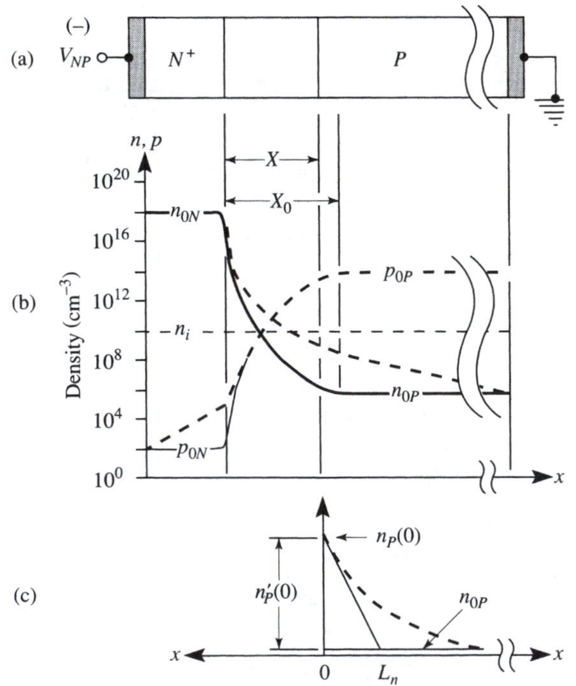

Figure F10 One-sided (N^+P) step junction under forward bias. (a) Physical representation of the junction sample. (b) Log-linear carrier-density profiles. (c) Linear presentation of the injected-carrier profile in the neutral *P*-type region.

threshold-of-inversion plane. It is evident in Figure F10 and is explicitly labeled in Figure F8. Other ways of stating the threshold condition are (assuming it is on the right-hand side of Figure F2),

$$U = -U_{02}, \tag{F39}$$

and equivalently,

$$W = 2U_{02}. \tag{F40}$$

It is because an inversion layer is a thin, dense sheet (nearly) of charge that electric field departs from the universal curve in Figure F3 at the threshold plane. Recall that Poisson's equation dictates a field slope proportional to charge density. Refer to Figure F4 to observe that the departure point of the inversion-layer curve is at $W = 20$ for $U_{02} = 10$, at $W = 40$ for $U_{02} = 20$, and so forth, consistent with Equation F40. The inversion-layer representation in Figure F8 does not appear steep because the positional scale has been grossly expanded. A quantitative look at this will be provided in an analytic problem.

Inversion layers also lend themselves to approximate-analytic modeling [F15, F16]. The expressions are more complex than those above, but are proving useful [F17]. These approximate-analytic functions describe the family of curves in Figure F4 to the left of the curve for $U_{02} = 0$. This "dividing-line" case of course represents a sample that is extrinsic on the left and intrinsic on the right, a special case that can be treated exactly [F18]. Moving to the right of the $U_{02} = 0$ curve brings us into the "accumulation set," or the case of a high–low junction that therefore incorporates an accumulation layer. Because the phenomenon of depletion is now no longer possible, these cases quickly converge on another universal curve. That is, the indicated curve is valid for any right-hand side bulk potential $|U_{02}| > {\sim}2$.

Thus, an inversion layer that exists on the lightly doped side of a grossly asymmetric junction is a thin, dense layer of carriers of the type that constitute majority carriers on the heavily doped side. It exists within a thicker layer of ionic charge of the same polarity. In a high–low junction, there is an accumulation layer on the low side, while the ionic charge (depletion layer) is totally absent there.

Finally, let us note that the existence of simple analytic expressions describing the "universal" portions of functions such as potential and electric field versus position (a universality achieved by appropriate normalization) is a consequence of the prudent choice of spatial origin. To illustrate the nontrivial nature of origin selection, we can point out that the essence of the shift from an obscure Ptolemaic model of the solar system to the simple and elegant Copernican model was the matter of spatial-origin choice. We offer this observation here, even at the risk of making too grand a comparison.

References

F1. W. Shockley, "The Theory of *p-n* Junctions in Semiconductors and *p-n* Junction Transistors," *Bell Syst. Tech. J.* **28,** 435 (1949).

F2. W. L. Brown, "*n*-Type Surface Conductivity on *p*-Type Germanium," *Phys. Rev.* **91,** 518 (1953).
F3. C. G. B. Garrett and W. H. Brattain, "Physical Theory of Semiconductor Surfaces," *Phys. Rev.* **99,** 376 (1955).
F4. R. H. Kingston and S. F. Neustadter, "Calculation of the Space Charge, Electric Field, and Free Carrier Concentration at the Surface of a Semiconductor," *J. Appl. Phys.* **26,** 718 (1955).
F5. P. Debye and E. Hückel, "Zur Theorie der Elektrolyte," *Z. Phys.* **24,** 185 (1923).
F6. P. Debye and E. Hückel, "Zur Theorie der Elektrolyte II," *Z. Phys.* **24,** 305 (1923).
F7. R. M. Warner, Jr., "Normalization in Semiconductor Problems," *Solid-State Electron.* **28,** 529 (1985).
F8. Ph. Passau and M. van Styvendael, "La Jonction *p-n* Abrupte dans le Cas Statique," *Proc. Int. Conf. Solid State Phys. in Electron. and Telecomm.*, Brussels, June 2–7, 1958, M. Desirant and J. F. Miciels (Eds.), *Int. Un. Pure and Appl. Phys.*, Vol. I, Academic Press, New York, 1960, p. 407.
F9. C. Goldberg, "Space Charge Regions in Semiconductors," *Solid-State Electron.* **7,** 593 (1964).
F10. R. P. Jindal and R. M. Warner, Jr., "A General Solution for Step Junctions with Infinite Extrinsic End Regions at Equilibrium," *IEEE Trans. Electron Devices* **28,** 348 (1981).
F11. R. M. Warner, Jr., R. P. Jindal, and B. L. Grung, "Field and Related Semiconductor-Surface and Equilibrium-Step-Junction Variables in Terms of the General Solution," *IEEE Trans. Electron Devices* **31,** 994 (1984).
F12. R. M. Warner, Jr., and B. L. Grung, *Transistors: Fundamentals for the Integrated-Circuit Engineer*, Wiley, New York, 1983; reprint edition, Krieger Publishing Company (P.O. Box 9542, Melbourne, FL 32902-9542), 1990, p. 360.
F13. R. M. Warner, Jr., and R. P. Jindal, "Replacing the Depletion Approximation," *Solid-State Electron.* **26,** 335 (1983).
F14. Warner and Grung, p. 346 and p. 697.
F15. J. R. Hauser and M. A. Littlejohn, "Approximations for Accumulation and Inversion Space-Charge Layers in Semiconductors," *Solid-State Electron.* **11,** 667 (1968).
F16. D.-H. Ju and R. M. Warner, Jr., "Modeling the Inversion Layer at Equilibrium," *Solid-State Electron.* **27,** 907 (1984).
F17. R. D. Schrimpf and R. M. Warner, Jr., "A Precise Scaling Length for Depleted Regions," *Solid-State Electron.* **28,** 779 (1985).
F18. R. P. Jindal, "Bulk and Surface Effects on Noise and Signal Behaviour of Semiconductor Devices," Ph.D. Thesis, University of Minnesota, Minneapolis, MN, 1981.

Appendix G
Depletion-Layer Capacitance of a *PN* Junction

An interface lying between a *P*-type region and an *N*-type region in a semiconductor single crystal is termed a *PN junction.* No matter how gradual the transition from *P*-type to *N*-type material, the junction surface is always sharply defined, the surface where net doping $(N_D - N_A)$ vanishes. To designate this mathematical surface in a specific way, the term *metallurgical junction* is often used, since *junction* frequently connotes the region of finite thickness that encompasses the space-charge or transition regions flanking the metallurgical junction, amounting typically to a few tenths of a micrometer overall. Within this still-thin layer, electron density on the *N*-type side makes a transition through many orders of magnitude to its low minority density on the *P*-type side, while hole density undergoes an analogous decline in the opposite direction.

Accompanying these extremely high-density gradients are electron and hole particle diffusion currents in opposite directions. Also, the density transitions "uncover" fixed ionic charge in the vicinity of the metallurgical junction, giving rise to space charge, of opposite sign on either side of the mathematical surface. In turn, the charges give rise to an internal electric field, and it in turn sets up particle drift currents of holes and electrons in opposite directions, precisely balancing the aforementioned diffusion currents at every position—a case of dynamic equilibrium. It is important to understand that the balance here is not a hole–electron balance, but rather one of hole diffusion against hole drift, with an analogous balance for electrons.

Integrating through the thin space-charge layer yields a voltage often termed the *built-in voltage*, which for a silicon junction is typically about 0.7 V. Applying a forward bias to a junction has the effect of reducing the magnitude of this voltage barrier and its accompanying electric field, upsetting the balance that existed in the equilibrium (zero-bias) case. Hence the diffusion currents now dominate, causing electrons to pour into the *P*-type side and vice versa, a phenomenon termed *minority-carrier storage*. In analogy to the familiar capacitor, such storage is used to define *diffusion capacitance*. While important in the context of the *PN*-junction diode, diffusion capacitance is not relevant in the MOS capacitor since the gate dielectric prohibits the conduction necessary for it. But we introduce the topic nonetheless because understanding the similarities and differences of the MOS and *PN*-junction "capacitors" is very important.

A phenomenon that the MOS and *PN*-junction devices share is *depletion-layer capacitance*, and to begin, it is helpful to compare the latter device with an ordinary parallel-plate capacitor. The primary difference is that in the conventional capacitor, the charge resides in planes of roughly atomic-dimension thickness in the metallic plates, while in *depletion-layer capacitance* double-layer charge is spread through regions that are thicker by many orders of magnitude, yet still typically a fraction of a micrometer overall, as noted above.

In the conventional capacitor governed by the law $Q = CV$, the quantity C is voltage-independent. Because C is constant, the capacitor is a *linear* device, and as a result, $(Q/V) = (dQ/dV)$. The reason that C is constant is that the planes of stored charge have a constant, voltage-independent spacing. Depletion-layer capacitance is by contrast nonconstant, and the corresponding "device" is thus nonlinear. The voltage dependence of depletion-layer thickness is the underlying physical reason for these facts.

We shall use the symbols C_t for depletion-layer capacitance and dQ_T for the relevant charge, exploiting the convenient subscript provided by the equivalent term *transition-region capacitance*. Using the small-signal technique for linearization, we write

$$C_t = \frac{dQ_T}{dV_{NP}}. \tag{G1}$$

The next step is to examine the stored charge Q_T. Because the parallel-plate capacitor is typically a symmetric device, let us first examine the symmetric step junction in order to enhance the similarity. And for simplest illustration, let us use the depletion approximation, which assumes that each of the two space-charge layers has a fully abrupt boundary with its corresponding adjacent neutral region. It will be shown later that this simplification can be dropped with no alteration of principles. Figure G1 depicts the sample to be analyzed, showing its depletion layer of sharply defined thickness X. Now note an elementary and important fact. When the charge on a capacitor is cited, what is meant is the charge on *one capacitor plate only.* The algebraic sum of the charges on the two plates is usually zero. Therefore, let us arbitrarily consider only the positive charge stored on the N-type side of the junction. It is evident that this charge can be written

$$Q_T = AqN\left(\frac{X}{2}\right). \tag{G2}$$

Combining this expression with the appropriate expression for depletion-layer thickness,

$$X = \left[\frac{4\epsilon(V_{NP} + \Delta\psi_0)}{qN}\right]^{1/2}, \tag{G3}$$

we have

$$Q_T = A\ [\epsilon qN(V_{NP} + \Delta\psi_0)]^{1/2}. \tag{G4}$$

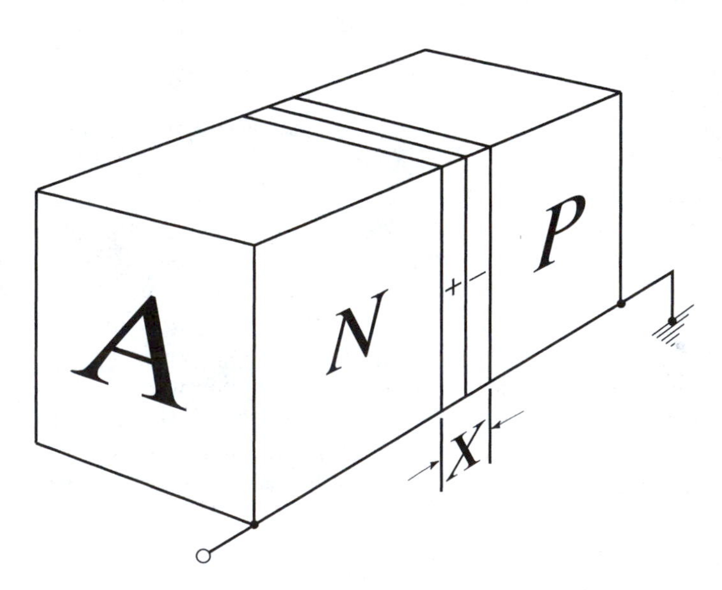

Figure G1 Sample chosen to illustrate calculation of depletion-layer capacitance.

Comparing this expression to the corresponding law for the conventional capacitor, $Q = CV$, we see the additional complexity associated with the nonlinear junction case. Equation G4 is plotted in Figure G2(a). While we could define a total capacitance for the junction at a point such as P_1 as the quotient Q_T/V_{NP} (the slope of the line from the point P_1 to the origin), it is a quantity that is rarely useful. On the other hand, C_t, the small-signal or incremental capacitance at the same point (represented by the slope of the tangent at P_1), has appreciable utility.

Inspection of Equation G4 shows that the curve in Figure G2(a) is a parabola with its vertex at point P_2. This point corresponds to a forward voltage (negative here) equal to the contact potential $\Delta\psi_0$. This point is not physically reachable because attempting to reduce the junction barrier to zero height causes enormous currents that destroy the device. The Q_T-axis intercept Q_{0T}, by contrast, is reachable and is of course a point corresponding to the device at equilibrium. The value of Q_{0T} is readily obtained from Equation G4 by setting V_{NP} equal to zero, and is the total charge "stored" on the capacitor at zero bias. However, SPICE analysis chooses to place the origin of the Q_T–V_{NP} plane at the point (Q_{0T}, 0) in Figure G2(a).

The desired symmetric-junction expression for C_t as a function of voltage can be obtained by differentiating Equation G4:

$$C_t = \frac{d}{dV_{NP}}\left[A\sqrt{\epsilon q N(V_{NP} + \Delta\psi_0)}\right] = \frac{A\sqrt{\epsilon q N}}{\qquad\qquad}. \tag{G5}$$

This expression is given in Table 1, and is plotted as the solid curve in Figure G2(b). It is a power-law relationship, but *not* a parabola. In fact, C_t does not increase without limit (as Equation G5 indicates) because the depletion approximation worsens with forward voltage, and fails for large values.

Observe that C_t is dominant for reverse bias and for small forward bias, a range wherein the depletion approximation is acceptable. The exponential dependence of C_s on forward bias causes it to sweep past C_t with its power-law dependence, and to dominate from there on, as can be seen (dashed curve) in Figure G2(b). For a typical junction this occurs in the upper half of the low-level forward-bias range, and well below the upper limit of that range.

Interestingly, through normalization it is possible to construct a universal curve for C_t as a function of total potential difference between the two sides of the junction. It is valid for any step junction. For this purpose we choose the most general kind of step junction, the asymmetric case, whose C_t expression is given in Table 1 along with expressions for other cases. Multiplying numerator and denominator of this expression, inside the radical, by kT/q gives us C_t in terms of normalized voltage:

$$C_t = A_q\left[\frac{\epsilon/kT}{(1/N_1) + (1/N_2)}\right]^{1/2} (\Delta U_0 + U_{NP})^{-1/2}. \tag{G6}$$

With the definition

$$K_t = Aq\left[\frac{\epsilon/kT}{(1/N_1) + (1/N_2)}\right]^{1/2}, \tag{G7}$$

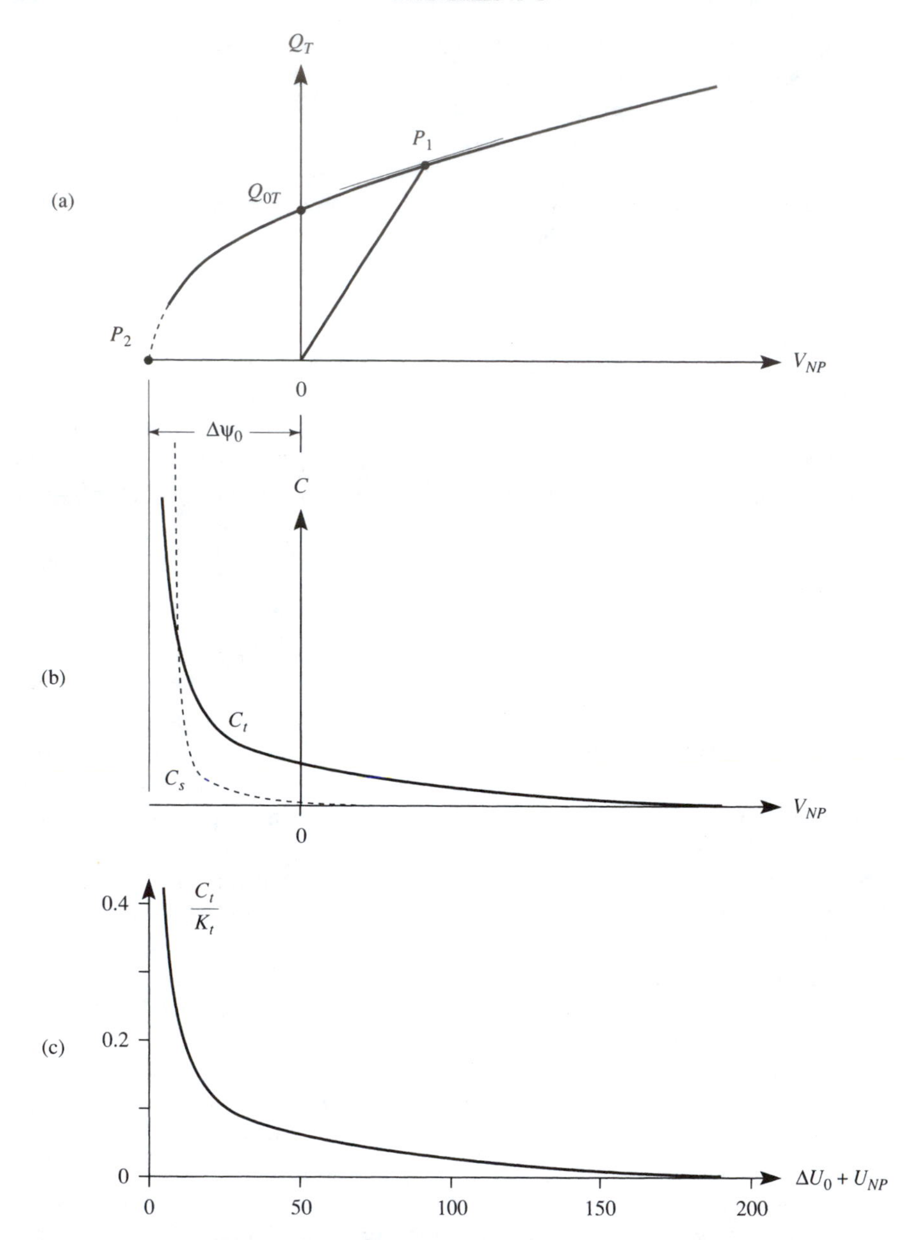

Figure G2 Functions related to depletion-layer capacitance C_t. (a) Total charge Q_T versus applied voltage V_{NP} for a step-junction sample. (b) Comparing depletion-layer capacitance C_t versus voltage (solid line) to diffusion capacitance C_s versus voltage (dashed line). (c) Normalized depletion-layer capacitance versus normalized potential difference through the junction, valid for any step junction.

Equation G6 becomes

$$\frac{C_t}{K_t} = (\Delta U_0 + U_{NP})^{-1/2}, \tag{G8}$$

which is plotted in Figure G2(c).

Now return to Equation G5, which gives C_t for the symmetric step junction. Notice that it can be rewritten with the aid of Equation G1 as

$$C_t = \frac{A\epsilon}{\sqrt{4\epsilon(V_{NP} + \Delta\psi_0)/qN}} = \frac{A\epsilon}{X}. \tag{G9}$$

Comparing this expression with the analogous expression for the ordinary capacitor,

$$C = \frac{A\epsilon}{d}, \tag{G10}$$

where d is plate spacing, leads to a seeming paradox. Why should *total* depletion-layer thickness X correspond to the plate spacing of the ordinary capacitor in view of the fact that the charge in the junction is distributed through the entire volume of thickness X? The matter is resolved by noting that a small-signal voltage variation superimposed on a fixed dc bias causes the storage and recovery of charge at the *boundaries* of the depletion layer. The physical situation is depicted in Figure G3(a), which shows the $\rho_v(x)$ profile for a symmetric junction in depletion-approximation terms: To the charge Q_T stored on one side, an incremental charge dQ_T has been added by the application of an increment of voltage. (The incremental charge may be positive or negative, depending on the sign of the voltage increment.) The main point is that the physical locations of the equal and opposite charge increments are spaced apart by the dimension X, while in a conventional capacitor the spacing is d, thus explaining the congruence of Equations G9 and G10.

The mechanism of charge storage at the boundary of a depletion layer has a binary interpretation. To be specific, again focus on the positive ionic charge Q_T stored on the N-type side of the sample. A positive voltage increment dV_{NP} will cause depletion-layer expansion and the storage of a positive increment dQ_T. This positive increment can be viewed as an incremental layer of donor ions "uncovered" by the application of dV_{NP}. Alternatively, it can be viewed as a *reduction* in the *negative* charge present on the N-type side, because majority electrons are withdrawn from there when dV_{NP} is applied.

Now turn to the more realistic depiction of the $\rho_v(x)$ profile in Figure G3(b). The charge increment dQ_T is now distributed through an appreciably greater volume than is acknowledged in the depletion-approximation representation, Figure G3(a). But because the true distribution of dQ_T has its center very nearly at the position defined by the depletion approximation, Equation G9 constitutes an excellent approximate expression for depletion-layer capacitance. Experimental observations have shown repeatedly that the depletion-approximation expression for C_t is in error by no more than a few percent for reverse bias and low forward bias.

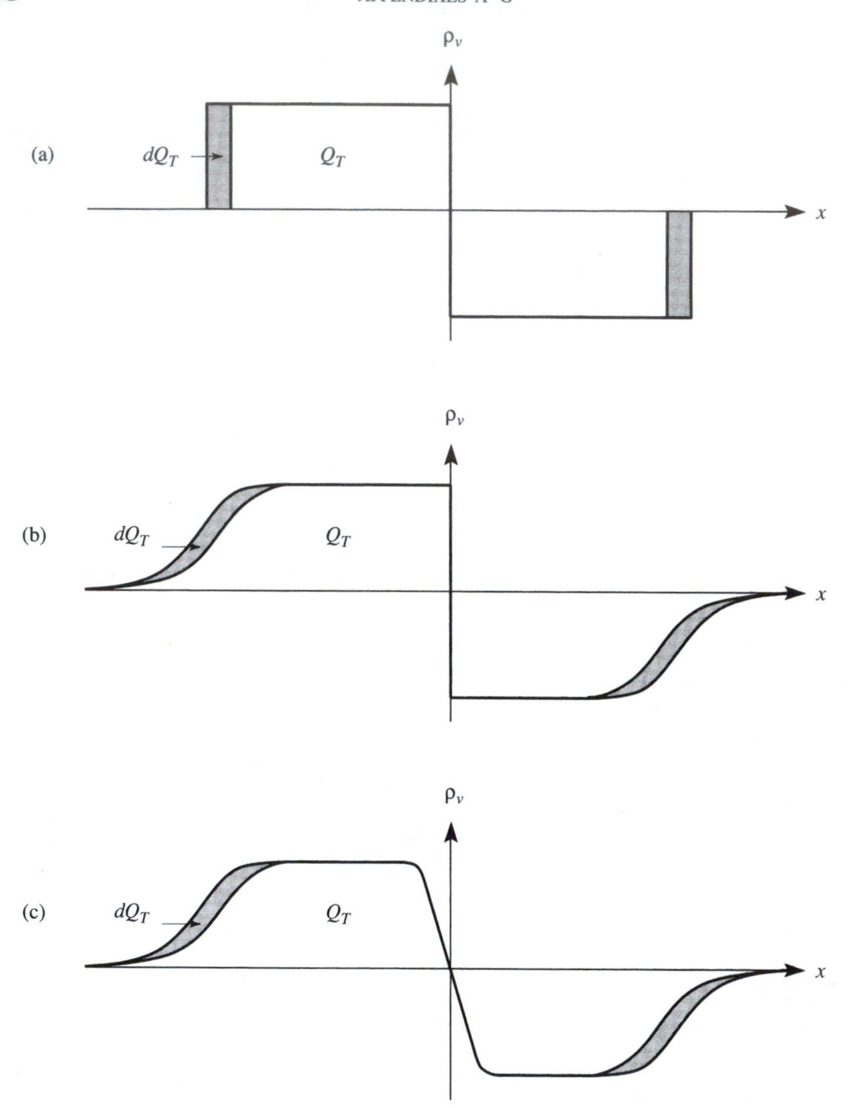

Figure G3 Charge-density profiles for symmetric step junctions. (a) Total and incremental charges for an ideally abrupt junction with the depletion approximation assumed. (b) Charges for an ideally abrupt junction with realistic charge profiles. (c) Charges for a realistically less-than-abrupt junction with realistic charge profiles.

While today's technology permits one to fabricate step junctions of near-ideal abruptness in the doping profile, it is nonetheless commonplace to observe a measure of "grading" or gradualness in the doping transition. Figure G3(c) illustrates the effect of such a departure from the ideal doping profile on the charge-density profile $\rho_v(x)$. Again, the effect on the positions of the increments dQ_T at a given bias voltage is small, and the effect on the field profile is simply to round the peak a bit. Thus the capacitance expression derived from the apparently simplistic depletion approximation is very useful in the real world through a wide voltage range.

Tables

Table 1. Depletion-Approximation Equations for Junctions*

Junction Type	Depletion-Layer Thickness	Peak (Maximum) Field	Depletion-Layer Capacitance
1. Symmetric step	$X = \left[\frac{4\epsilon(V_{NP} + \Delta\psi_0)}{qN}\right]^{1/2}$	$E_M = \left[\frac{qN}{\epsilon}(V_{NP} + \Delta\psi_0)\right]^{1/2}$	$C_T = \frac{A}{2}\left[\frac{\epsilon qN}{V_{NP} + \Delta\psi_0}\right]^{1/2}$
2. Asymmetric step	$X = \left[\frac{2\epsilon(V_{NP} + \Delta\psi_0)}{q}\left(\frac{1}{N_1} + \frac{1}{N_2}\right)\right]^{1/2}$	$E_M = \left[\frac{2q(V_{NP} + \Delta\psi_0)}{\epsilon[(1/N_1) + (1/N_2)]}\right]^{1/2}$	$C_T = A\left[\frac{\epsilon q}{2(V_{NP} + \Delta\psi_0)[(1/N_1) + (1/N_2)]}\right]^{1/2}$
	$X_1 = \left[\frac{2\epsilon}{q}\frac{N_2}{N_1}\left(\frac{V_{NP} + \Delta\psi_0}{N_1 + N_2}\right)\right]^{1/2}$		
	$X_2 = \left[\frac{2\epsilon}{q}\frac{N_1}{N_2}\left(\frac{V_{NP} + \Delta\psi_0}{N_1 + N_2}\right)\right]^{1/2}$		
3. One-sided step (N_2 is light-side doping)	$X = \left[\frac{2\epsilon(V_{NP} + \Delta\psi_0)}{qN_2}\right]^{1/2}$	$E_M = \left[\frac{2qN_2}{\epsilon}(V_{NP} + \Delta\psi_0)\right]^{1/2}$	$C_T = A\left[\frac{\epsilon qN_2}{2(V_{NP} + \Delta\psi_0)}\right]^{1/2}$
4. P^+IN^+	$X =$ constant	$E_M = \frac{V_{NP} + \Delta\psi_0}{X}$	$C_T = \frac{A\epsilon}{X}$
5. Linearly graded ($a \equiv$ gradient)	$X = \left[\frac{12\epsilon}{qa}(V_{NP} + \Delta\psi_0)\right]^{1/3}$	$E_M = \frac{qaX_0^2}{8\epsilon}$	$C_T = A\left[\frac{\epsilon^2 qa}{12(V_{NP} + \Delta\psi_0)}\right]^{1/3}$

*For cases 1–4, the contact potential (built-in voltage) is given by $\Delta\psi_0 = (kT/q)\ln(N_1N_2/n_i^2)$, and for case 5, $\Delta\psi_0 = (qa/\epsilon)(X_0^3/12) = (kT/q)\ln(aX_0/2n_i)^2$, which requires iterative solution.

Table 2. Experimentally Determined Ionization Energies for Various Impurities in Silicon*

Periodic Column							
I	**II**	**III**	**V**	**VI**	**VII**	**VIII**	**Rare Earths**
Ionization Energy Measured from Conduction-Band Edge							
Li(D)0.033	Mg(D)0.11		Sb(D)0.039	S(D)0.11	Mn(D)0.53	Ni(A)0.35	Na(D)0.25
Ag(A)0.22	Mg(D)0.25		N(D)0.04	Te(D)0.14		Co(A)0.53	Tm(D)0.29
Au(A)0.54	Hg(A)0.31		P(D)0.044	O(D)0.16		Fe(D)0.55	
	Hg(A)0.36		As(D)0.049	S(D)0.18			
	Zn(A)0.55		Bi(D)0.069	W(A)0.22			
				W(A)0.30			
				Mo(D)0.33			
				S(D)0.37			
				W(A)0.37			
				O(A)0.38			
				S(D)0.61			
Ionization Energy Measured from Valence-Band Edge							
		Tl(A)0.26		O(A)0.35		Fe(D)0.40	
Cu(A)0.49		In(A)0.16		W(D)0.35		Pt(D)0.37	
Au(D)0.35	Hg(D)0.33	Ga(A)0.065		Mo(D)0.34		Co(A)0.35	
Ag(D)0.32	Zn(A)0.31	Al(A)0.057		W(D)0.31		Pt(D)0.31	
Cu(D)0.24	Hg(D)0.25	B(A)0.045		Mo(D)0.30		Ni(A)0.23	

*The symbols (D) and (A) designate donors and acceptors, respectively. (After Neuberger and Welles, *Silicon*, Electronic Properties Information Center, Hughes Aircraft Company, Culver City, CA, October 1969, with permission.)

Table 3. Approximate Expressions for Normalized Electric Field $|dW/d(x/L_D)|$ as a Function of Normalized Potential W*

Range of $\lvert U_B \rvert$	Range of W	Approximate Expression	Maximum Error within Indicated Ranges of $\lvert U_B \rvert$, W
A. For depletion and inversion			
2 to ∞	0 to ∞	$\sqrt{2[(e^{-W}+W-1)+e^{-2\lvert U_B\rvert}(e^{W}-W-1)]}$	0.91%
A′. At threshold			
2 to ∞	$W=2\lvert U_B\rvert$	$2\sqrt{\lvert U_B\rvert(1-e^{-2\lvert U_B\rvert})}$	0.91%
B. For depletion and inversion			
3 to ∞	0.5 to ∞	$\sqrt{2(e^{-W}+W-1+e^{W-2\lvert U_B\rvert})}$	1.86%
B′. At threshold			
2 to ∞	$W=2\lvert U_B\rvert$	$2\sqrt{\lvert U_B\rvert}$	1.85%
C. For depletion and inversion			
2 to ∞	3 to ∞	$\sqrt{2(W-1+e^{W-2\lvert U_B\rvert})}$	1.42%
D. For extreme inversion			
3 to ∞	$4\lvert U_B\rvert$ to ∞	$2e^{(W/2)-\lvert U_B\rvert}$	1.22%
E. For depletion (ionic component only)			
3 to ∞	0 to $(2\lvert U_B\rvert-2)$	$\sqrt{2(e^{-W}+W-1)}$	0.13%
F. For depletion (ionic component only)			
4 to ∞	3 to $(2\lvert U_B\rvert-2)$	$\sqrt{2(W-1)}$	1.21%
G. For depletion or accumulation			
0 to ∞	0 to 0.1	W	0.042%
H. For accumulation			
2 to ∞	0 to ∞	$\sqrt{2(e^{W}-W-1)}$	1.2%
J. For accumulation			
2 to ∞	6 to ∞	$\sqrt{2e^{W/2}}$	1.7%

*Expressions are accurate within 2% in indicated ranges. Correct algebraic sign can be incorporated herein by using the coefficient $-U_B/\lvert U_B\rvert$, where U_B is normalized bulk potential in the semiconductor portion of the MOS capacitor (using Fermi level as reference). A range limit of ∞ is intended to indicate the largest value that does not violate underlying approximations and assumptions.

Table 4. Exact Expressions for Normalized Electric Field $|dW/d(x/L_D)|$ as a Function of Normalized Potential W*

For depletion and inversion	$\sqrt{2\left[\dfrac{e^{\lvert U_B\rvert}(e^{-W}+W-1)+e^{-\lvert U_B\rvert}(e^{W}-W-1)}{e^{\lvert U_B\rvert}+e^{-\lvert U_B\rvert}}\right]}$
For accumulation	$\sqrt{2\left[\dfrac{e^{-\lvert U_B\rvert}(e^{-W}+W-1)+e^{\lvert U_B\rvert}(e^{W}-W-1)}{e^{-\lvert U_B\rvert}+e^{\lvert U_B\rvert}}\right]}$
For right-hand side intrinsic	$\sqrt{e^{W}+e^{-W}-2}$

*Correct algebraic sign can be incorporated herein by using the coefficient $-U_B/|U_B|$, where U_B is normalized bulk potential in the semiconductor portion of the MOS capacitor (using Fermi level as reference) [55, 56].

Table 5. Symbol Conversion Chart for Basic and Physical Quantities in SPICE Level-2 MOSFET Model

SPICE Symbol	SPICE Description	Default Value	Unit	Present Symbol	Present Description
	Basic set: X_{OX}, μ_n, U_{EXP}, E_{CRIT}, L_D^*, C_n/q, λ				
TOX	Thin oxide thickness	1.0	m	X_{OX}	Oxide thickness
UO	Surface mobility	600	$cm^2/V{\cdot}s$	μ_n	Surface mobility
UEXP	Exponential coefficient for mobility	0.0	—	U_{EXP}	Exponent for mobility adjustment
UCRIT	Critical electric field for mobility	10^4	V/cm	E_{CRIT}	Critical electric field for mobility adjustment
LD	Lateral diffusion distance	0.0	m	L_D^*	Lateral diffusion distance
NFS	Surface-fast state density	0.0	$F/C{\cdot}cm^2$	C_n/q	Subthreshold fitting parameter
LAMBDA	Channel-length modulation	0.0	V^{-1}	λ	Channel-length modulation
	Physical set: N_A, T_G, N_f^*, R_{SH}, J_0				
NSUB	Substrate doping	0.0	cm^{-3}	N_A	*P*-substrate doping
TPG	Type of gate	1	—	T_G	Type of gate
NSS	Surface-state density	0.0	cm^{-2}	N_f^*	Effective oxide fixed-charge density
RSH	Source and drain diffusion sheet resistance	0.0	ohm/m^2	R_{SH}	Source and drain diffusion sheet resistance
JS	Bulk-junction reverse saturation current density	10^{-8}	A/m^2	J_0	Saturation-current density

Table 6. Symbol Conversion Chart for Capacitive Quantities in SPICE Level-2 MOSFET Model

SPICE Symbol	SPICE Description	Default Value	Unit	Present Symbol	Present Description
	Capacitance effects: $\Delta\psi_0$, $C_{0.B}$, M_B, $C_{0.SW}$, M_{SW}, C_{0GS}, C_{0GD}, C_{0GB}				
VJ	BS junction potential	0.8	V	$\Delta\psi_0$	BS contact potential
CJ	Zero-bias BS junction bottom capacitance	0.0	F/cm^2	$C_{0.B}$	Zero-bias BS junction-bottom depletion-layer capacitance
MJ	BS grading coefficient, bottom	0.5	—	M_B	BS grading exponent, bottom
CJSW	Zero-bias BS junction-sidewall capacitance	0.0	F/cm^2	$C_{0.SW}$	Zero-bias BS junction-sidewall depletion-layer capacitance
MJSW	BS grading coefficient, sidewall	0.5	—	M_{SW}	BS grading exponent, sidewall
CGSO	GS overlap capacitance per meter of channel width	0.0	F/m	C_{0GS}	GS overlap capacitance per meter of channel width
CGDO	GD overlap capacitance per meter of channel width	0.0	F/m	C_{0GD}	GD overlap capacitance per meter of channel width
CGBO	GB overlap capacitance per meter of channel length	0.0	F/m	C_{0GB}	GB extrinsic capacitance per meter of channel length

Table 7. Representative Set of SPICE Parameter Values

Parameter	Value	Parameter	Value
X_{OX}	1.139×10^{-7} m	N_A	10^{16} cm^{-3}
μ_n	879.31 cm/s	T_G	1
U_{EXP}	0.26742	N_f^*	6.5279×10^{10}/cm^2
E_{CRIT}	10,000 cm/s	R_{SH}	80 ohm
L_D^*	0.175×10^{-6} m	J_0	3.125×10^{-7} A/m^2
C_N	10^{11} F/cm^2	C_{0GS}	400×10^{-12} F/m
λ	0.022980/V	C_{0GD}	400×10^{-12} F/m
$\Delta\psi_0$	0.8 V	C_{0GB}	200×10^{-12} F/m
M_B	0.5	$C_{0.B}$	3.125×10^{-4} F/m^2
M_{SW}	0.5	$C_{0.SW}$	2.0×10^{-10} F/m
Z	20×10^{-6} m	L	4×10^{-6} m
N_{RS}	0.5	N_{RD}	0.5
A_S	320×10^{-12} m^2	A_D	320×10^{-12} m^2
P_S	50×10^{-6} m	P_D	50×10^{-6} m

Table 8. Symbol Conversion Chart for Electrical Quantities in SPICE Level-2 MOSFET Model

SPICE Symbol	SPICE Description	Default Value	Unit	Present Symbol	Present Description
		Basic set: V_T, K, γ, ψ_B, R_S, R_D, I_0			
VTO	Zero-bias threshold voltage	1.0	V	V_T	Threshold voltage
KP	Transconductance parameter	0.0	A/V^2	$2K$	Twice the current coefficient
GAMMA	Body-effect parameter	0.0	—	γ	Bulk-charge parameter
PHI	Surface inversion potential	0.6	V	$-2\psi_B$	Band bending at threshold
RS	Source ohmic resistance	0.0	ohm	R_S	Source ohmic resistance
RD	Drain ohmic resistance	0.0	ohm	R_D	Drain ohmic resistance
IS	Bulk–junction saturation current	10^{-14}	A	I_0	Bulk–source–junction saturation current

Table 9. Equivalent Sets of SPICE Parameter Values

Physical Parameter	Value	Electrical Parameter	Value
N_A	10^{16} cm^{-3}	V_T	1.03 V
T_G	1	$2KL^*/Z$	2.67×10^{-5} A/V^2
N_f^*	6.5279×10^{10}/cm^2	γ $-2\psi_B$	1.90 V$^{1/2}$ 0.695 V
R_{SH}	80 ohm	R_S and R_D	40 ohm
J_0	3.125×10^{-7} A/m^2	I_0	10^{-16} A

Table 10. Operating-Point Information Corresponding to the SPICE Parameter Set in Table 9

$I_D = 0.055$ mA	$g_m = 0.069$ mS
$V_{GS.P} = +2.471$ V	$g_o = 0.001$ mS
$V_{BS.P} = +0.000$ V	$g_{mbs} = 0.062$ mS
$V_{DS.P} = +2.471$ V	$f_T = 0.5$ GHz
	$A_V = 69$

Table 11. Calculated dc Values, Numerical Example

Symbol	Unit	*A* (OFF)	*B* (Bias)	*C* (ON)
$V_{GS.P}$	V	0.356	2.471	5.000
$V_{BS.P}$	V	0.000	0.000	0.000
$V_{DS.P}$	V	5.000	2.471	0.356
I_D	mA	3.90×10^{-9}	0.055	0.096
I_G	mA	0.0	0.0	0.0
I_B	mA	0.0	0.0	0.0
V_{GS}	V	0.356	2.469	4.996
V_{BS}	V	-1.55×10^{-10}	−0.002	−0.004
V_{DS}	V	5.000	2.467	0.348
C_{gs}	fF	8.000	22.754	19.658
q_{gs}	fC	0.0	0.0	0.0
C_{gd}	fF	8.000	8.000	18.422
q_{gd}	fC	0.0	0.0	0.0
C_{gb}	fF	22.862	0.730	0.730
q_{gb}	fC	0.0	0.0	0.0
C_{bs}	fF	110.00	109.84	109.73
q_{bs}	fC	-1.70×10^{-8}	−0.244	−0.424
C_{bd}	fF	40.853	54.415	91.667
q_{bd}	fC	-2.98×10^{2}	-1.80×10^{2}	−35.200
$g_m \approx g_{m.P}$	mS	5.62×10^{-8}	0.069	0.020
$g_o \approx g_{o.P}$	mS	1.01×10^{-10}	0.001	0.253
$g_{mbs} \approx g_{mbs.P}$	mS	4.85×10^{-8}	0.062	0.020

Table 12. Calculated Transient Values, Large-Signal Numerical Example

Symbol	Unit	Particular Times $t^{m=1}$ (1)	$t^{m=2}$ (2)	$t^{m=3}$ (3)	$t^{m=4}$ (4)
t	ns	10.0	17.68	20.0	91.0
$v_{GS.P}(t)$	V	0.356	3.924	5.000	5.000
$v_{BS.P}(t)$	V	0.000	0.000	0.000	0.000
$v_{DS.P}(t)$	V	5.000	2.355	0.536	0.356
$i_D(t)$	mA	-8.47×10^{-7}	0.119	0.108	0.096
$i_G(t)$	mA	1.39×10^{-7}	0.023	0.022	5.43×10^{-5}
$i_B(t)$	mA	7.12×10^{-7}	0.060	0.018	2.74×10^{-4}
$v_{GS}(t)$	V	0.356	3.916	4.994	4.996
$v_{BS}(t)$	V	-1.55×10^{-10}	-0.008	-0.006	-0.004
$v_{DS}(t)$	V	5.000	2.342	0.525	0.348
$g_m \approx g_{m.P}$	mS	5.62×10^{-8}	0.122	0.031	0.020
$g_o \approx g_{o.P}$	mS	1.01×10^{-10}	0.005	0.228	0.253
$g_{mbs} \approx g_{mbs.P}$	mS	4.85×10^{-8}	0.092	0.029	0.020
$C_{GS}(t)$	fF	8.000	22.754	20.008	19.658
$q_{GS}(t)$	fC	2.847	75.492	98.618	98.659
$C_{GD}(t)$	fF	8.000	8.000	17.985	18.422
$q_{GD}(t)$	fC	-37.153	12.596	49.977	53.246
$C_{GB}(t)$	fF	22.862	0.730	0.730	0.730
$q_{GB}(t)$	fC	8.135	19.513	20.298	20.298
$C_{BS}(t)$	fF	1.10×10^2	1.09×10^2	1.10×10^2	1.10×10^2
$q_{BS}(t)$	fC	-1.71×10^{-8}	-0.891	-0.652	-0.424
$C_{BD}(t)$	fF	40.853	55.437	85.269	91.667
$q_{BD}(t)$	fC	-2.98×10^2	-1.73×10^2	-51.045	-35.200
$i_D^R(t)$	mA	-3.88×10^{-9}	0.193	0.139	0.096
$i_G^{CS}(t)$	mA	-6.27×10^{-18}	0.011	0.009	2.56×10^{-5}
$i_G^{CD}(t)$	mA	1.39×10^{-7}	0.013	0.012	2.87×10^{-5}
$i_G^{CB}(t)$	mA	-2.83×10^{-18}	3.39×10^{-4}	3.39×10^{-4}	-9.95×10^{-17}
$i_B^{CD}(t)$	mA	7.12×10^{-7}	0.061	0.018	1.37×10^{-4}
$i_B^{CS}(t)$	mA	6.41×10^{-17}	-1.68×10^{-4}	1.16×10^{-4}	1.36×10^{-4}

Table 13. Representative Values Chosen for Evaluating the Quotient between BJT and MOSFET Transconductance-Normalizing Quantities in Section 6-4

$\frac{q}{kT}\frac{q}{\epsilon_{OX}} = 1.853 \times 10^{-5}$ cm	$\frac{A_E}{A_B} = 1$
$\frac{\mu_{nb}}{\mu_{ns}} = 3$	$n_{0B} = 10^3/\text{cm}^3$
$\frac{X_{OX}}{X_B} = \frac{1}{3}$	$L^2 = 10^{-8}$ cm^2

Table 14. Classification of SPICE MOSFET Models

Model Generation	Major Characteristics	Model	Major Developer	Standard References
First	Emphasis on analytical description of device behavior with a small number of model parameters	Level 1	U. of California/Berkeley	[74, 86–92]
		Level 2	U. of California/Berkeley	[74, 86–92]
		Level 3	U. of California/Berkeley	[74, 86–92]
Second	Emphasis on efficient circuit simulation with a large number of parameters that are difficult to extract	BSIM	U. of California/Berkeley	[86, 93]
		BSIM2	U. of California/Berkeley	[86, 94]
		Level13	MetaSoft	[92]
		Level28	MetaSoft	[86, 92]
Third	Emphasis on analytical description of device behavior and on efficient circuit simulation with a moderate number of parameters that are somewhat easy to extract	BSIM3	U. of California/Berkeley	[86, 95–96]
		Level49	MetaSoft	[86, 92]
		Level9	Philip Electronics	[86, 97]
		Power-Lane	Power-Lane	[86–98]
		PCIM	Digital Equipment Corp.	[86, 99]
		EKU	Enz-Krummenacher-Vittoz	[86, 100]

References

1. J. E. Lilienfeld, U.S. Patent 1,745,175, filed October 8, 1926 (October 22, 1925, in Canada), and issued January 28, 1930.
2. O. Heil, British Patent 439,457, issued 1939.
3. W. Shockley, *Electrons and Holes in Semiconductors*, Van Nostrand, New York, 1950, p. 30.
4. L. H. Hoddeson, "Multidisciplinary Research in Mission-Oriented Laboratories; The Evolution of Bell Laboratories' Program in Basic Solid-State Physics Culminating in the Discovery of the Transistor, 1935–1948," Department of Physics, University of Illinois at Urbana–Champaign, November 1978. Also, "The Discovery of the Point Contact Transistor," *Historical Studies in the Physical Sciences* **12,** 41 (1981).
5. J. Bardeen and W. H. Brattain, "The Transistor—A Semiconductor Triode," *Phys. Rev.* **74,** 230 (1948).
6. W. Shockley, U.S. Patent 2,569,347, filed June 26, 1948, and issued September 25, 1951.
7. R. M. Warner, Jr., and B. L. Grung, *Transistors: Fundamentals for the Integrated-Circuit Engineer*, Wiley, New York, 1983; reprint edition, Krieger Publishing Company (P. O. Box 9542, Melbourne, FL 32902-9542), 1990, p. 30.
8. W. Shockley, "A Unipolar Field-Effect Transistor," *Proc. IRE* **40,** 1365 (1952).
9. G. C. Dacey and I. M. Ross, "Unipolar Field-Effect Transistor," *Proc. IRE* **41,** 970 (1953).
10. H. Christensen and G. K. Teal, U.S. Patent 2,692,839, filed April 7, 1951, and issued October 26, 1954.
11. R. M. Warner, Jr., G. C. Onodera, and W. J. Corrigan, U.S. Patent 3,223,904, filed February 19, 1962, and issued December 14, 1965.
12. W. Shockley, U.S. Patent 2,787,564, filed October 28, 1954, and issued April 2, 1957.
13. R. H. Crawford, "Implanted Depletion Loads Boost MOS Array Performance," *Electronics* **45,** 85 (1972).
14. Warner and Grung, p. 64.
15. D. Kahng and M. M. Atalla, "Silicon–Silicon Dioxide Field Induced Surface Devices," *IRE-AIEE Solid-State Device Research Conference*, Carnegie Institute of Technology, Pittsburgh, PA, June 1960.
16. M. M. Atalla, E. Tannenbaum, and E. J. Scheibner, "Stabilization of Silicon Surfaces by Thermally Grown Oxides," *Bell Syst. Tech. J.* **38,** 749 (1959).
17. C. A. Mead, "Schottky Barrier Gate Field-Effect Transistor," *Proc. IEEE* **54,** 307 (1966).
18. W. W. Hooper and W. I. Lehrer, "An Epitaxial GaAs Field-Effect Transistor," *Proc. IEEE* **55,** 1237 (1967).
19. R. Dingle, H. L. Stormer, A. C. Gossard, and W. Wiegmann, "Electron Mobilities in

Modulation-Doped Semiconductor Heterojunction Superlattices," *Appl. Phys. Lett.* **37,** 805 (1978).

20. T. Mimura, S. Hizamizu, T. Fujii, and K. Nanbu, "A New Field Effect Transistor with Selectively Doped GaAs/n-$Al_xGa_{1-x}As$ Heterojunctions," *Jpn. J. Appl. Phys.* **19,** L225 (1980).
21. N. C. Cirillo, M. Shur, P. J. Vold, J. K. Abrokwah, R. R. Daniels, and O. N. Tufte, "Insulated Gate Field Effect Transistor," *IDEM Tech. Digest*, 317 (1985).
22. T. Mizutani, S. Fujita, and Y. Yanagawa, "Complementary Circuit with AlGaAs/GaAs Heterostructure MISFETs Employing High-Mobility Two-Dimensional Electron and Hole Gases," *Electron. Lett.* **21,** 1116 (1985).
23. M. Shur, J. K. Abrokwah, R. R. Daniels, and D. K. Arch, "Mobility Enhancement in Highly Doped GaAs Quantum Wells," *J. Appl. Phys.* **61,** 1643 (1987).
24. M. Sweeny, J. Xu, and M. Shur, "Hole Subbands in One-Dimensional Quantum-Well Wires," *Superlattices and Microstructures* **4,** 623 (1988).
25. F. G. Allen and G. W. Gobeli, "Work Function, Photoelectric Threshold, and Surface States of Atomically Clean Silicon," *Phys. Rev.* **127,** 150 (1962); "Comparison of the Photoelectric Properties of Cleaved, Heated, and Sputtered Silicon Surfaces," *J. Appl. Phys.* **35,** 597 (1964).
26. B. E. Deal and E. H. Snow, "Barrier Energies in Metal–Silicon Dioxide–Silicon Structures," *J. Phys. Chem. Solids* **27,** 1873 (1966).
27. W. M. Werner, "The Work Function Difference of the MOS-System with Aluminum Field Plates and Polycrystalline Silicon Field Plates," *Solid-State Electron.* **17,** 769 (1974).
28. J. W. Gibbs, "Thermodynamics," *Collected Works*, Vol. 1, Yale University Press, New Haven, CT, 1948, p. 219.
29. I. E. Tamm, "Über eine mögliche Art der Electronenbindung an Kristalloberflächen," *Phys. Z. Sowjetunion* **1,** 733 (1932).
30. W. Shockley, "On the Surface States Associated with a Periodic Potential," *Phys. Rev.* **56,** 317 (1939).
31. A. Many, Y. Goldstein, and N. B. Grover, *Semiconductor Surfaces*, North-Holland, Amsterdam, 1965, p. 165.
32. A. S. Grove, *Physics and Technology of Semiconductor Devices*, Wiley, New York, 1967, p. 144.
33. A. van der Ziel, "Flicker Noise in Electronic Devices," *Advances in Electronics and Electron Devices*, Vol. 49, Academic Press, New York, 1979, p. 225.
34. B. E. Deal, "Standardized Terminology for Oxide Charges Associated with Thermally Oxidized Silicon," *IEEE Trans. Electron Devices* **27,** 606 (1980).
35. J. Bardeen, "Surface States and Rectification at a Metal Semiconductor Contact," *Phys. Rev.* **71,** 717 (1947).
36. E. H. Nicollian and A. Goetzberger, "The Si–SiO_2 Interface—Electrical Properties as Determined by the MIS Conductance Technique," *Bell Syst. Tech. J.* **46,** 1055 (1967).
37. R. D. Schrimpf, private communication.
38. J. R. Schwank, D. M. Fleetwood, P. S. Winokur, P. V. Dressendorfer, D. C. Turpin, and D. T. Sanders, "The Role of Hydrogen in Radiation-Induced Defect Formation in Polysilicon Gate MOS Devices," *IEEE Trans. Nuclear Sci.* **NS-34,** 1152 (1987).
39. B. E. Deal, M. Sklar, A. S. Grove, and E. H. Snow, "Characteristics of the Surface-State Charge (Q_{ss}) of Thermally Oxidized Silicon," *J. Electrochem. Soc.* **114,** 266 (1967).

40. S. I. Raider and A. Berman, "On the Nature of Fixed Oxide Charge," *J. Electrochem. Soc.* **125,** 629 (1978).
41. S. I. Raider and R. Flitsch, "X-Ray Photoelectron Spectroscopy of SiO_2–Si Interfacial Regions: Ultra Thin Oxide Films," *IBM J. Res. Dev.* **22,** 294 (1978).
42. O. L. Krivanek, T. T. Sheng, and D. C. Tsui, "A High-Resolution Electron Microscopy Study of the Si–SiO_2 Interface," *Appl. Phys. Lett.* **32,** 437 (1978).
43. D. Frohman-Bentchkowsky, "FAMOS—A New Semiconductor Charge Storage Device," *Solid-State Electron.* **17,** 517 (1974).
44. A. van der Ziel, private communication.
45. W. L. Brown, "*n*-Type Surface Conductivity on *p*-Type Germanium," *Phys. Rev.* **91,** 518 (1953).
46. C. G. B. Garrett and W. H. Brattain, "Physical Theory of Semiconductor Surfaces," *Phys. Rev.* **99,** 376 (1955).
47. R. H. Kingston and S. F. Neustadter, "Calculation of the Space Charge, Electric Field, and Free Carrier Concentration at the Surface of a Semiconductor," *J. Appl. Phys.* **26,** 718 (1955).
48. C. E. Young, "Extended Curves of the Space Charge, Electric Field, and Free Carrier Concentration at the Surface of a Semiconductor, and Curves of the Electrostatic Potential Inside a Semiconductor," *J. Appl. Phys.* **32,** 329 (1961).
49. A. S. Grove and D. J. Fitzgerald, "Surface Effects on *P–N* Junctions—Characteristics of Surface Space-Charge Regions under Non-Equilibrium Conditions," *Solid-State Electron.* **9,** 783 (1966).
50. Ph. Passau and M. van Styvendael, "La Jonction *p–n* Abrupte dans le Cas Statique," *Proc. Int. Conf. Solid State Phys. in Electron. and Telecomm.*, Brussels, June 2–7, 1958, M. Desirant and J. F. Miciels (eds.), International Union for Pure and Applied Physics, Vol. I, Academic Press, New York, 1960, p. 407.
51. C. Goldberg, "Space Charge Regions in Semiconductors," *Solid-State Electron.* **7,** 593 (1964).
52. R. P. Jindal and R. M. Warner, Jr., "A General Solution for Step Junctions with Infinite Extrinsic End Regions at Equilibrium," *IEEE Trans. Electron Devices* **28,** 348 (1981).
53. R. M. Warner, Jr., and R. P. Jindal, "Replacing the Depletion Approximation," *Solid-State Electron.* **26,** 335 (1983).
54. H. C. Pao and C. T. Sah, "Effects of Diffusion Current on Characteristics of Metal-Oxide (Insulator)-Semiconductor Transistors," *Solid-State Electron.* **9,** 927 (1966).
55. Warner and Grung, pp. 372–377.
56. R. M. Warner, Jr., R. P. Jindal, and B. L. Grung, "Field and Related Semiconductor-Surface and Equilibrium-Step-Junction Variables in Terms of the General Solution," *IEEE Trans. Electron Devices* **31,** 994 (1984).
57. W. Shockley, "The Theory of *p–n* Junctions in Semiconductors and *p–n* Junction Transistors," *Bell Syst. Tech. J.* **28,** 435 (1949).
58. R. M. Warner, Jr., R. D. Schrimpf, and P. D. Wang, "Explaining the Saturation of Potential Drop on the High Side of a Grossly Asymmetric Junction," *J. Appl. Phys.* **57,** 1239 (1985).
59. A. Goetzberger, "Ideal MOS Curves for Silicon," *Bell Syst. Tech. J.* **45,** 1097 (1966).
60. W. S. Boyle and G. E. Smith, "Charge Coupled Semiconductor Devices," *Bell Syst. Tech. J.* **49,** 587 (1970).
61. A. S. Grove, E. H. Snow, B. E. Deal, and C. T. Sah, "Simple Physical Model for the Space-

Charge Capacitance of Metal-Oxide-Semiconductor Structures," *J. Appl. Phys.* **35,** 2458 (1964).
62. C. N. Berglund, "Surface States at Steam-Grown Silicon–Silicon Dioxide Interfaces," *IEEE Trans. Electron Devices* **13,** 701 (1966).
63. R. Castagné, "Détermination de la Densité d'États Lents d'une Capacité Métal-Isolant-Semiconducteur par l'Étude de la Charge sous une Tension Croissant Linéairement," *C. R. Acad. Sci. (Paris)* **267,** 866 (1968).
64. D. R. Kerr, "M.I.S. Measurement Techniques Utilizing Slow Voltage Ramps," J. Bonel (ed.), *Trans. Int. Conf. on Properties and Use of MIS Structures*, Grenoble, France, 1969, p. 303.
65. M. Kuhn, "A Quasi-Static Technique for MOS *C–V* and Surface State Measurements," *Solid-State Electron.* **13,** 873 (1970).
66. A. Goetzberger, E. Klausmann, and M. J. Schulz, "Interface States on Semiconductor/Insulator Surfaces," *CRC Crit. Rev. Solid-State Sci.* **6,** 1 (1976).
67. A. S. Grove, B. E. Deal, E. H. Snow, and C. T. Sah, "Investigation of Thermally Oxidised Silicon Surfaces Using Metal-Oxide-Semiconductor Structures," *Solid-State Electron.* **8,** 145 (1965).
68. R. M. Warner, Jr., "A Note on Equality of Bulk and Inversion-Layer Capacitance at Threshold," *Solid-State Electron.* **30,** 181 (1987).
69. M. C. Tobey, Jr., and N. Gordon, "Concerning the Onset of Heavy Inversion in MIS Devices," *IEEE Trans. Electron Devices* **ED-21,** 649 (1974).
70. Y. Tsividis, "Moderate Inversion in MOS Devices," *Solid-State Electron.* **25,** 1099 (1982).
71. R. D. Schrimpf, D.-H. Ju, and R. M. Warner, Jr., "Subthreshold Transconductance in the Long-Channel MOSFET," *Solid-State Electron.* **30,** 1043 (1987).
72. Warner and Grung, Chapters 9 and 10.
73. H. K. J. Ihantola and J. L. Moll, "Design Theory of a Surface Field-Effect Transistor," *Solid-State Electron.* **7,** 423 (1964).
74. A. Vladimirescu, A. R. Newton, and D. O. Pederson, *Spice Version 2G.2 User's Guide*, Electronics Research Laboratory, University of California, Berkeley, 1981.
75. J. R. Brews, "Physics of the MOS Transistor," *Applied Solid State Science, Supplement 2A*, D. Kahng (ed.), Academic Press, New York, 1981.
76. Warner and Grung, Fig. 6-10(b), p. 421.
77. W. M. Penney and L. Lau (eds.), *MOS Integrated Circuits*, Van Nostrand Reinhold, New York, 1972, p. 92.
78. J. R. Brews, "A Charge-Sheet Model of the MOSFET," *Solid-State Electron.* **21,** 345 (1978).
79. G. Baccarani, M. Rudan, and G. Spadini, "Analytical IGFET Model Including Drift and Diffusion Currents," *Solid-State Electron Devices* **2,** 62 (1978).
80. D.-H. Ju and R. M. Warner, Jr., "Modeling the Inversion Layer at Equilibrium," *Solid-State Electron.* **27,** 907 (1984).
81. L. M. Dang, "A One-Dimensional Theory on the Effects of Diffusion Current and Carrier Velocity Saturation on E-type IGFET Current–Voltage Characteristics," *Solid-State Electron.* **20,** 781 (1977).
82. H. Shichman and D. A. Hodges, "Modeling and Simulation of Insulated-Gate Field-Effect Transistor Switching Circuits," *IEEE J. Solid-State Circuits* **SC-3,** 285 (1968).
83. J. E. Meyer, "MOS Models and Circuit Simulation," *RCA Rev.* **32,** 42 (1971).

84. D. Frohman-Betchkowsky and A. S. Grove, "On the Effect of Mobility Variation on MOS Device Characteristics," *Proc. IEEE* **56,** 217 (1968).
85. R. J. Gravrok, private communication.
86. D. P. Foty, *MOSFET Modeling with SPICE*, Prentice-Hall, Englewood Cliffs, NJ, 1997.
87. L. Nagel and R. Rohrer, "Computer Analysis of Nonlinear Circuits, Excluding Radiation (CANCER)," *IEEE J. Solid-State Circuits* **SC-6,** 166 (1971).
88. L. Nagel and D. Pederson, "Simulation Program with Integrated Circuit Emphasis," University of California/Berkeley, Electronics Research Laboratory Memorandum UCB/ERL M352, 1973.
89. L. Nagel, "SPICE2: A Computer Program to Simulate Semiconductor Circuits," University of California/Berkeley, Electronics Research Laboratory Memorandum UCB/ERL M520, 1975.
90. PSPICE User's Guide, MicroSim, Inc., Irvine, CA, 1992.
91. G. Massobrio and P. Antognetti, *Semiconductor Device Modeling with SPICE*, 2nd ed., McGraw-Hill, New York, 1993.
92. HSPICE User's Manual, Meta-Software, Inc., Campbell, CA, 1993.
93. B. Sheu, "MOS Transistor Modeling and Characterization for Circuit Simulation," University of California/Berkeley, Electronics Research Laboratory Memorandum UCB/ERL M85/85, 1985.
94. B. Sheu, D. Scharfetter, P. Ko, and M. Jeng, "BSIM: Berkeley Short Channel IGFET Model for MOS Transistors," *IEEE J. Solid-State Circuits* **22,** 558 (1987).
95. J. Huang et al., "BSIM3 Manual" (version 2.0), University of California/Berkeley, 1994.
96. Y. Cheng et al., "BSIM3 Version 3.0 Manual," University of California/Berkeley, 1995.
97. R. Velge, D. Klaasen, and F. Klaasen, "MOS Model 9," Philips Electronics N.V., Unclassified Report NL-UR 003/94, 1994.
98. J. Power and W. Lane, "An Enhanced SPICE MOSFET Model Suitable for Analog Applications," *IEEE Trans. Computer-Aided Design* **11,** 1418 (1992).
99. N. Arora, R. Rios, C. Huange, and K. Raol, "PCIM: A Physically Based Continuous Short Channel IGFET Model for Circuit Simulation," *IEEE Trans. Electron Devices* **ED-41,** 988 (1994).
100. C. Enz, F. Krummenacher, and E. Vittoz, "An Analytical MOS Transistor Model Valid in All Regions of Operation and Dedicated to Low Voltage and Low Current Applications," *Analog Integ. Circuit Signal Process.* **8,** 83 (1995).
101. Y. Tsividis and K. Suyama, "MOSFET Modelling for Analog Circuits CAD: Problems and Prospects," *IEEE J. Solid-State Circuits* **SC-29,** 210 (1994).
102. R. M. Warner, Jr., "Comparing MOS and Bipolar Integrated Circuits," *IEEE Spectrum* **4,** 50 (1967).
103. P. L. Hower, "Bipolar vs. MOSFET: Seeing where the Power Lies," *Electronics* **53,** 106 (Dec. 18, 1980).
104. Y. Tsividis, "Moderate Inversion in MOS Devices," *Solid-State Electron.* **25,** 1099 (1982).
105. E. O. Johnson, "The Insulated-Gate Field-Effect Transistor—A Bipolar Transistor in Disguise," *RCA Rev.* **34,** 80 (1973).
106. C. T. Sah and H. C. Pao, "The Effects of Fixed Bulk Charge on the Characteristics of Metal–Oxide Semiconductor Transistors," *IEEE Trans. Electron Devices* **ED-13,** 393 (1966).

107. F. Van de Wiele, "A Long-Channel MOSFET Model," *Solid-State Electron.* **22,** 991 (1979).
108. A. G. Milnes, *Semiconductor Devices and Integrated Electronics*, Van Nostrand Reinhold, New York, 1980, p. 410.
109. J. Evans and K. A. Pullen, Jr., "Limitation of Properties of Field-Effect Transistors," *Proc. IEEE* **54,** 82 (1966).
110. R. M. Warner, Jr., and R. D. Schrimpf, "BJT-MOSFET Transconductance Comparisons," *IEEE Trans. Electron Devices* **34,** 1061 (1987).
111. H. K. Gummel and H. C. Poon, "An Integral Charge Control Model of Bipolar Transistors," *Bell Syst. Tech. J.* **49,** 827 (1970).
112. R. H. Crawford, *MOSFET in Circuit Design*, McGraw-Hill, New York, 1967, p. 53.
113. M. B. Barron, "Low-Level Currents in Insulated Gate Field Effect Transistors," *Solid-State Electron.* **15,** 293 (1972).
114. R. F. van Overstraeten, G. Declerck, and G. F. Broux, "Inadequacy of the Classical Theory of the MOS Transistor Operating in Weak Inversion," *IEEE Trans. Electron Devices* **ED-20,** 1150 (1973).
115. R. R. Troutman, "Subthreshold Design Considerations for Insulated Gate Field-Effect Transistors," *IEEE J. Solid-State Circuits* **SC-9,** 55 (1974).
116. W. Fichtner and H. W. Pötzl, "MOS Modelling by Analytical Approximations. I. Subthreshold Current and Threshold Voltage," *Int. J. Electron.* **46,** 33 (1979).
117. L. D. Yau, "A Simple Theory to Predict the Threshold Voltage of Short-Channel IGFETs," *Solid-State Electron.* **17,** 1059 (1974).
118. Warner and Grung, pp. 454–455.
119. R. M. Warner, Jr., D.-H. Ju, and B. L. Grung, "Electron-Velocity Saturation at a BJT Collector Junction under Low-Level Conditions," *IEEE Trans. Electron Devices* **ED-30,** 230 (1983).

Problems

Topics for Review

R1. Why does the MOSFET make more severe demands on the knowledge and control of silicon surfaces than does the BJT?

R2. Explain the acronym *MOSFET.*

R3. Identify materials most commonly used in the three regions of the MOS capacitor.

R4. Name and describe briefly at least two members of the FET family in addition to the MOSFET.

R5. Explain the terms *source* and *drain.*

R6. What bias polarity on the gate of a MOSFET produces useful results? Why? Specify the substrate type and channel type you have assumed in your answer.

R7. Explain the term *flatband voltage.*

R8. What is *subthreshold current* and what accounts for it?

R9. Explain, with examples, the terms *unipolar* and *bipolar.*

R10. Why was a field-effect transistor the first solid-state device to be conceived, but nearly the last to be realized?

R11. Make a four-way comparison of mobility values, *N*-channel versus *P*-channel, and surface (channel) versus bulk.

R12. Define *threshold of weak inversion.*

R13. Define *threshold of strong inversion.*

R14. What is meant by *sheet resistance*? Why is it a useful concept?

R15. How do the dielectric permittivities of silicon and silicon dioxide compare?

R16. Why does the surface depletion layer essentially stop growing at the threshold (of strong inversion) voltage?

R17. In what respect does the rudimentary analysis of the MOSFET yield a square-law model?

R18. What major simplifying assumption is made in the rudimentary model?

R19. Name the two major operating regimes and indicate where they reside.

R20. Give a "zeroth-order" explanation for why drain current is approximately independent of drain–source voltage in saturation.

R21. Explain the terms *E-mode*, *D-mode*, and *E–D circuits*. What are the advantages and disadvantages of the last?

R22. How do the electrical properties of a D-mode MOSFET differ from those of a JFET (which is also a D-mode device)?

R23. How does MOSFET transconductance depend on input voltage?

R24. How can one realize a linear voltage amplifier using MOSFETs?

R25. Explain the term *saturated load.* What are its advantages and disadvantages?

R26. Compare and contrast the E-mode MOSFET *transfer characteristic* and the *locus of points* separating the curved and the saturation regimes.
R27. Compare and contrast the E-mode and D-mode MOSFET transfer characteristics.
R28. How can both the E-mode and the D-mode MOSFET have characteristic currents?
R29. Explain the acronym *CMOS* and describe the circuitry it identifies. What are its advantages and disadvantages?
R30. What primarily fixes switching delay in MOS circuits?
R31. Does choice of load device mainly affect the *turn-on* or *turn-off* properties of the inverter? Why?
R32. What is meant by the term *discrete device*?
R33. What is *totempole current*?
R34. Why did computerized circuit design get an earlier and faster start in MOS technology than in bipolar technology?
R35. Compare the step-junction problem and the silicon-surface (as encountered in the MOS capacitor) problem.
R36. Describe the oxide–silicon boundary conditions for an ideal MOS capacitor.
R37. Describe the field profile in an ideal MOS capacitor in depletion, but well below threshold. Explain why its major features exist.
R38. What principle determines field magnitude on the two sides of the oxide–silicon interface?
R39. How do an ordinary parallel-plate capacitor and the MOS capacitor differ?
R40. Explain the meaning of the symbol ψ_{SB}.
R41. What is meant by *band bending*?
R42. Explain *vacuum level* and *work function*.
R43. Explain *photoelectric threshold* and *electron affinity*.
R44. Name advantages of a *polysilicon* gate electrode over a metal electrode.
R45. Cite the several kinds of parasitic charge that can affect the oxide–silicon system. Describe the characteristics of each, and state its accepted symbol.
R46. Why is *interface trapped charge* (in "surface states") particularly damaging?
R47. Explain the comparison of surface states with an electrostatic shield. What was their historical role?
R48. Mobile charge is almost as intolerable as interface trapped charge. Why?
R49. Identify the species primarily responsible for mobile charge in SiO_2.
R50. Explain the term *dangling bonds*.
R51. Given a band diagram for an MOS capacitor, identify the *surface potential*.
R52. What is meant by the *accumulation* condition in an MOS capacitor? What bias polarity produces it?
R53. What does the acronym *EPROM* stand for?
R54. What do the terms *volatile* and *nonvolatile* mean when applied to an electronic memory?
R55. Why does interface trapped charge affect the mobility of inversion-layer carriers?
R56. Why is inversion-layer mobility only a fraction of bulk mobility, even in the absence of parasitic charge?
R57. What are the meanings and causes of *mobility fluctuations* and *density fluctuations*?
R58. What property of a MOSFET is affected by such fluctuations?
R59. What technology has made it possible to adjust doping and charge conditions near a silicon surface with great precision?

R60. Describe the equivalent-circuit model appropriate to each operating regime for the MOS capacitor.
R61. At what voltage do the *C–V* profiles for the inversion-layer and depletion-layer capacitances intersect? Why?
R62. Compare the junction-capacitance equivalent circuits with those for the MOS capacitor.
R63. What capacitance component is effectively absent in each case and why?
R64. Does *capacitance crossover* exist for a junction as well as for the MOS capacitor? At what approximate voltage does it occur?
R65. Give a physical explanation for approximate MOS capacitance crossover at the voltage stated in response to R61.
R66. Why has *C–V* assessment been so important in MOS technology?
R67. How does small-signal frequency affect a (combined or overall) *C–V* profile for the MOS capacitor?
R68. Describe the quasistatic technique for observing a *C–V* profile. What is its advantage over the small-signal method?
R69. What feature of a *C–V* profile reveals the presence of "fast surface states," capable of harboring interface trapped charge?
R70. Explain *long-channel approximation.*
R71. Also explain *gradual-channel approximation* and relate the two terms. State an algebraic condition defining one or both.
R72. What major simplification exists in a long-channel model as compared to a short-channel model?
R73. Identify the Level-1, -2, and -3 SPICE models. How do they differ? What are their major features?
R74. What is meant by *body effect*?
R75. Explain how body effect alters threshold voltage.
R76. Describe a sample situation where body effect exists.
R77. How is a D-mode load device affected by body effect? Why?
R78. Describe the modeling approaches taken by Pao and Sah, and by Brews.
R79. Name and describe the operating regimes identified in the SPICE program.
R80. How does SPICE modeling predict nonconstant current in the saturation regime?
R81. Compare the gate terminal and bulk region as control electrodes.
R82. How can the bulk region act as a "gate?"
R83. Does it make sense to talk about the current gain of a MOSFET? Why?
R84. What is f_T for a small-signal MOSFET amplifier?
R85. What parasitic elements are dominant in determining f_T?
R86. What small-signal and large-signal characterizing quantities have a product that equals a constant for a given circuit?
R87. What is *noise margin*?
R88. Why does noise margin matter?
R89. What parasitic capacitance is most important in large-signal applications? in small-signal applications?
R90. List three ways of comparing the transconductance properties of BJT and MOSFET.
R91. Discuss qualitatively the findings in each case.
R92. Why is transconductance important?

R93. In view of the large transconductance deficit of the MOSFET under many circumstances, why is it the dominant device in the world today?

R94. Comment on the device-dimension dependencies of transconductance in the BJT and MOSFET.

R95. Describe the properties of the source–channel junction as a function of gate bias.

R96. Describe channel transport well above threshold; below threshold.

R97. In view of your answer to the last question, why does the MOSFET not exhibit a diffusion capacitance, as does the BJT?

R98. Why does subthreshold transconductance in the MOSFET approach the theoretical limit of qI_D/kT?

Analytic Problems

A1. A certain E-mode MOSFET has a gate capacitance of 0.02 pF, $V_T = 1$ V, $\mu_n = 700$ cm^2/V·s, $X_{OX} = 0.05$ μm, and $L = 3$ μm.

a. Calculate its gate width Z in μm.
b. Calculate its transconductance g_m at $V_{GS} = 3$ V in the saturation regime.

A2. Consider an N-channel E-mode MOSFET in the curved (nonsaturation) regime.

a. Derive an expression for the transconductance g_m.
b. Rewrite the expression for I_D in terms of normalized voltage, that is, in terms of V_{GS}/V_T and V_{DS}/V_T.
c. From the result in part b, write an approximate expression for I_D that is valid in the linear regime, which is to say, *very* near the origin of the output plane. Use this expression to derive an expression for channel resistance r (source–drain resistance) for the MOSFET in the linear regime.
d. Use the result in part c to find r for the MOSFET of the previous problem when it has $V_{GS} = 3$ V.

A3. The MOSFET of Problem A1 is placed in the circuit shown.

a. Find I_D.
b. Find V_{DS}.
c. Sketch an accurate loadline diagram, plotting an I–V curve for the MOSFET and another for the resistor R_D. On the diagram indicate V_{DD} and the numerical coordinates of the quiescent point, (I_D, V_{DS}).

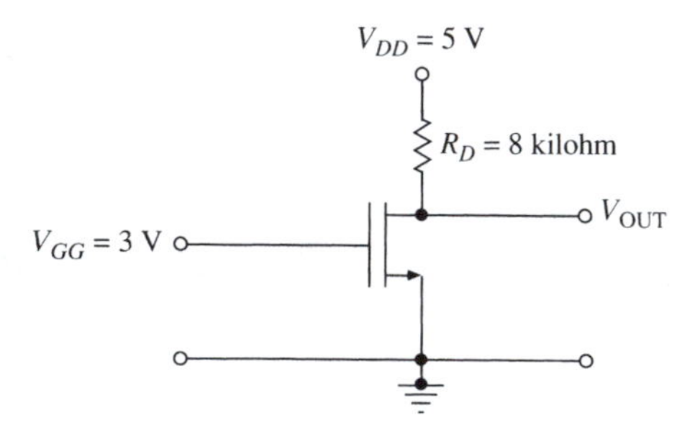

A4. For a certain N-channel E-mode MOSFET the curved-regime characteristics are described by

$$I_D = K[2(V_{GS} - V_T)V_{DS} - V_{DS}^2],$$

and the saturation-regime characteristics are described by

$$I_D = K(V_{GS} - V_T)^2,$$

where $K = 1$ mA/V^2 and $V_T = 1$ V. The MOSFET is placed in this circuit:

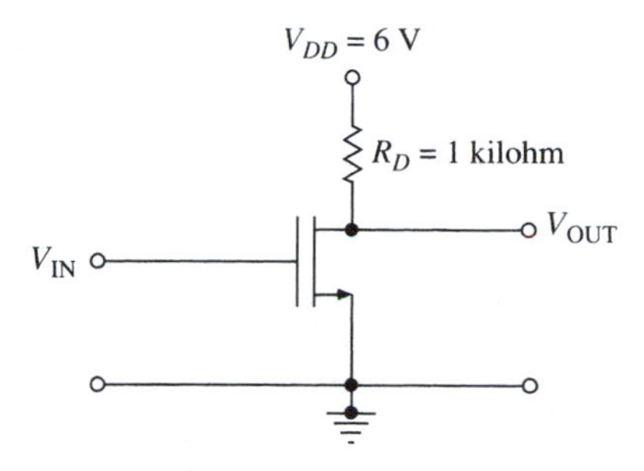

V_{IN}	V_{OUT}
0 V	
1 V	
2 V	
3 V	

a. Complete the chart at the right of the diagram.
b. Plot the points whose coordinates are in the chart above on the axes below, after calibrating the axes appropriately. Draw a smooth curve through the points you have plotted. This curve is known as the *voltage transfer curve.*

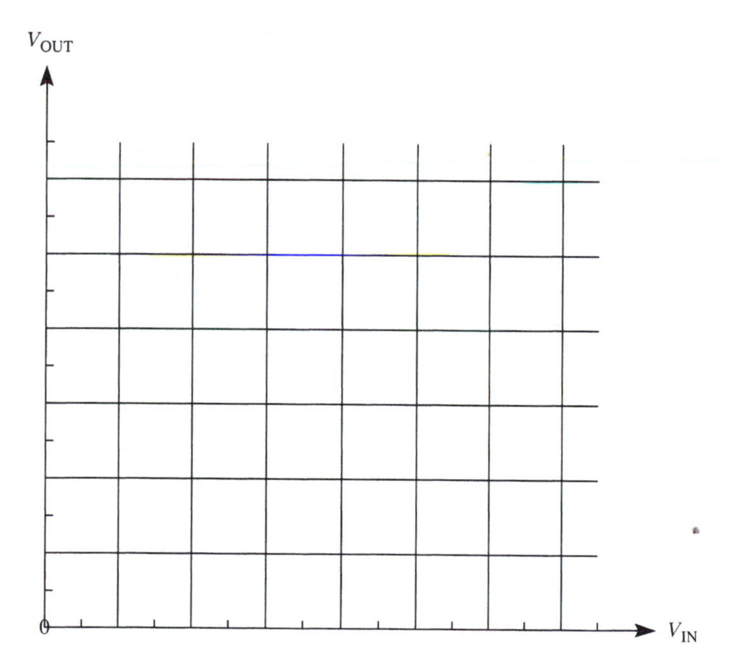

c. What is the significance of the derivative (dV_{OUT}/dV_{IN})?
d. Estimate the value of the derivative at $V_{IN} = 2$ V by using graphical or analytic means.

A5. Given here is one output characteristic for a particular N-channel E-mode MOSFET. Its saturation-regime output-current equation is $I_{D1} = K_1(V_{GS} - V_T)^2$.

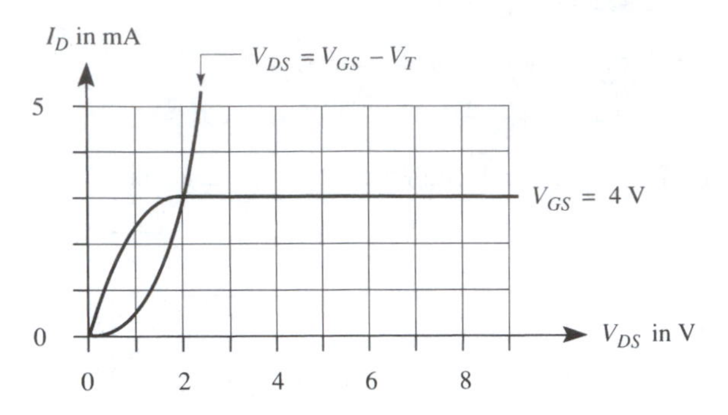

a. Calculate V_T.
b. Calculate K_1. Include units.
c. A second MOSFET is identical in every respect, except that its channel length is $L_2 = (1.5)L_1$, where L_1 is channel length in the first device. On the output plane above, accurately sketch the I–V characteristic for the second device at $V_{GS} = 4$ V.

A6. In Exercise 1-8, Section 1-4, we noted that the boundary between the saturation and curved regimes is the locus of points for which $V_{DS} = V_{GS} - V_T$ is valid, and that this voltage expression is not the "equation for" the boundary curve. Derive the equation for the boundary curve plotted in Figure 1-6(a).

A7. Shown here is a diode-connected E-mode MOSFET, a configuration used in some circuits as a quasi-resistor. Determine whether this device is operating in the saturation regime, the curved regime, some combination of the two, or on the boundary between them.

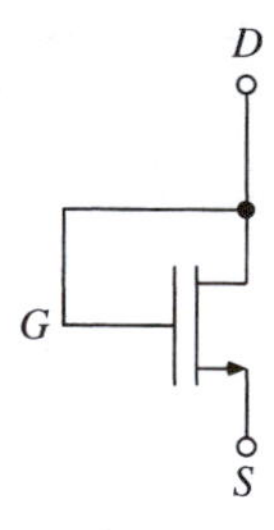

A8. The current–voltage equation for the diode-connected E-mode MOSFET was derived in Problem A7. Comment on the relationship of the resulting expression to that for the appropriate transfer characteristic.

A9. Two discrete MOSFETs are interconnected in the manner shown here to create an inverting amplifier stage. The two devices are identical in every respect except in the lateral dimensions Z and L. (See also Problem D2.)

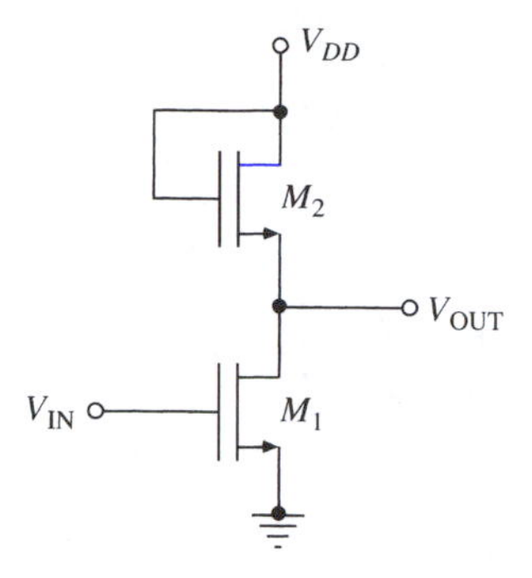

a. Draw an output-plane diagram and apply the principles of loadline-diagram construction to it so that the interaction of the two devices can be visualized.
b. Demonstrate that the given circuit constitutes a linear voltage amplifier by deriving an equation for its small-signal voltage gain, $A_V \equiv dV_{OUT}/dV_{IN}$.
c. Explain graphically the linearity of this amplifier circuit by means of the loadline diagram constructed above.

A10. The cross-sectional diagram in Figure 1-4 is for an *N*-channel MOSFET operating in the curved regime, well out of saturation.

a. Draw a similar diagram, but for a long-channel MOSFET operating well inside the saturation regime. On your diagram, indicate clearly the depletion-layer boundaries in the *P*-type region.
b. On the same or a similar diagram, sketch several lines of force (with correct direction indicated) in each depletion layer.
c. The device being considered exhibits the extremely high dynamic output resistance depicted in the shaded portion of Figure 1-5(b). The fact that I_D is essentially independent of V_{DS} is a consequence of Ohm's law. Using the diagram of part a or b, or a similar diagram, explain the constancy of $I_D(V_{DS})$. The only additional item of information you need is that when an MOS capacitor is brought to the threshold condition, with the voltage V_T from gate to bulk, there also exists a voltage V_{TOX} from gate to the oxide–silicon interface (voltage drop through the oxide) that is just as clearly defined by the structure of the capacitor as is V_T. Obviously $V_{TOX} < V_T$.
d. Relate the explanation given in part c to the regimes of operation depicted in Figure 1-5.
e. Construct a voltage-scale diagram, indicating the voltages and voltage intervals cited in part c, for operation at the boundary of the curved and saturation regimes.
f. In part e, the voltage at the drain terminal is lower than the voltage at the point where the channel vanishes (because $V_T > V_{TOX}$). How can the electrons make their way from the channel end to the drain?
g. By considering the same phenomena that enter into part c, and by drawing another diagram, explain why a short-channel MOSFET exhibits much lower dynamic output resistance in the saturation regime than does a long-channel device that is identical in every respect except channel length. Name an analogous BJT effect.

A11.

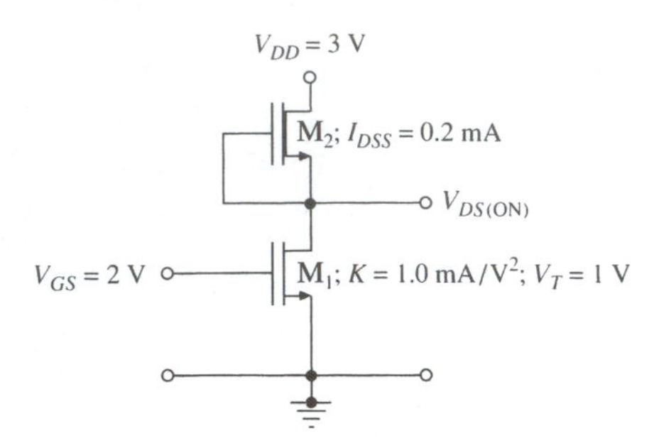

Given above is the schematic diagram for a MOSFET enhancement–depletion (E–D) inverter under bias. On the pair of axes in the next figure is plotted the output characteristic for M_1 for its given bias condition.

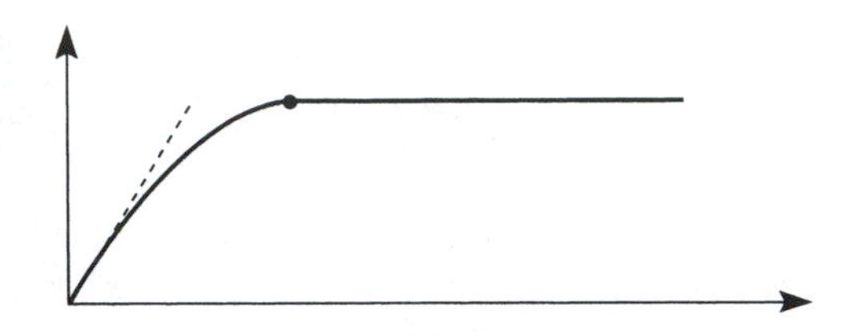

a. Label the axes using correct symbols.
b. Calibrate both axes. [*Note*: For *each* axis you must carry out a calculation in order to calibrate it correctly.]
c. On the same diagram, label the voltages $V_{DS(ON)}$ and V_{DD} very clearly.
d. Using symbols only (no numbers), derive an approximate expression for $V_{DS(ON)}$. [*Hint*: Because $V_{DS(ON)}$ is small, use a linear approximation (as shown by the dashed line in the figure) to the curved-regime characteristic.]
e. Evaluate the expression derived in part d.

A12. One-sided junctions and high–low junctions have important similarities and important differences that enter the MOS-capacitor problem as well.

a. Construct linearly calibrated axes, and on them, using a solid line, sketch a reasonably accurate electric-field profile for an equilibrium N^+N high–low junction.
b. Using a dashed line, draw an equally accurate electric-field profile for the equilibrium N^+P one-sided junction of Figure 3-5. Carry out any necessary calculations and explain your reasoning. Comment qualitatively but accurately on the differences in the two profiles.

A13. For the junction of Problem A12b determine the amount of reverse bias needed to cause a rightward shift of the right-hand depletion-layer boundary by 1 μm. Draw a diagram showing the relationship of the new electric-field profile to the equilibrium profile for the N^+P junction.

A14. A certain MOS capacitor has the structure shown in Figure 3-4(a). It also has a gate measuring 2 μm $\times$ 2 μm, an oxide thickness of $X_{OX} = 400$ Å, and a net substrate doping of $N_A = 5 \times 10^{15}/\text{cm}^3$. Calculate the capacitance of the device when it is biased precisely at the threshold voltage.

A15. It is noted in Section 2-6 that the fixed charge Q_f at the oxide–silicon interface was found in the 1960s to be a function of crystal orientation, and to exhibit values following the same sequence as surface-atom density for those orientations. A substantial set of data drawn from the contemporaneous literature demonstrated a Q_f sequence having the ratios given in the following table for three different orientations. (Details of the experimental data as well as indications of the data sources can be found in [7, p. 730 and 749].)

	$\frac{Q_f(100)}{Q_f(100)}$	$\frac{Q_f(110)}{Q_f(100)}$	$\frac{Q_f(111)}{Q_f(100)}$
Median ratios, experimental data	1	1.45	3.15

Employing information on the silicon crystal given in [7] (page 730), calculate the corresponding ratios of surface-atom densities, and compare them with fixed-charge ratios observed experimentally. In the case of the (111) orientation, treat the two closely spaced planes of atoms as a single plane.

A16. Equation 3-15 gives a useful approximation for C_b for conditions well below threshold, where Q_b is taken from Equation 3-19 as

$$Q_b = \frac{\epsilon\sqrt{2}}{L_D}\frac{kT}{q}\sqrt{W_S - 1}.$$

a. How large an error results from using this approximate expression *at* threshold for a sample of net doping $N = 10^{15}/\text{cm}^3$?

b. If one makes the erroneous assumption that C_i can be obtained from $C_i = dQ_n/d\psi_S$, where Q_n is inferred from Equation 3-19 as

$$Q_n = \frac{\epsilon\sqrt{2}}{L_D}\frac{kT}{q}\sqrt{e^{W_S-2|U_B|}},$$

how serious is the resulting error at threshold for the same sample as in part a?

A17. Equations 3-22 and 3-23 are often written in terms of L_{Di}. Make the conversion.

A18. A certain long-channel MOSFET is biased at $V_{GS} = 2\ V_T$ and $V_{DS} = 5\ V_T$. For simplicity, assume that it is free of all parasitic effects. Of interest are the areal charge densities Q_n and Q_B at the one-third point of the channel, $y = (L/3) \equiv y_1$, and at the two-thirds point, $y = (2L/3) \equiv y_2$. Construct a chain of qualitative logic proving unequivocally that $Q_n(y_2) < Q_n(y_1)$ and $Q_b(y_2) > Q_b(y_1)$.

A19. Figure 5-7(d) presents profiles for both carrier populations in the bulk region of an MOS capacitor at threshold. Let us assume that its net doping is $N_A = 10^{16}/\text{cm}^3$. The density values are well known in the undepleted bulk region, and also at the oxide–silicon interface by virtue of the given conditions.

a. Using the depletion-approximation replacement (DAR), derive an expression for $p(x/L_D)$ valid for most of the region lying between these two extremes.

b. Calculate six or seven points in the intermediate region, and plot your expression on a semilog diagram (log p versus x/L_D).

c. Calculate and plot the electron profile also, $n(x/L_D)$.
d. For the given conditions, the oxide–silicon interface is sometimes described as the *strong-threshold plane.* In a parallel way, state the position of the *weak-threshold plane.* Calculate its position in terms of the DAR spatial origin.
e. Calculate accurately in micrometers the separation of the strong-threshold and weak-threshold (or intrinsic) planes.

A20. Use the same conditions and distance calibration as in Problem A19 to plot the following curves:

a. Generate curves of $p(x/L_D)$ and $n(x/L_D)$, but with linear density calibration this time, as in Figure 3-7(d).
b. Plot a normalized profile of total volumetric charge density in the silicon.

A21. Derive the last expression of Table 4.

A22. A certain MOS capacitor of 1966 was "quasi-ideal" in the sense that H_D and Q_f precisely compensated Q_{it} through a certain C–V range. Also, $Q_m = Q_{ot} = 0$. The experimental C–V curve was as shown below:

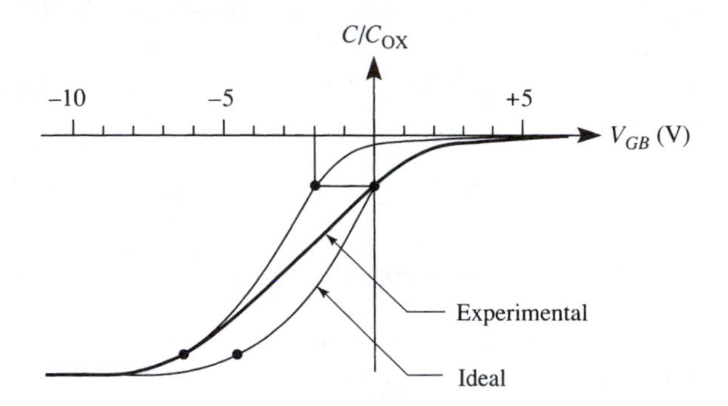

a. Determine the bulk conductivity type.
b. Determine the sign(s) and identity(ies) of possible contributors to Q_s in the operating regime relevant to this problem.
c. Given that $\psi_B = \pm 0.4$ V, sketch a band diagram for the bulk region in the flat-band condition and indicate on it the energy positions of the states responsible for interface trapped charge, Q_{it} ("fast surface states").
d. Determine whether the states named in part c are donorlike or acceptorlike and explain your reasoning.

A23. Carry out the following steps for the device of Problem A22.

a. Calculate N_{it}, the areal density of "fast surface states."
b. Calculate Q_f/C_{OX}, given $H_D = -0.2$ V.
c. Calculate the net doping N.
d. Calculate an approximate value for $X_{D(MAX)}$, the greatest depletion-layer thickness at the surface in the absence of drain–source bias.

A24. The MOSFET inverter that is repeated here was also given in Problem A9. There we showed that the voltage gain A_V is (1) constant (meaning that the resulting amplifier is linear), and (2) designable. An analogous BJT inverter is shown below in the same diagram.

a. Analyze the BJT inverter to determine its gain properties, and compare them with those of the MOSFET inverter.

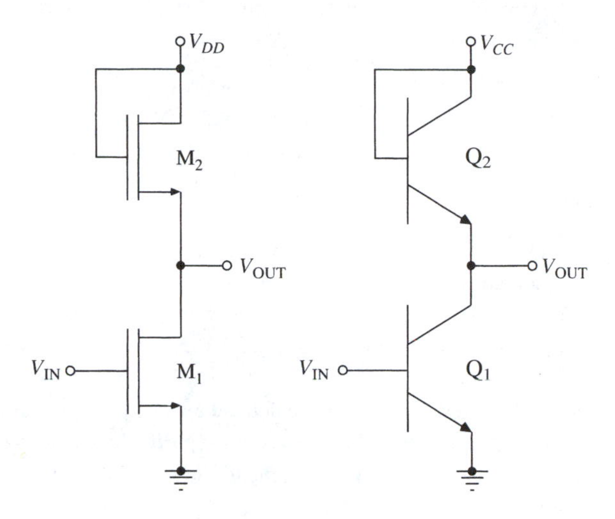

b. What feature(s) of the diagrams in Section 6-4 is (are) related to the comparison between these two kinds of inverters? Comment on the origin of the differing gain properties of the two inverters.

A25. Verify Equation 4-18.

A26. Use the analysis of Chapter 1, the Level-1 model, to follow these instructions:

a. Derive an expression for conductance g that is valid throughout the curved regime of the output plane.
b. Derive an expression for conductance in the linear regime $g\,(\sim 0)$ very near the origin of the output plane.
c. Derive an expression for $I_D\,(\sim 0)$, drain current in the linear regime.

A27. Use the analysis of Section 4-2, the Level-2 model, to follow these instructions:

a. Derive an expression for $I_D\,(\sim 0)$, drain current in the linear regime.
b. Compare the expression just derived with that obtained in Problem A26c and explain any similarity or difference.

A28. Give a physical interpretation of Equation 4-22.

A29. Equation 4-13 can be rewritten as

$$I_D = \mu_n C_{OX}\frac{Z}{L}\left\{\left(V_{GS} - V_{FB} + 2\psi_B - \frac{V_{DS}}{2}\right)V_{DS} - \frac{2}{3}\frac{\sqrt{2\epsilon q N_A}}{}(-2\psi_B)^{3/2}\left[\left(1 + \frac{V_{DS}}{(-2\psi_B)}\right)^{3/2} - 1\right]\right\}.$$

Using the three-term binomial-series expansion for $(1 + \lambda)^n$, where $n = 3/2$ and $\lambda \ll 1$, derive a version of this equation valid in the linear regime and the adjacent portion of the curved regime, but not valid throughout the curved regime as is Equation 4-13. In other words, the expression to be derived will be intermediate between Equation 4-13 and the expression written in Problem A26a.

A30. Calculate the value of g_{mbs} given as Equation 5-44, using the MOSFET data given in Section 4-4.

Note: In Problems A31 through A35 you are to compare the SPICE results with your calculations. For this comparison, use the RMKSA system of units and the following values from SPICE:

$$\epsilon = 1.03594 \times 10^{-10} \text{ F/m} \qquad (kT/q) = 0.02586 \text{ V}$$

$$\epsilon_{OX} = 3.4531 \times 10^{-11} \text{ F/m} \qquad n_i = 1.45 \times 10^{16}/\text{m}^3$$

$$q = 1.602 \times 10^{-19} \text{ C}$$

A31. Prove that the right-hand column of parameters given in Table 9 can be derived from the left-hand column. Omit consideration of the parameters R_S and R_D.

A32. Assuming that $\mu_n = 1200 \text{ cm}^2/\text{V·s}$, $N_A = 10^{16}/\text{cm}^3$, and $\tau = 3.605 \times 10^{-7}$ s, calculate a value for J_0 and compare the result with the value of 3.125×10^{-7} A/m^2 used in the SPICE calculations.

A33. Assuming that $\Delta\psi_0 = 0.8$ V and $N_A = 10^{16}/\text{cm}^3$, calculate a value for $C_{0.B}$ and compare the result with the value of 3.125×10^{-4} F/m^2 used in the SPICE calculations.

A34. Using the values given in Tables 7, 9, and 11, calculate the MOSFET on voltage V_{ON} and compare the result with V_T.

A35. Using the values given in Tables 7, 9, and 11, calculate the small-signal capacitance values given in the equivalent circuit of Figure 5-10.

A36. Give a qualitative explanation for the impact of the body effect on the *I–V* characteristic of a D-mode load, displayed in Figure 5-15.

A37. For the MOSFET of Figure 5-1, evaluate the bulk-charge parameter γ using the SPICE data, and also using the values given in Table 14 for $T = 297.8$ K; determine the percentage difference.

A38. The large-signal SPICE calculation in Section 5-4 employs an inverter with an E-mode driver and a D-mode load. The layout of the driver is given here.

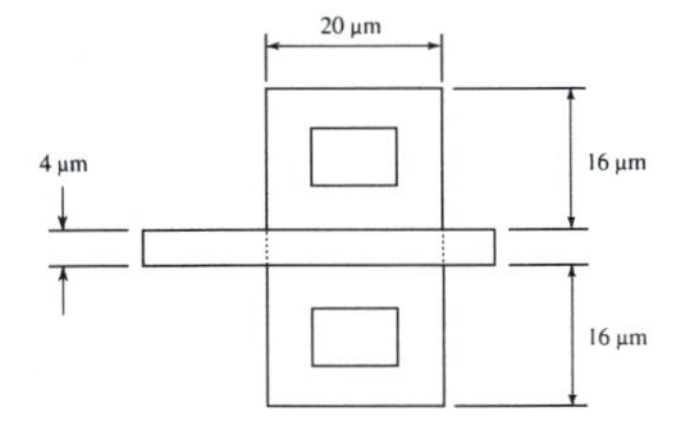

In Table 7 a set of SPICE parameters is given, including

$$A_S = A_D = 320 \times 10^{-12} \text{ m}^2,$$

$$P_S = P_D = 50 \times 10^{-6} \text{ m},$$

and

$$N_{RS} = N_{RD} = 0.5.$$

Show that these values are consistent with the layout.

A39. Using the information given in the previous problem, indicate clearly the various areas corresponding to the overlap capacitances. Calculate their values for the given layout, and determine the overlap capacitances in terms of picofarads per meter of gate width. Determine the limiting values of gate–bulk overlap capacitance for widely varying operating conditions, particularly in V_{GS}.

A40. Starting with the small-signal equivalent circuit given in Figure 5-10, interchange the bulk and the gate terminals so that the amplifier operates in the inverted mode. Construct an approximate equivalent circuit similar to that in Figure 5-11. Calculate the resulting voltage gain and short-circuit current-gain cutoff frequency and compare these values with those for the normal mode.

A41. The configuration shown here is a useful two-terminal current regulator ("current source"). The current level can be tailored in the range $0 < I_D < I_{DSS}$ by adjusting R, which varies V_{GS}.

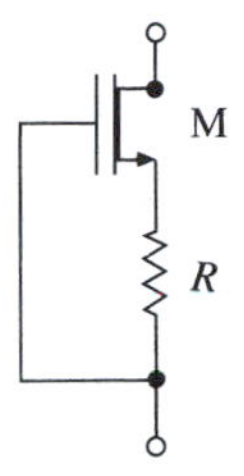

A certain manufacturer purchases discrete N-channel D-mode MOSFETs at high volume for use in this application in his product to deliver the current $I_D = 1$ mA $\pm$ 0.05 mA. He wishes to use a fixed-value resistor (that you may assume has an exact value) rather than selecting resistors with values that compensate for MOSFET variations. The D-mode MOSFET in saturation is well described by the expression

$$I_D = I_{DSS}\left[1 - \frac{V_{GS}}{V_P}\right]^2.$$

Also, you may assume that

$$I_{DSS} = (-V_P H)^{3/2},$$

where H is a constant.

a. Write an expression for the value of R needed as a function of I_{DSS}, V_P, and the selected value of I_D.
b. Derive an expression for I_D as a function of R, I_{DSS}, and V_P.
c. Derive an expression for the relative change in I_D as a function of R, I_{DSS}, and the relative change in I_{DSS}.
d. Calculate H, R, and RH for the nominal values $I_{DSS} = 1.5$ mA and $V_P = -1$ V. Invert the expression derived in part c, and calculate the relative change in I_{DSS} that can be tolerated for $(dI_D/I_D) = 0.05$.
e. Specify the upper and lower limits for the range of I_{DSS} that can be accepted, consistent with the specification $I_D = 1$ mA $\pm$ 0.05 mA.
f. Name the most straightforward way for the MOSFET manufacturer to adjust H, and indicate the direction of the change in H for an increase in the variable you have identified. [This problem is based upon a question posed by Larry McNichols.]

Computer Problems

C1. Consider the circuit shown in Figure 5-8. SPICE parameter values for this transistor are given in Table 7, and dc operating-point information is given in Table 11.

a. Using the SPICE program, calculate the ac voltage gain in dB as a function of frequency for values of frequency from 1 kHz to 10 GHz. Plot the results along with the phase shift in degrees. Use the .OP command in SPICE to print out the values of the small-signal hybrid-pi parameters and compare them with those given in Figure 5-10. The SPICE model is:

```
.MODEL ENHD NMOS (LEVEL=2 TOX=1.139E-07
+UO=879.31 UEXP=0.26742 UCRIT=10K
+LAMBDA=0.022980 NFS=1E11
+NSUB=1E16 NSS=6.5279E10 TPG=1
+RSH=80 LD=0.175U JS=3.125E-7
+CJ=3.125E-4
+MJ=0.5 PB=0.8 FC=0.5
+CJSW=2.0E-10 MJSW=0.5
+CGSO=400P CGDO=400P CGBO=200P)
```

The following format should be used:

```
MQ1 3 7 0 0 ENHD W=20U L=4U AD=320P AS=320P
+                    NRD=0.5 NRS=0.5 PS=50U PD=50U
```

b. Replace the physical parameters by the electrical parameters, as given in Table 9. Then repeat the calculation and show that the results are the same. The SPICE model is:

```
.MODEL ENHD NMOS (LEVEL=2 TOX=1.139E-07
+UO=879.31 UEXP=0.26742 UCRIT=10K
+LAMBDA=0.022980 NFS=1E11
+VTO  =1.03
+KP  =2.6658E-005
+GAMMA  =1.900
+PHI  =0.6954
+CJ=3.125E-4 LD=0.175U
+MJ=0.5 PB=0.8 FC=0.5
+CJSW=2.0E-10 MJSW=0.5
+CGSO=400P CGDO=400P CGBO=200P)
```

c. Replace the transistor in parts a and b by a small-signal circuit that uses resistors, capacitors, and voltage-controlled current sources. Repeat the calculation of ac voltage gain and show that the results are identical to those already obtained in parts a and b.

C2. Each of the four identical inverters in the diagram below employs an E-mode MOSFET in combination with a D-mode load.

a. Using SPICE and the transistor-parameter values given in Table 7, calculate the transient response of the circuit shown in the figure below for an input pulse that is identical to the curve given in Figure 5-18(a). Show that the output waveform at OUTA is identical to the curve in Figure 5-18(b). The SPICE models are:

```
.MODEL ENHD NMOS (LEVEL=2 TOX=1.139E-07
+UO=879.31 UEXP=0.26742 UCRIT=10K
+LAMBDA=0.022980 NFS=1E11
+NSUB=1E16 NSS=6.5279E10 TPG=1
```

```
+RSH=80 LD=0.175U JS=3.125E−7
+CJ=3.125E−4
+MJ=0.5 PB= 0.8 FC=0.5
+CJSW=2.0E−10 MJSW=0.5
+CGSO=400P CGDO=400P CGBO=200P)
```

and

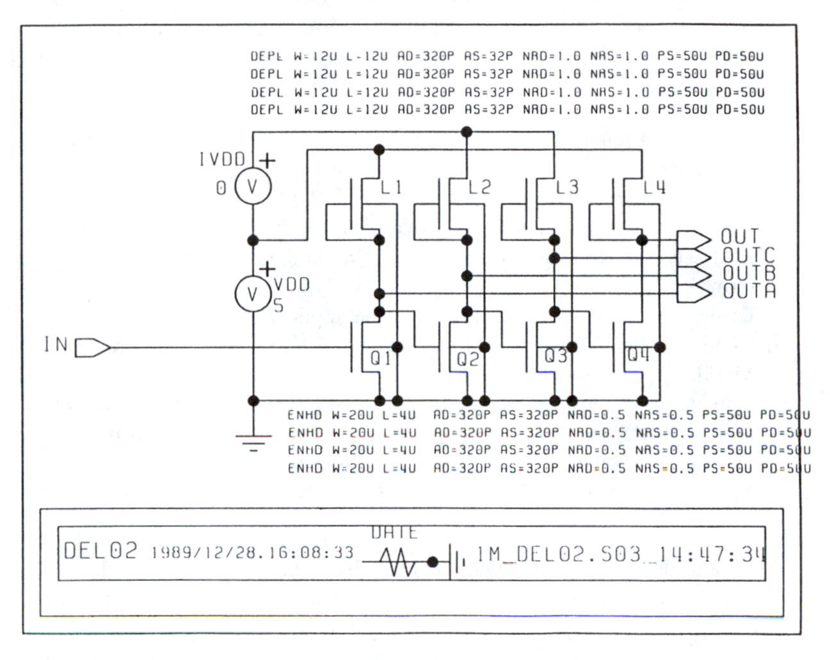


```
MODEL DEPL NMOS (LEVEL=2 TOX=1.139E−07
+UO=879.31 UEXP=0.26742 UCRIT=10K
+LAMBDA=0.022980 NFS=1E11
+NSUB=1E16 NSS=1.2062E12 TPG=1
+RSH=80 LD=0.175U JS=3.125E−7
+CJ=3.125E−4
+MJ=0.5 PB=0.8 FC=0.5
+CJSW=2.0E−10 MJSW=0.5
+CGSO=400P CGDO=400P CGB0=200P)
```

Also, the following formats should be used:

```
V$D2 3 0 PULSE(3.55853E−01 5.0 1.0E−08 1.0E−08
+      1.0E−08 7.1E−08 1.6E−07)
ML1 6 3 3 0 DEPL W=12U L=12U AD=320P AS=32P
+      NRD=1.0 NRS=1.0 PS=50U PD=50U
MQ1 3 7 00 ENHD W=20U L=4U AD=320P AS=320P
+      NRD=0.5 NRS=0.5 PS=50U PD=50U
.OP
.OPTION RELTOL=0.0001
.OPTION GMIN=1E−18
```

b. Find the delay time between stages.

Design Problems

D1. You are to design a MOSFET for use as a voltage-controlled resistor at low voltages. You have access to a technology that delivers a threshold voltage of $V_{T(\text{MIN})} = 1$ V at a gate-oxide thickness of $X_{\text{OX(MIN)}} = 0.04$ μm, and for reasons of technological compatibility, you must use these values. The minimum effective channel length is $L^* = 1$ μm, and gate-oxide breakdown strength is $E_B = 1.0 \times 10^7$ V/cm. Channel-electron mobility declines with V_{GS}, and is 150 cm²/V·s. If any other MOSFET data are needed, take them from Tables 5 through 12.

a. The desired tuning range is 10 ohms to 1000 ohms, and the maximum voltage the device will experience is $V_{D(\text{MAX})} = 100$ mV. Determine the maximum current it will carry, $I_{D(\text{MAX})}$.
b. Devise a criterion that is directly related to the linearity of the resistor. Can linearity be improved without limit by pushing in the direction indicated by this criterion? Why?
c. Determine necessary Z/L and V_{GS} for the application.
d. Choose a channel length, and make a case for your choice in terms of parasitic MOSFET properties.
e. You are given the specification that in the $r_{\text{MIN}}(0)$ condition, the current $I_{D(\text{MAX})}$ is to depart from that in a truly linear 10-ohm resistor by no more than 1%. Determine whether your design meets this spec.
f. Make a plan-view sketch of your design, clearly indicating terminal identities and channel dimensions.

D2. You are to design a linear MOSFET amplifier stage like that of Problem A9. All of the technological information given in Problem D1 applies here as well. The desired voltage gain is $A_V = -64$, and the bandwidth should be as large as possible. Before proceeding with the design, you must deal qualitatively with several issues that may affect your design:

a. The MOSFET is not precisely a square-law device. How will this fact affect performance of the linear amplifier?
b. What is the impact of body effect on the I–V characteristic of the saturated load? What, if anything, can be done to ameliorate it?
c. Comment qualitatively on the effect of absolute channel length upon amplifier properties. Explain what properties are primarily affected and why.
d. Sketch a compact (plan-view) layout for your design, indicating all dimensions and terminal identities clearly. Comment on possible variations.

D3. You are to design a MOSFET for use in a digital integrated circuit. It requires a transconductance of $g_m = 0.69$ mS. Starting with the same process as used for the small-signal example in Section 5-3, what size MOSFET would you use? For a voltage gain of $A_V = -3$ and a supply voltage of $V_{DD} = 5$ V, design a minimum-size load resistor using a thin-film CrSi resistor process (that provides a linear device) with a sheet resistance of $R_{\text{SH}} = 200$ ohms and a minimum width of $Z = 10$ μm.

D4. You are to design a MOSFET for use in a digital integrated circuit.

a. The MOSFET must drive an interconnect line of length $L = 1000$ μm and a width of $Z = 2$ μm, having a capacitance of 0.2 fF/μm of length. Starting with the same process as used for the large-signal example in Section 5-4, what is the minimum-size MOSFET that will produce a rise time of $t_{RISE} = 20$ ns?

b. What static power will be required for this MOSFET?

D5. You are to design a 21-stage ring oscillator using the depletion-load inverter analyzed in Section 5-4. This oscillator is formed by connecting the inverters into a continuous cascade, with the output of one inverter driving the input of the next. What is the frequency of oscillation and what is the total power? Using a capacitor process with a capacitance of 1 fF/μm^2, design a ring oscillator with a frequency of 50 kHz.

D6. Starting with the D-mode-load inverter analyzed in Section 5-4, design and lay out a two-input NAND gate having the same logic levels as the inverter. The layout of the inverter driver is given in Problem A38. Schematic diagrams of the inverter and NAND circuits are given here:

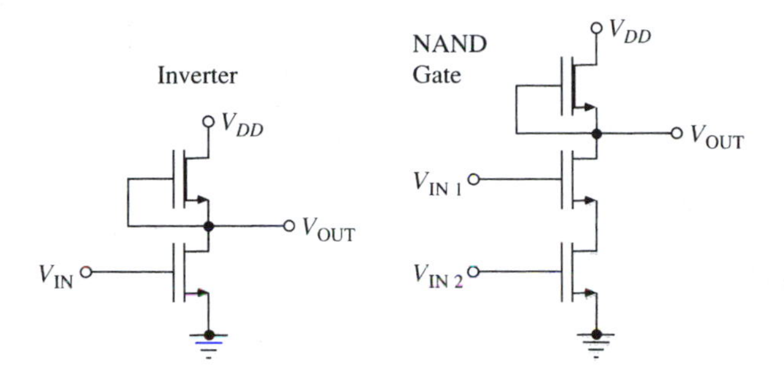

Symbol Index

Numbers following the symbols listed here indicate pages where each symbol is explained and (or) introduced. SPICE symbols are found here and also in Tables 5 through 12, and 14.

Subject Index

Italicized page numbers cite figures; tables are cited by table number; "(ex.)" means exercise.